Sustainable Catalytic Processes

Sustainable Catalytic Processes

Edited By

Basudeb Saha
Department of Chemistry, Purdue University, West Lafayette, Indiana, USA; Laboratory of Catalysis, Department of Chemistry, University of Delhi, Delhi, India

Maohong Fan
Department of Chemical and Petroleum Engineering, University of Wyoming, Laramie, WY, USA

Jianji Wang
School of Chemistry and Chemical Engineering, Key Laboratory of Green Chemical Media and Reactions, Ministry of Education, Henan Normal University, Xinxiang, Henan, PR China

ELSEVIER

AMSTERDAM • BOSTON • HEIDELBERG • LONDON • NEW YORK • OXFORD
PARIS • SAN DIEGO • SAN FRANCISCO • SINGAPORE • SYDNEY • TOKYO

Elsevier
Radarweg 29, PO Box 211, 1000 AE Amsterdam, Netherlands
The Boulevard, Langford Lane, Kidlington, Oxford OX5 1GB, UK
225 Wyman Street, Waltham, MA 02451, USA

ISBN: 978-0-444-59567-6

British Library Cataloguing in Publication Data
A catalogue record for this book is available from the British Library

Library of Congress Cataloging-in-Publication Data
A catalog record for this book is available from the Library of Congress

For Information on all Elsevier publications
visit our website at http://store.elsevier.com/

Contents

List of Contributors ..... ***ix***

Preface ..... ***xi***

Chapter 1: Role of Meso/Microporous Molecular Sieve Composite Materials on Various Catalytic Transformations ***1***

A. Sakthivel, A.K. Singh

1. Introduction 1
2. Zeolite and Zeolite-Like Molecular Sieves 2
3. Mesoporous Molecular Sieves 4
4. Progress in Mesoporous Molecular Sieves Assembled from Microporous Precursors 6
5. Meso/Microaluminosilicate Composite Materials as Potential Catalysts 8
6. Titanium-Based Meso/Microcomposite as Sustainable Catalysts 13
7. Meso/Microcomposite of Aluminophosphate-Based Materials as Potential Catalysts 14
8. Summary 18

Acknowledgements 18

References 18

Chapter 2: Functionalized Mesoporous Materials as Sustainable Catalyst in Liquid Phase Catalytic Transformations ***23***

Nabanita Pal, Asim Bhaumik

1. Introduction 23
2. Synthesis and Types of Functionalized Mesoporous Materials 25
3. Different Organic Transformations over Functionalized Mesoporous Catalysts 28
   - 3.1 Acid–Base Catalysed Reactions on Functionalized Mesoporous Materials 29
   - 3.2 Redox Reactions Catalysed by Functionalized Mesoporous Materials 36
   - 3.3 Different Coupling Reactions over Functionalized Mesoporous Materials 43
   - 3.4 Polymerization Reactions 52
   - 3.5 Role of Functionalization over Ordered Mesoporous Silicas 52
4. Summary 54

List of Abbreviations 54

References 54

Chapter 3: Sustainable Catalysis Systems Based on Ionic Liquids 61
Jianji Wang, Anlian Zhu, Lingjun Li

1. Introduction 61
2. Physical Properties of Ionic Liquids and Their Potentials in Sustainable Catalysis 62
   2.1 Liquid Range 62
   2.2 Density and Viscosity 64
   2.3 Polarity 64
3. The Utilization of Ionic Liquids in Sustainable Catalysis Procedures 65
   3.1 The Utilization of Ionic Liquids in Transition Metal Catalysis Reactions 66
   3.2 Enzymatic Reactions in Ionic Liquids 70
   3.3 The Dissolution and Conversion of Biomass in Ionic Liquids 74
   3.4 The Transformation of CO_2 to Cyclocarbonates Using Ionic Liquids 80
   3.5 The Utilization of Functionalized Ionic Liquids in Sustainable Catalysis 83
4. Outlook and Summary 88

References 89

Chapter 4: Catalysis for the Production of Sustainable Chemicals and Fuels from Biomass 99
Md Imteyaz Alam, Basudeb Saha

1. Introduction 99
2. Biomass and Biomass Compositions 99
3. Strategy for Biomass Conversion 100
4. Biomass to Value-Added Chemicals 101
5. 5-Hydroxymethyl Furfural 102
   5.1 Aspects of HMF Synthesis 103
   5.2 By-products in HMF Production 105
6. Catalyst for Biomass Conversion 105
7. Significance of 5-Hydroxymethyl Furfural as a Platform Chemical 107
   7.1 Oxidation of HMF to FDCA 107
   7.2 Oxidation of HMF to DFF 107
   7.3 HMF to LA 107
8. Second-Generation Biofuels from Biomass 108
   8.1 Cellulosic Ethanol 109
   8.2 5-Ethoxymethyl Furfural 110
   8.3 2,5-Dimethylfuran 111
   8.4 Long Chain Hydrocarbons 114
9. Summary 120

References 120

Chapter 5: Lignin Deconstruction: Chemical and Biological Approaches.............. 125
Saikat Dutta

1. Introduction 125
2. Lignin Valorization Techniques 128
   2.1 Chemical Techniques 128
3. Enzymatic Techniques 138
4. Carbonization 144
   4.1 Carbon Fibres 145
5. Summary and Outlook 151

References 153

Chapter 6: Integrated Bio- and Chemocatalytic Processing for Biorenewable Chemicals and Fuels.. 157
Md Imteyaz Alam, Shelaka Gupta, Ejaz Ahmad, M. Ali Haider

1. Introduction 157
2. Biocatalytic Transformation to Platform Molecules 159
3. Challenges Related to the Nature of Biogenic Impurities 165
4. Strategies for Catalytic Transformation of Platform Molecules 168
5. Conclusions 174

References 174

Chapter 7: Catalytic Coal Gasification 179
Xin Huang, Fan Zhang, Maohong Fan, Yonggang Wang

1. Introduction 179
2. Catalysts for Coal Gasification 181
   2.1 Alkali Metal Compounds 181
   2.2 Alkaline Earth Metal Compounds 185
   2.3 Transition Metal Compounds 189
   2.4 Composite Catalysts 191
   2.5 Invalidation and Recovery of Catalyst 194
3. Conclusion and Outlook 196

References 196

Index 201

List of Contributors

Ejaz Ahmad Renewable Energy and Chemicals Laboratory, Department of Chemical Engineering, Indian Institute of Technology, Delhi, India

Md Imteyaz Alam Renewable Energy and Chemicals Laboratory, Department of Chemical Engineering, Indian Institute of Technology, Delhi, India

Asim Bhaumik Department of Materials Science, Indian Association for the Cultivation of Science, Kolkata, West Bengal, India

Saikat Dutta Department of Chemical Engineering, National Taiwan University, Taipei, Taiwan

Maohong Fan Department of Chemical and Petroleum Engineering, University of Wyoming, Laramie, WY, USA

Shelaka Gupta Renewable Energy and Chemicals Laboratory, Department of Chemical Engineering, Indian Institute of Technology, Delhi, India

M. Ali Haider Renewable Energy and Chemicals Laboratory, Department of Chemical Engineering, Indian Institute of Technology, Delhi, India

Xin Huang School of Chemical and Environmental Engineering, China University of Mining and Technology, Beijing, PR China; Department of Chemical and Petroleum Engineering, University of Wyoming, Laramie, WY, USA

Lingjun Li School of Chemistry and Chemical Engineering, Key Laboratory of Green Chemical Media and Reactions, Ministry of Education, Henan Normal University, Xinxiang, Henan, PR China

Nabanita Pal Department of Materials Science, Indian Association for the Cultivation of Science, Kolkata, West Bengal, India; Surface Physics and Materials Science Division, Saha Institute of Nuclear Physics, Kolkata, West Bengal, India

Basudeb Saha Department of Chemistry, Purdue University, West Lafayette, Indiana, USA; Laboratory of Catalysis, Department of Chemistry, University of Delhi, Delhi, India

A. Sakthivel Department of Chemistry, Inorganic Materials and Catalysis Laboratory, University of Delhi (North Campus), Delhi, India

A.K. Singh Department of Chemistry, Inorganic Materials and Catalysis Laboratory, University of Delhi (North Campus), Delhi, India

Jianji Wang School of Chemistry and Chemical Engineering, Key Laboratory of Green Chemical Media and Reactions, Ministry of Education, Henan Normal University, Xinxiang, Henan, PR China

Yonggang Wang School of Chemical and Environmental Engineering, China University of Mining and Technology, Beijing, PR China

Fan Zhang School of Chemical and Environmental Engineering, China University of Mining and Technology, Beijing, PR China; Department of Chemical and Petroleum Engineering, University of Wyoming, Laramie, WY, USA

Anlian Zhu School of Chemistry and Chemical Engineering, Key Laboratory of Green Chemical Media and Reactions, Ministry of Education, Henan Normal University, Xinxiang, Henan, PR China

Preface

It was a fine spring day when my good friend Professor Maohong Fan and the co-editor of this book proposed me to edit a research-based book with Elsevier in the area of catalysis. The proposal was great, and I thought hard about a possible title for the book. Being a teacher and a researcher in the area of sustainable chemistry and catalysis with a reputed university and a former scientist of an international chemical company, I found a reference book summarizing contemporary research outcome in sustainable catalytic chemistry is important. So I agreed to edit this book with my co-editors, Professor Maohong Fan and Professor Jianji Wang, which can serve as a reference to researchers in industrial and academic settings to meet their growing interests in conducting research on the development of environmentally and economically viable chemical processes with catalyst participation.

During 2005−2015, there has been an increasing importance in the development of catalytic processes for the production of chemicals and fuels from nonconventional renewable sources to control rapid depletion of conventional fossil resources and prevent greenhouse gas emissions. Among several renewable sources, biomass has emerged as a preferred and sustainable resource for the production of chemical precursors, commodity and speciality chemicals and biofuels because of its abundancy and accessibility. However, deconstruction of biomass, especially second-generation lignocellulose, to access the constituents cellulose, hemicellulose and lignin for upgrading poses a significant challenge, which necessitates the utilization of heterogeneous catalytic substances of appropriate properties. Therefore, two chapters discussing various catalytic properties, preparation and characterization techniques and applications are included in this book. Further the utilization of these materials, along with several homogeneous catalytic substances for biomass deconstruction and upgradation of cellulose and hemicellulose into chemicals and biofuels, is summarized in two new chapters.

Another important component of biomass is lignin. Lignin is a natural biopolymer of important phenolic subunits and constitutes about 40% of carbon and energy values of biomass. While in 2005−2015, there has been significant attention given to the development of catalytic processes for the conversions of cellulose and hemicellulose, the valorization of lignin constituent phenolic monomers into high value chemicals for speciality applications such as ingredients for aroma, performance and agricultural products is important to improve the economic viability of biorefinery processes and to ensure the utilization of complete carbon value of biomass. By keeping this in mind, one chapter is dedicated to the discussion of the catalytic processes for selective upgrading of lignin to the valuable products and the associated mechanistic insights.

An integrated biological and catalytic conversion process for upgrading the entire biomass into a variety of commodity and speciality chemicals is seen as a viable strategy for the production of diverse products. A limitation of this strategy is the deactivation of catalytic substances by biogenic impurities that are formed in the biological conversion step. Therefore, a chapter is presented discussing a detailed overview of the available knowledge on the mechanism leading to the deactivation of the catalyst surface by biogenic impurities. This will help in the better designing of catalysts and purification methods for the development of effective integrated processes for biomass conversions.

Coal gasification is an important process in the purview of sustainable catalytic processes. The efficiency of the coal gasification process in producing syngas with high carbon value and optimized carbon monoxide to hydrogen ratio is largely dependent on the effectiveness of catalysts. Therefore, a chapter is compiled on the latest development of catalysts for coal gasification processes. Besides discussing the benefits and reactivity of earth abundant and inexpensive alkali, alkaline earth and transition metals for the said process, the mechanisms and the recovery protocols of the aforementioned catalytic substances are presented with fundamental concepts and future outlook.

I am greatly thankful to my co-editors Professor Maohong Fan and Professor Jianji Wang for their valuable time in contributing and reviewing their chapters and helping me in designing appropriate chapter contents for the book. Their sincere support and courage are enormous.

I am also equally indebted to all of my outstanding colleagues and friends who authored the chapters in this book and gave their time to thoroughly review the chapters. This book will remain incomplete without the continuous support and help of the people of Elsevier. My sincere appreciation goes to Dr Kostas Marinakis – Senior Acquisition Editor, Sarah Jane Watson – Editorial Project Manager, Paul Prasad Chandramohan – Senior Project Manager and Tharangini Sakthivel – Contracts Coordinator with Global Rights Department for their help throughout the publication process as well as overseeing proof composition and corrections.

With my co-editors, I hope this book will become an ideal reference to students, scientists, academicians and industrialists of all areas of chemistry, especially sustainable, catalysis and environmental chemistry. I apologize in advance for any unforeseen errors in the composition, and would appreciate your sharing such mistakes along with any advice for the future editions.

Editor: Basudeb Saha
Co-editors: Maohong Fan and Jianji Wang

CHAPTER 1

Role of Meso/Microporous Molecular Sieve Composite Materials on Various Catalytic Transformations

A. Sakthivel, A.K. Singh
Department of Chemistry, Inorganic Materials and Catalysis Laboratory, University of Delhi (North Campus), Delhi, India

1. Introduction

The development of inorganic materials possessing a framework structure with channels, voids, cavities and appropriate pore dimensional accesses (voids) is a prime area of materials sciences owing to their sorption, catalytic and molecular sieve properties [1]. In this regard, the synthesis and design of open framework molecular sieves, with tailored pore size and controllable framework topology, have attracted great interest among materials researchers. Notably, the field of heterogeneous catalysis has witnessed the major application of such engineered materials in petrochemical and fine chemical processes, specifically for adsorption and support [2,3]. The basic criteria that enable their extensive application are the inherent porosity and high surface area. The International Union of Pure and Applied Chemistry has classified these porous materials into three classes based on pore size: microporous ($d < 2$ nm), mesoporous (2 nm $< d < 50$ nm) and macroporous ($d > 50$ nm) [4]. In scientific applications, it is more precise to use the term 'molecular sieves' in place of porous materials. These materials possess selective sieving/adsorption properties at the molecular level owing to their exclusive shape, size and molecular polarity [5]. A material is considered a molecular sieve when it can selectively separate particular molecules or components from a mixture based on shape and size. The term molecular sieve was coined by McBain to explain the sieving properties of certain microporous charcoal and natural zeolites [5]. Scheme 1 summarizes the classification of molecular sieves, different types of molecular sieves materials and their importance in various applications.

Sustainable Catalytic Processes. http://dx.doi.org/10.1016/B978-0-444-59567-6.00001-7

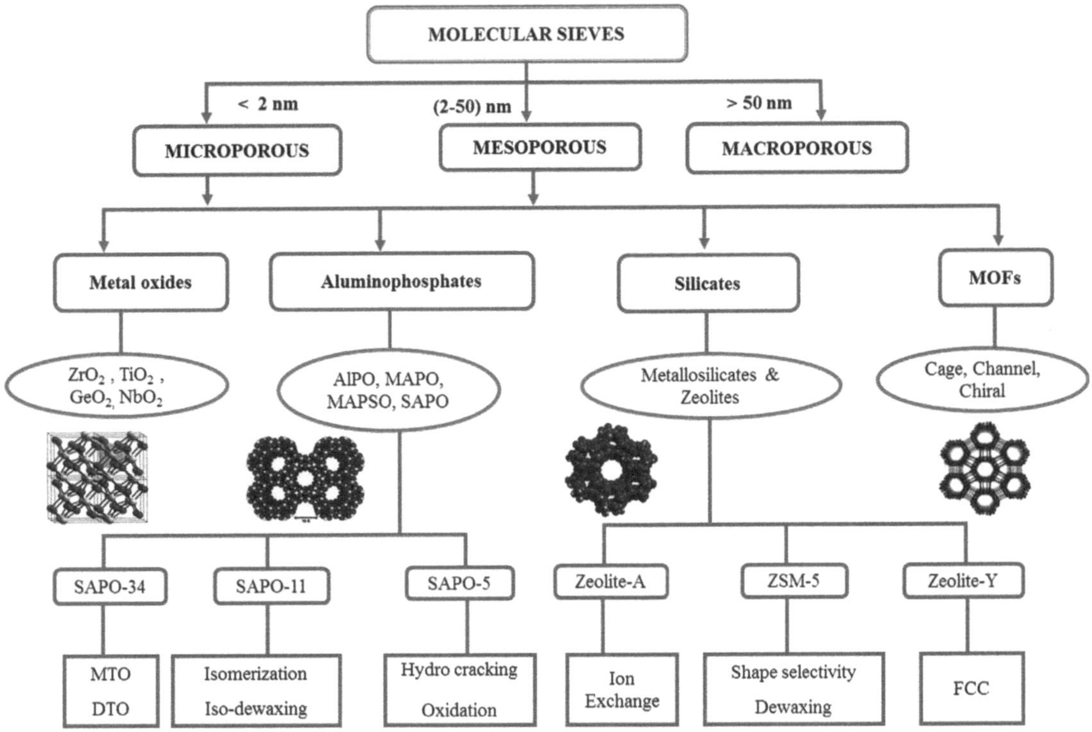

Scheme 1
Various types of molecular sieve classifications and their important applications.

2. Zeolite and Zeolite-Like Molecular Sieves

Zeolite and zeolite-like molecular sieves are one of the most important framework molecular sieve materials extensively used as catalysts and adsorbents in several petroleum and fine chemical processes [6–9]. Their elemental composition can be regulated by the introduction of heteroatoms (B, Ti, V, Cr, Fe, Co, Ga, etc.) either in the framework or extra-framework and subsequently alter the active sites of materials. By definition, zeolites are aluminosilicates framework materials that is based on a three-dimensional network of $(SiO_4)^{4-}$ and $(AlO_4)^{5-}$ tetrahedra [5,10–13]. The extended arrangements of connected primary tetrahedra may be defined as secondary building units (SBU), such as single four rings, six rings and more complex units like double four rings and double six rings, etc., which are further linked to form a structural sub unit (SSU) (Figure 1) [5,10,14]. Aluminophosphate molecular sieves are another class of porous crystalline inorganic solid materials that are formed from alternating AlO_4^{5-} and PO_4^{3-} tetrahedra, which were first synthesized by Wilson and co-workers [15,16]. This was the first family of framework oxide molecular sieves, which is free from silicate. Their general formula can be expressed as $[(AlO_2)x(PO_2)x]\cdot y\ H_2O$, indicating that, contrary to most of the zeolites, aluminophosphate molecular sieves are ordered with an Al/P ratio that is always unity.

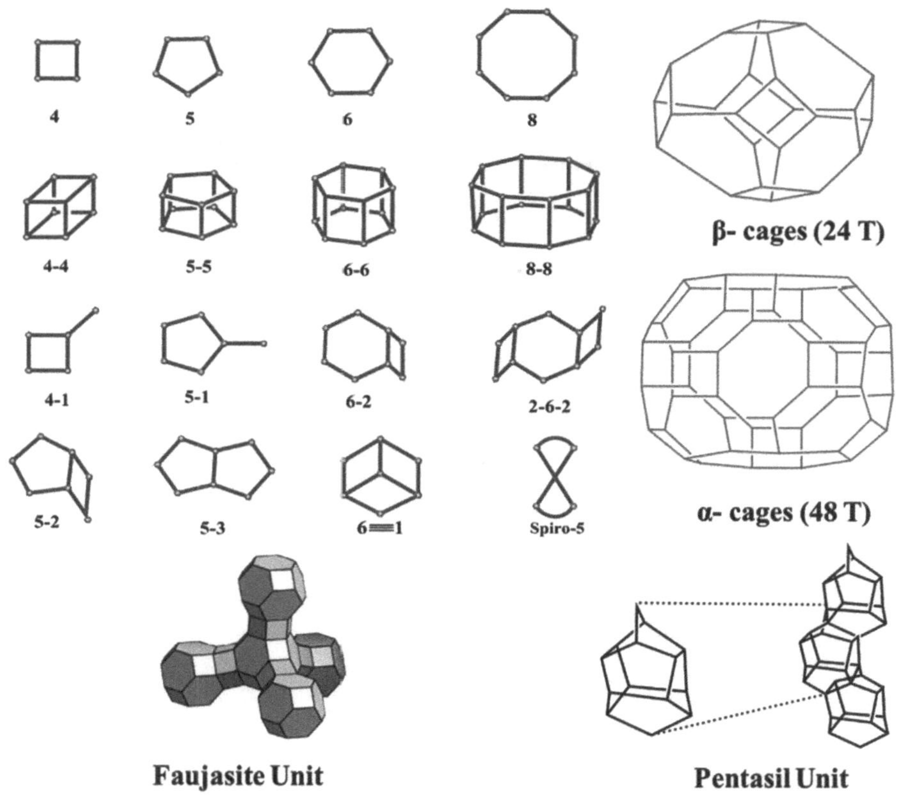

Figure 1
Secondary building unit and structural sub units of zeolite molecular sieves [5].

Conventionally, zeolites and zeolite-like molecular sieve materials are synthesized under hydrothermal conditions using silicon and aluminium sources in the presence of alkali metal cations or organic amines/ammonium cations, which act as templates or structure-directing agents (SDAs). The primary Td units ($[SiO_4]^{4-}$ and $[AlO_4]^{5-}$) combine to form SBUs, which grow around the template of different framework structures. The properties of the resulting zeolite are influenced by several factors, including the composition and pH of the reaction mixture, crystallization temperature, duration of the reaction and choice of SDAs. Figure 2 represents the several stages involved in the synthesis of different zeolitic systems [17].

The intrinsic properties of these molecular sieves, namely strong acidity, large internal surface area, pore volume and unique framework topology facilitate as possible applications in catalysis and adsorption processes. Further, researchers have focused on developing new zeolitic materials with improved textural properties. The major drawback of zeolites is their inability to diffuse bulkier reactant molecules due to their small pore openings and channel size (<0.8 nm) [18].

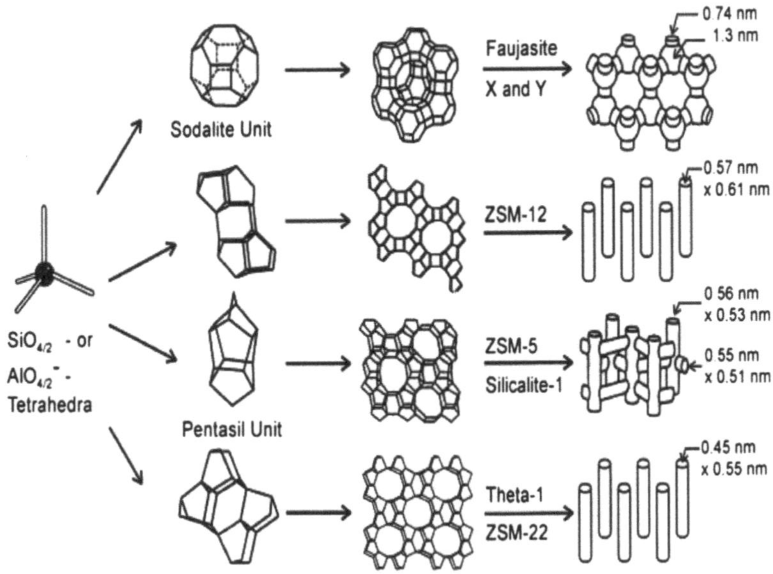

Figure 2

Structures of four selected zeolites (from top to bottom: (1) faujasite or zeolites X, Y; (2) zeolite ZSM-12; (3) zeolite ZSM-5 or silicalite-1; (4) zeolite Theta-1 or ZSM-22) and their micropore systems and dimensions. *Reproduced from Ref. [17].*

3. Mesoporous Molecular Sieves

The limitations of microporous zeolite were addressed, to some extent, in 1992, when researchers at Mobil synthesized M41S molecular sieves, the first family of mesoporous silicates with a larger pore size (1.5–10 nm) [19]. A liquid crystal templating (LCT) mechanism, in which surfactant media serve as the organizing agents, was used to assemble hexagonal, cubic and lamellar mesophases of MCM-41, MCM-48 and MCM-50, respectively (Figure 3). These materials possess a high surface area (>1000 m^2/g), and their pore size can be tailored by varying the reactant stoichiometry and nature of the

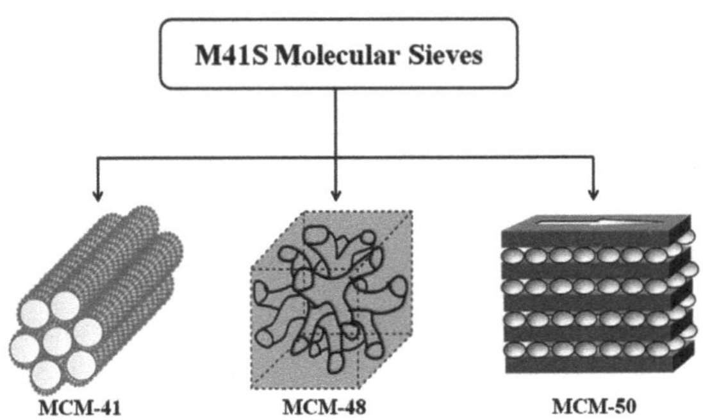

Figure 3

Structures of MCM-41, MCM-48 and MCM-50.

surfactant (cationic/anionic, alkyl chain length), or by post-synthetic techniques. This strategy has successfully led to the synthesis of other mesophases, for example, SBA-15, SBA-1 and HMS [20].

The assembly of these different mesophases depends upon the surfactant organization in the micellar liquid crystalline phase, which serves as a template. The inorganic silicates are cast over the surface of the template by electrostatic interaction. Beck et al. believe that the liquid crystal arrangement is solely credited to surfactants, and the silicate condensation is not the dominant factor in forming the structure [19–21]. Later, Stucky and co-workers proposed another mechanism, in which a cooperative interaction between the surfactant and the silicate precursors (cooperative self-assembly (CSA) mechanism) leads to different mesophases [22]. In general, the LCT mechanism is favoured at high surfactant concentrations, while the CSA mechanism operates at low surfactant concentrations. The CSA mechanism has been widely accepted and extensively evidenced by different advanced characterization techniques [23]. Figure 4 presents the two different mechanisms proposed for the synthesis of mesoporous materials [24].

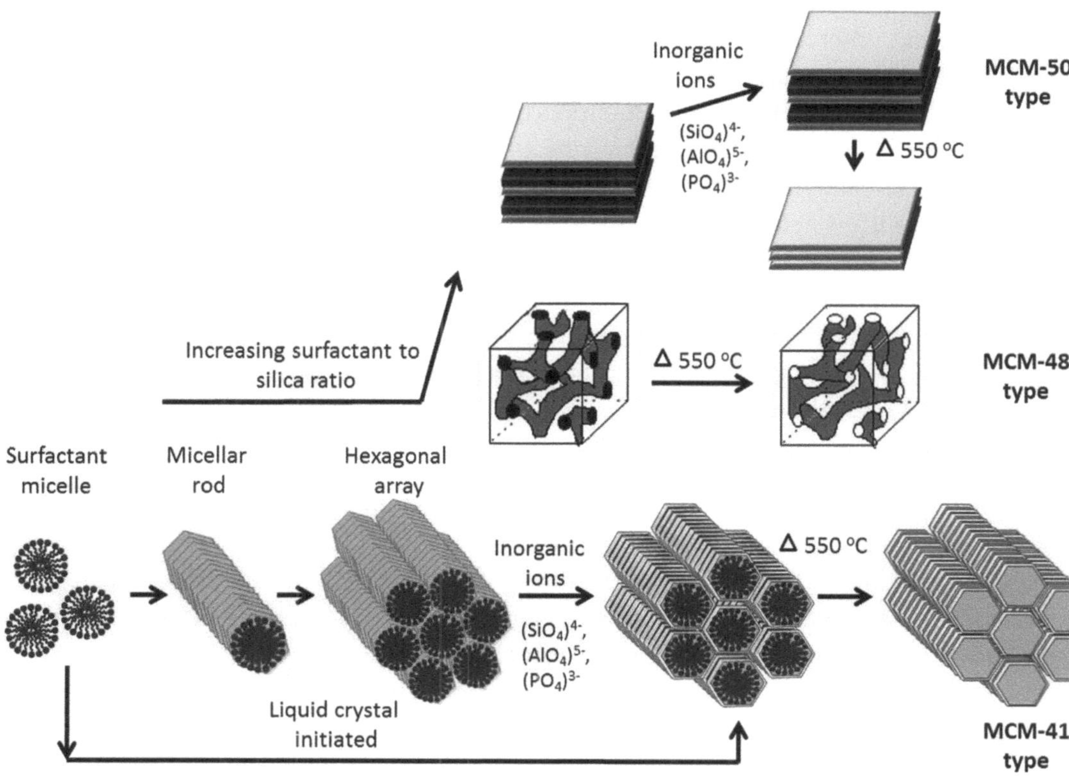

Figure 4

Schematic illustration of (1) the liquid crystal templating mechanism and (2) the cooperative self-assembly mechanism for the formation of ordered mesoporous materials [24].

The trivalent metal ion-substituted mesoporous molecular sieve showed promising catalysts for various organic transformations [25–27]. On the other hand, the introduction of transition metal ions results in redox properties and shown as potential oxidation catalysts [28–30]. However, the amorphous wall properties of these materials limit their application at high temperatures and reactions involving water owing to their poor hydrothermal stability.

4. Progress in Mesoporous Molecular Sieves Assembled from Microporous Precursors

The quest to synthesize thermally and mechanically strong mesoporous materials has driven researchers to adopt various synthetic routes and in situ and ex situ methods. During 1992–till present, numerous studies have focused on the synthesis of efficient heterogeneous molecular sieve catalysts with superior characteristics to those of conventional ones. In this regard, the development of hierarchical porous and meso–microcomposite materials is a prospective field for heterogeneous catalysis. Hierarchical or meso/microcomposite materials are generally associated with multidimensional, interconnected pores in an ordered structure, with a high internal surface area and pore volume [31]. Furthermore, these composite materials overcome the diffusion limitation that arises in microporous systems. The interest in such materials is rapidly growing due to their versatile applications in various fields (e.g. nanoscience, catalysis, separation, electronics, optics, optoelectronics, energy, environment and life sciences) [32]. Several synthetic procedures have been adopted to fabricate hybrid materials. The various strategies involve (1) post-synthetic silylation, followed by crystallization; (2) improving the framework wall thickness by aluminium grafting; (3) introducing inorganic salts to improve condensation and cross-linkages in the framework; (4) post-coating of amorphous walls by additional templating agents; (5) developing meso/microcomposites in single- or multistep synthesis or (6) recrystallizing zeolites [33–45].

Broadly, the various routes have been categorized as 'bottom-up' or 'top-down' methods. In the bottom-up approach, researchers utilize multistep 'soft templating' or 'hard templating' methods. In soft templating methods, meso/microcomposite materials are prepared either by a direct synthesis route, using a mixture of templates and surfactants for the formation of cubic and hexagonal mesoporous arrays of aluminosilicates [46–49], or by synthesis of mesoporous materials followed by coating of the pore walls with zeolite SBUs [34–38]. In hard templating methods, zeolite seeds are first prepared, and then used to assemble the mesoporous phase [39–45]. On the other hand, in top-down processes, certain framework elements are removed from the zeolitic species (dealumination or desilication) in order to generate mesopores in them. These processes

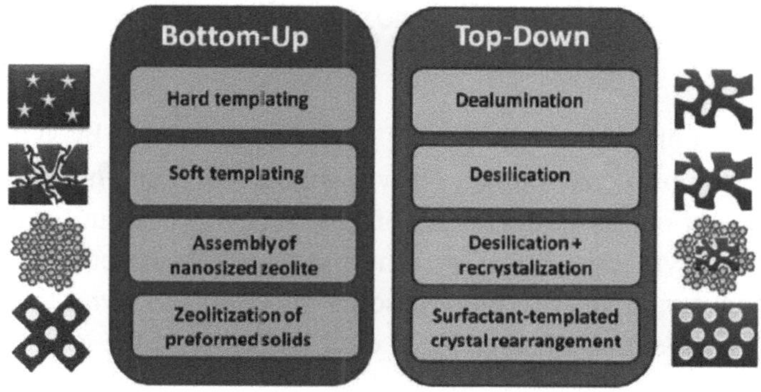

Figure 5
Different synthetic approaches to prepare hierarchical zeolites [32].

generally do not involve the use of surfactants. Figure 5 shows a schematic representation of the different synthetic approaches for preparing mesoporous hierarchical zeolites [32].

The reported meso/microcomposites and hierarchical materials have emerged as an important class of catalytic materials, since they possess significant advantages over both zeolites and ordered mesoporous materials [50]. Compared to their microporous counterparts, the meso/microcomposites have the following advantages [50]:

1. Mesopores offer better catalytic efficiency resulting from the improved diffusion ability of reactant molecules.
2. Hierarchical materials lead to a decrease in path length and contact time, which favours improved selectivity for the chosen reaction.
3. The reduced path length minimizes coke formation, blocking of active sites and catalyst deactivation.

Similarly, composite materials have advantages over their mesoporous analogues, which are as follows [50]:

1. Microporous zeolitic building units on the walls of mesoporous channels impart high thermal and mechanical stability, which makes them suitable for vigorous reaction conditions.
2. High surface acidity derived from microporous building units widens their catalytic applications.

The following section of this chapter summarizes the use of such meso/microcomposite materials as important catalysts for various organic transformations.

5. Meso/Microaluminosilicate Composite Materials as Potential Catalysts

The first reported example of the synthesis of such composite was by synthetic deposition of MCM-41 on faujasite-type zeolite, and it was studied for cracking of vacuum gas oil [51]. The Na_2O content on the synthesis gel of zeolite-Y has been utilized for the coating of MCM-41 by in situ synthesis. It was evident that the MCM-41-coated zeolite-Y showed about 10% more conversion and 3% more gasoline product selectivity than did the parent pure MCM-41 and zeolite-Y samples [51]. Li et al. synthesized the MCM-41/ZSM-5 composite using a two-step approach [36], by first assembling mesoporous MCM-41 in the presence of a surfactant by a self-assembling method, followed by recrystallization of the amorphous wall into ZSM-5 using tetrapropylammonium bromide. The resultant materials showed an increase of about 9% in catalytic activity for the cracking of *n*-dodecane compared to the physical mixture of MCM-41 and ZSM-5. The observed higher catalytic activity of composite prepared by the in situ method can be explained based on the interconnection of micropores present on the walls of the mesoporous channel [36].

Xiao et al. [39] assembled mesoporous aluminosilicate (MAS-5) possessing a high thermal stability and strong acidity using a preformed precursor containing a zeolite beta structural unit. The materials were found to be promising for the catalytic cracking of trialkyl benzene [39]. Pure mesoporous MCM-41 showed a 65.8% conversion of 1,3,5-triisopropyl benzene, whereas MAS-5 containing the zeolitic building unit showed a 95% conversion. Similarly, the alkylation of butane with 2-butene, which is an important reaction to get a low-vapour-pressure gasoline blend component, was studied on both H-MCM-41 and MAS-5 materials having similar SiO_2/Al_2O_3 ratios. It was evident from this study that mesoporous aluminosilicates assembled from microporous precursor (Beta) result in excellent conversion in the range of about 98% to 100% and the activity remain constant for a prolonged period of time, owing to less coke formation and intrinsic acidic sites derived from the microporous building units [39].

Sakthivel et al. prepared meso—microcomposite materials having an MCM-48-type mesoporous structure with ZSM-5 wall properties (RMM-1 and MMSH) by two methods and studied the Friedel—Graft alkylation of ethylbenzene with ethanol in the vapour phase. It was found that the composite materials yield better catalytic conversion than the parent pure mesoporous aluminosilicate molecular sieves (Table 1). Further, it is interesting to note from their study that both the composite catalysts showed a slight decrease in activity in the first hour due to coke formation; however, the activity stabilized after 90 min. About 90% of diethyl benzene selectivity was achieved in both the composite materials, which was found to be much better than even microporous ZSM-5 catalyst [52].

Table 1: Catalytic Data Obtained from Various Samples During Ethylbenzene Alkylation Reaction.[a]

Catalysts	MMSH	RMM-1	Al-MCM-48	H-ZSM-5
Ethylbenzene conversion (%)	20.2	29.9	11.9	44.1
Diethylbenzene selectivity (%)	89.6	86.6	88.3	69.0
Others[b]	10.4	13.4	11.7	31.0

[a]Reaction conditions: temperature = 623 K; WHSV = 4.6 h^{-1}; ethylbenzene: ethanol = 1:1 (molar ratio); Time on Stream (TOS) = 1.5 h.
[b]Other products including benzene, cracking products and trace amount of trialkyl benzene.
Reproduced from Ref. [52].

The core-shell composites of ZSM-5-MesoSiO_2 were prepared by Zhao et al. [53] and studied for catalytic cracking of *n*-dodecane. It was found that ZSM-5-MesoSiO_2 showed a conversion level of 90% as compared to 83% conversion by conventional ZSM-5. The observed higher conversion on core–shell ZSM-5-MesoSiO_2 was explained based on the graded acidity distribution, from weak in the exterior mesopores to strong in the inner micropores, in addition to which the presence of hierarchical mesopores favours the diffusion of larger molecules which facilitate more cracking [53].

A series of meso/microcomposite materials were derived from Mordenite zeolite (Re-MOR1, Re-MOR-2 and Re-MOR-3) by dealumination followed by recrystallization [54]. The catalytic properties of Re-MOR were studied in the trans-alkylation of biphenyl with diisopropyl benzene. Conversion levels of 69% and 25% were obtained for biphenyl and diisopropyl benzene, which was far superior to microporous dealuminated Mordenite (25% and 4%) and pure mesoporous MCM-41 (24% and 15%). The selectivity of diisopropyl biphenyl was improved significantly to about 47% in comparison to the other system, where the selectivity was found to be in the range of 3% to 18%. The improved catalytic activity and selectivity were explained based on the combined properties of zeolitic acidity and molecular diffusion properties on mesoporous molecular sieves [54].

Aluminium and zirconium substituted meso–microcomposite (AlZrMMSH) has been shown to be a promising catalyst for esterification of glycerol with lauric acid under a supercritical carbon dioxide (SCC) medium [55]. The use of the SCC medium compared to a conventional medium favoured the improvement of catalytic performance by high diffusivities of reactant, products and coke precursors [55]. Similarly, AlMMSH and calcium-containing AlMMSH were utilized for amidation of palmitic acid with *N*-hexadecylamine [56]. The conversion was found to be in the range of 5% to 70%. The author explained that the incorporation of calcium in the framework of silicate results in weak acidity that has the effect of reducing strong acidic sites, which enhances catalytic performance [56]. A report [57] also addressed the direct synthesis of zeolite-β,

Table 2: Nitration of Toluene over SBA-15, Zeolite-β and ZBS-15 Catalysts.[a]

Catalysts	Conversion (%)	*o*-Nitrotoluene	*p*-Nitrotoluene	*m*-Nitrotoluene
SBA-15	33	35	61	4
Zeolite-β	53	39	57	4
ZBS-15	100	39	56	5
No catalyst	13	45	51	4

[a]Reaction conditions: Catalyst = 20 mg; toluene (1 mmol), HNO_3 (2 mmol) temperature = 373 K; time = 4 h.
Reproduced from Ref. [57].

mesoporous SBA-15 molecular sieve composite (ZBS-15) and utilized this for the industrially important nitration process under solvent-free conditions. The superior activity evident on ZBS-15 (Table 2) was explained based on the strong acidity derived from walls having a zeolite-β unit and larger mesopores that provide a diffusion-free environment [57]. The activity of the catalyst ZBS-15 was maintained even after four recycles.

Laha et al. [58] developed a three-dimensional MCM-48 type molecular sieve in situ, assembled in the presence of a zeolite-β SBU (Al-MCM-48-BEA) and utilized for the alkylation–condensation of isophytol with trimethylhydroquinone to get α-tocopherol (vitamin E). The use of Al-MCM-48-BEA showed about 95% yield (Figure 6) in comparison to conventional Al-MCM-48, which showed only a 16% yield, and microporous zeolite-Y, which showed only about 10% yield. The lower yield in microporous materials was explained based on the diffusion limitation, and the observed high activity on Al-MCM-48-BEA was demonstrated based on the available strong acidic sites that are derived from zeolitic SBUs present on the wall of the materials [58].

Microporous zeolite-L and mesoporous SBA-15 composites (LSBA-15; LS) were in situ synthesized and used as support additives to prepare a hydro-upgrading catalyst for fluid catalytic cracking (FCC) of gasoline [59]. The corresponding CoMo-supported catalysts

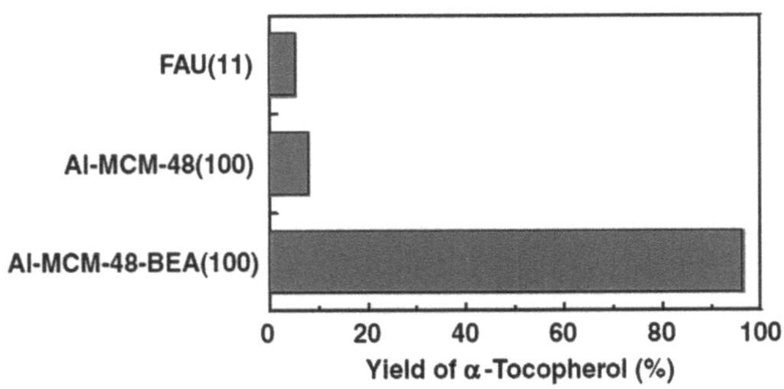

Figure 6
Yield of α-tocopherol by alkylation condensation over various catalysts under supercritical CO_2 medium. *Reproduced from Ref. [58].*

($CoMo/LS$-γ-Al_2O_3; CoMo/LSA) were prepared by the two-step incipient-wetness impregnation of ammonium molybdate (MoO_3 10 wt%) and cobalt nitrate (CoO 5 wt%). It was identified that CoMo/LSA possessed the advantages of the mesoporous SBA-15 structural properties and acidity (both Brönsted and Lewis) similar to that of zeolite L. The presence of meso/microcomposite properties reduces the diffusion limitation and avoids unwanted side reactions such as cracking and coke formation [59]. The CoMo/LSA showed a hydrodesulphurization (HDS) efficiency of 95.1% with highly efficient conversion of olefins to *i*-paraffins with a high research octane number of 90.9. It was explained that the outstanding physicochemical properties of LS micro/mesoporous composites made it a promising material for FCC gasoline hydro-upgrading [59].

Meso-structure aluminosilicate (MAS) assembled from ZSM-5 zeolite nanocrystals at different hydrothermal durations (4, 16 and 24; represented as MAS-4, MAS-16 and MAS-24, respectively) and subsequently modified with NiMo (MAS-16) was used for HDS of dibenzothiophene (DBT) as well as HDS and hydrodenitrogenation (HDN) of light gas oil (LGO) [60]. The hydrotreating activity of MAS-16-supported NiMo catalyst was found to be better than that of AlSBA-15-NiMo catalyst. A good dispersion of active sites with excellent textural properties and the presence of strong Brönsted acidic sites favoured the superior activity observed. The HDS and HDN activities of NiMo/MAS-16 are reported to be better than that of NiMo supported on commercial γ-Al_2O_3. The naphtha product distributions based on the boiling point (BP) range are displayed in Figure 7 (reproduced

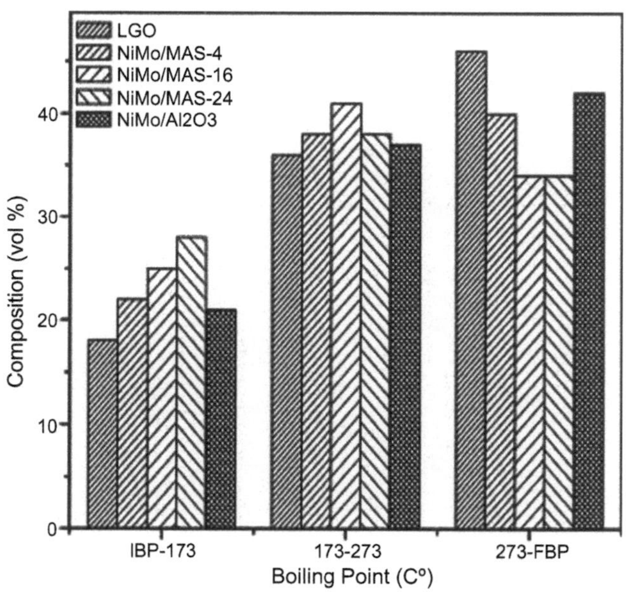

Figure 7
LGO hydrotreating product distribution based on boiling point over different catalysts. *Reproduced from Ref. [60].*

from Ref. [60]), and it was evident from the results that the formation of lighter naphtha (BP range of 90–170 °C) in NiMo/MAS-16 is greater than that in commercial NiMo/γ-Al_2O_3 catalyst. The activity was explained based on the large pores and narrow pore size distribution in NiMo/MAS-16, which eliminates the diffusion limitation and facilitate the better catalytic performance.

Hybrid zeolitic-ordered mesoporous materials (HZM) and hierarchical ZSM-5 were utilized for 1,2-epoxyoctane rearrangement and anisole acylation [61]. The presence of strong Lewis acidity and the best accessibility of active sites present on HZM facilitate the highest conversion (~100%) for 1,2-epoxyoctane isomerization. The microporous ZSM-5 showed about 20% conversion, with the formation of 1-octaldehyde and octenols as products with equal distribution. The formation of a heavier aldol condensation product is evident around 58 mol % on HZM, due to the presence of more Lewis acidic sites and the absence of steric and diffusion limitations. The HZM was further investigated for anisole acylation to get industrially important *p*-methoxyacetophenone (*p*-MAP). The conventional ZSM-5 yielded a remarkably lower conversion of only 10%, obtained owing to the small pore openings that limit the conversion. The hierarchical ZSM-5 having 5% and 12% enhances a strong Lewis acidity and the pore accessibility facilitates the improvement in conversion (Figure 8). The HZM showed a far superior conversion to both microporous ZSM-5 and mesoporous Al-MCM-41 (Figure 8) with a *p*-MAP yield of about 95%.

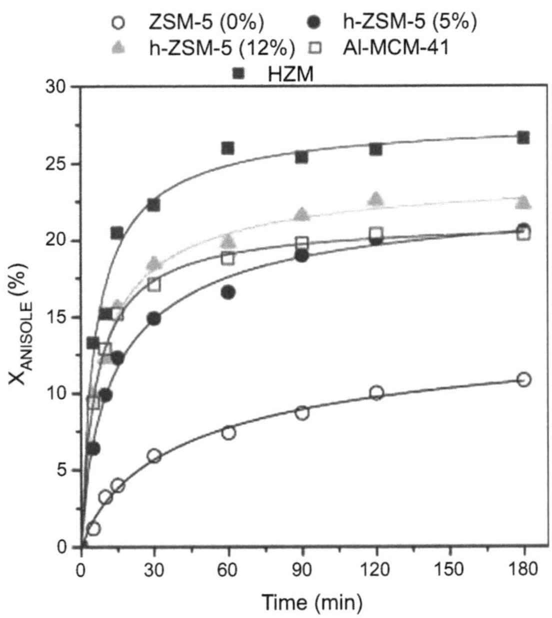

Figure 8
Acylation of anisole using different hierarchical ZSM-5. *Reproduced from Ref. [61].*

6. Titanium-Based Meso/Microcomposite as Sustainable Catalysts

Landry et al. [62] synthesized titanium-containing micro-mesoporous materials (Ti-MMM-1) and studied them for the oxidation of cyclohexane and *n*-octane. It was evident from their studies that more cyclohexane conversion is obtained from Ti-MMM-1 compared to that from TiMCM-41. Further, the authors observed the selective formation of desired products such as cyclohexanone and cyclohexanol (Table 3). Similar results were also evident on linear alkane, namely, *n*-octane. The desired product of the corresponding alcohol was obtained as the predominant product. It was explained that the presence of smaller and more strained microporous crystallites in Ti-MMM-1 was responsible for its enhanced reactivity. Hexagonal mesoporous solids obtained using microporous titanosilicate (Mime-TS-1) were utilized for epoxidation of cyclohexene, and it was found that the conversion greatly improved compared to TiMCM-41 [63]. The formation of both epoxide and allylic oxidized products was evident, while Mime-1h showed a more allylic product than epoxide [63]. It was observed that the conversion over Mime-TS-1 increased twofold compared with that over microporous TS-1 when the authors used *tert*-Butyl hydroperoxide (TBHP) as the oxidant. The larger selectivity of allylic products evident on Mime-TS-1 was explained based on the abundance of surface OH groups [63].

Li et al. utilized the TiHMS/TS-1 composite for oxidative desulphurization of thiophene (TH), benzothiophene (BT) and DBT, which are model compounds present in the fuel. The results are reproduced in Figure 9 [64]. It was reported that microporous TS-1 did not show any catalytic activity, owing to the difficulty of bulk molecules diffusion inside the channels of microporous molecular sieves. For the oxidation of DBT and BT, both pure mesoporous TiHMS and TiHMS/TS-1 composite showed excellent catalytic activity. On the other hand, for TH, a highly stable sulphur compound, the oxidative removal rate on TiHMS was found to be only 33%, whereas the composite showed 91%, which is very close to the catalytic behaviour of microporous TS-1.

Table 3: Oxidation of Cyclohexane and *n*-Octane over Ti-MMM-1.

	Conversion (mol%)	Ketone(s) (mol%)	Alcohol(s) (mol%)
Cyclohexane			
Ti-MMM-1	9.2	35.1	54.7
Microporous TS-1	4.2	26.4	27.6
Mesoporous TiMCM-41	1.9	9.8	17.0
n-Octane			
Ti-MMM-1	19.8	14.5	80.8
Microporous TS-1	13.3	10.3	80.3
Mesoporous TiMCM-41	2.9	21.5	52.7

Reproduced from Ref. [62].

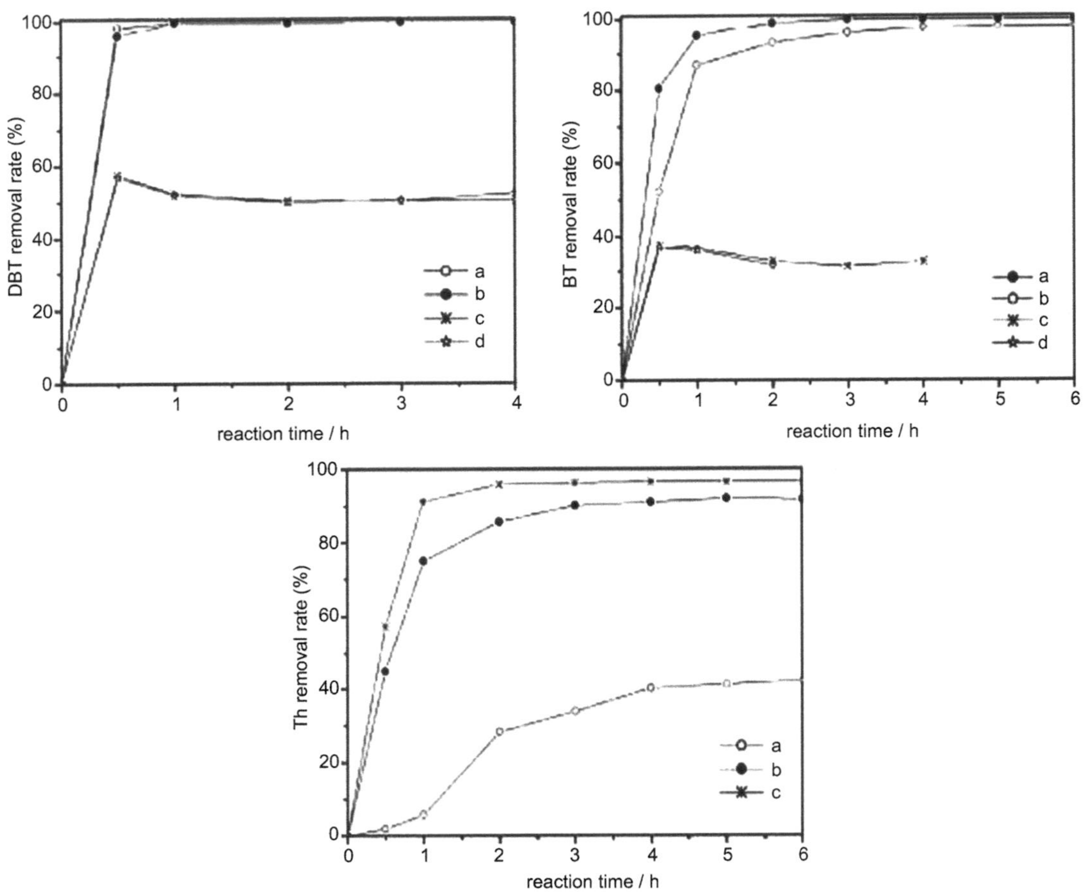

Figure 9
Catalytic performance on oxidative desulphurization of various sulphur compounds using (a) TiHMS, (b) TiHMS/TS-1, (c) TS-1 and (d) blank. *Reproduced from Ref. [64].*

7. Meso/Microcomposite of Aluminophosphate-Based Materials as Potential Catalysts

Aluminophosphate materials are another class of materials analogous to zeolite, and this was the first family of non-silica-based framework oxide molecular sieves, first reported by Wilson et al. in the 1980s [65]. Aluminophosphate molecular sieves (AlPO-*n*; *n* refers to a distinct structure type) are composed of alternating AlO_4 and PO_4 tetrahedra attached by bridging oxygen atoms, leading to a neutral open framework structure. The framework of AlPO is flexible compared to the rigid zeolite framework. The structural flexibility of AlPO results in structural and compositional diversity, as aluminium and/or phosphorus in the lattice framework can be isomorphously substituted by ions of a similar charge and size. The incorporation of silicon or any other metal cations into the framework leads to the formation of microporous silicoaluminophosphate (SAPO-*n*) or metalloaluminophosphate/metallo-silicoaluminophosphate (MAPO-*n*/MAPSO-*n*) molecular sieves, respectively [66]. The successful development of mesoporous composites of the aluminosilicate family and the demonstration of their importance in catalytic and adsorption studies of bulk molecules have motivated researchers to develop similar composite framework materials. Regardless of the challenges regarding the synthesis and structural stability of aluminophosphate-based materials, the vast structural diversity, framework flexibility and catalytic demand prompted researchers to espouse innovative synthetic routes and explore the mechanistic aspects of SAPO-based materials. In the following section, we briefly summarize the recent developments on meso/microcomposites of aluminophosphate-based molecular sieves and their applications.

The first stable X-ray mesoporous silicoaluminophosphate was assembled from a microporous SAPO-34 precursor using a multistep method by Sakthivel et al. [67] and was utilized for the liquid phase synthesis of phenoxy propanol using phenol and propylene oxide. A comparable conversion of phenol of about 23% was evident on Meso-SAPO-34, which was found to be better than on mesoporous Al-MCM-41 and microporous SAPO-34 materials (Table 4) [67]. The presence of strong acidic sites along with the mesoporous channels present on the Meso-SAPO-34 yield better conversion and the desired 1-phenoxy-2-propanol selectivity (80%).

Subsequently, series of Meso-SAPO (Meso-SAPO-37, Meso-SAPO-34 and Meso-SAPO-5) have been prepared from different microporous precursors (SAPO-37, SAPO-34 and SAPO-37) and utilized for the petrochemically important hydroisomerization of 1-octene, and the results are reproduced here [68]. The schematic representation of the assembly of Meso-SAPO-*n* and its application for hydroisomerization of olefin is represented in Scheme 2 [68].

Table 4: Phenoxylation of Propylene on MESO-SAPO-34.[a]

Catalysts	Conversion of Phenol (wt%)	Selectivity of 1-phenoxy-2-propanol (wt%)
MESO-SAPO-34	23.3	81.3
First recycled	22.7	74.8
Second recycled	21.9	70.0
Without catalysts	4.1	30.3
SAPO-34	7.0	56.8
Al-MCM-41	11.5	84.5

[a]Reaction conditions: catalysts = 0.25 g; phenol = 1 g; propylene oxide = 0.93 g (1:1.5 mol ratio); time = 8 h; temperature 423 K.

Reproduced from Ref. [67].

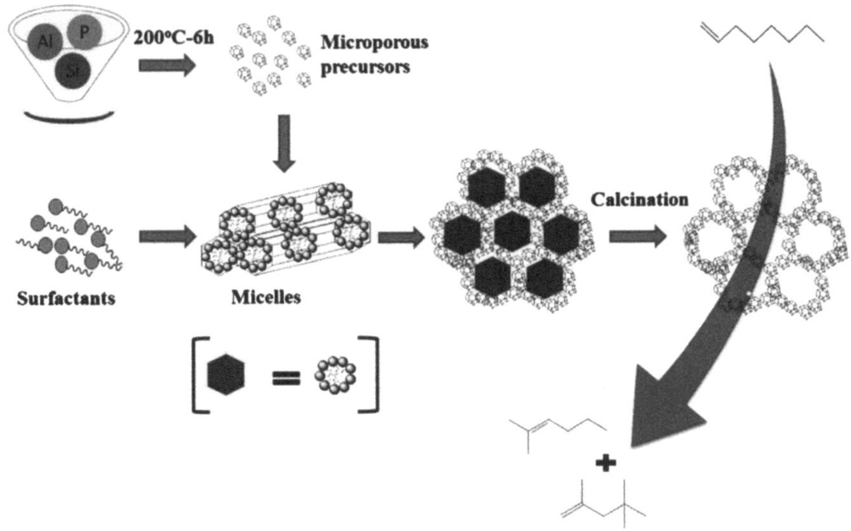

Scheme 2
Schematic representation for the formation of MESO-SAPO-*n* and its application on hydro-isomerization of 1-octene. *Reproduced from Ref. [68].*

The catalytic activities of various mesoporous analogues were compared by maintaining the temperature (400 °C) and Weight Hour Space Velocity (WHSV) ($8\ h^{-1}$) under identical conditions; the results are shown in Figure 8. The conversion of 1-octene decreased in the order Meso-SAPO-37 > Meso-SAPO-5 > Meso-SAPO-34. The selectivity of branched to linear isomerization products (Figure 10) decreased in the order Meso-SAPO-34 ~ Meso-SAPO-37 > Meso-SAPO-5. The observed branched isomer selectivity and better catalytic conversion on Meso-SAPO-37 was explained based on strong Lewis and Brönsted acidic sites present together on the materials.

The catalytic activity was further compared to that of microporous SAPO analogues, and the results are shown in Table 5. The microporous system having small pore openings (SAPO-34) and a one-dimensional pore structure (SAPO-5) yielded less conversion and poor branched isomer selectivity. The observed branched isomer selectivity on faujasite-type SAPO-37 catalyst was demonstrated based on three-dimensional large pore openings of the framework. However, the presence of strong Lewis and Brönsted sites on MESO-SAPO-37 and mesoporous channels favoured superior conversion of 1-octene compared to other systems with comparable branched isomer selectivity.

The hydroisomerization activity over meso–microcomposites of SAPO-34 (Meso-SAPO-34) further improved when the uniformity of the mesoporous channels was improved by post-synthetic vapour phase treatment (VPT) to tether uncondensed inorganic ions [69]. The post-synthesis VPT-treated Meso-SAPO-34 (Meso-SAPO-34-VPT) using

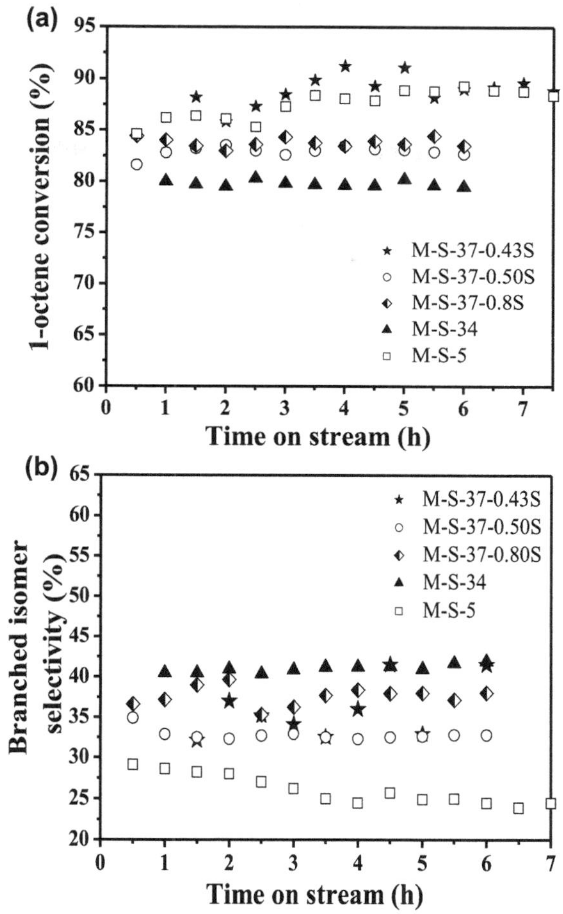

Figure 10

Role of different MESO-SAPO-*n* on hydroisomerization of octene at 400 °C with WHSV = 8 h^{-1}, (a) 1-octene conversion and (b) branched isomer selectivity, (Where M-S-34 = MESO-SAPO-34, M-S-5 = MESO-SAPO-5, M-S-37 = MESO-SAPO-37). *Reproduced from Ref. [68].*

Table 5: Catalytic Activity of Various Silicoaluminophosphate (SAPO) Materials on Hydroisomerization of 1-Octene.[a]

Sample Name	1-Octene Conversion (%)	Product Selectivity (%)				
		Linear Octenes			Branched Octenes	
		2-Octene	3-Octene	4-Octene	Methyl-heptenes	Dimethyl-hexenes
MESO-SAPO-37	88.5	12.3	28.8	24.6	23.5	10.6
MESO-SAPO-34	79.6	6.9	25.4	26.4	15.8	23.5
MESO-SAPO-5	87.4	20.3	16.1	40.1	20.9	2.6
SAPO-37	79.6	8.4	20.2	27.1	18.1	25.3
SAPO-34	71.7	4.5	25.2	8.5	2.8	28.1
SAPO-5	64.8	18.4	47.5	29.5	2.1	2.5

[a]Reaction condition: temperature = 400 °C, WHSV = 8 h^{-1} at TOS = 3 h.
Reproduced from Ref. [68].

tetraethylorthosilicate helps to improve the Brönsted acidity by the introduction of silicate species in the framework, and improving the uniformity of the mesoporous channels helps to enhance the 1-octene conversion level by approximately 95% while the desired branched isomer selectivity of 58% was achieved. It was explained that the uniform mesoporous channels and strong acidic sites derived from microporous building units facilitate the superior catalytic performance observed [69].

8. Summary

Numerous reports have shown the importance of developing mesoporous materials possessing microporous SSUs on the wall. These materials provide a new direction in molecular sieve-based sustainable catalysts for fine and petrochemical processes. The major advantages of such composite materials improve the structural stability, control the diffusion limitations of bulk molecules and enhance the surface active sites. The importance of such composites in aluminosilicate-based materials is well established and flexible framework aluminophosphate-based molecular sieves prove difficult to assemble owing to several inorganic anions that play an important role in assembly. The VPT method has recently been shown as an alternative route to improve composites' textural properties and stability. Overall, composite molecular sieve materials have been shown to be promising sustainable future materials for bulk molecule transformations.

Acknowledgements

The authors express their sincere thanks to UGC (No. 41-237/2012/(SR)) and University of Delhi R&D grant (DRCH/R&D/2013-14/4155) for financial support.

References

[1] Barton TJ, Bull LM, Klemperer WG, Loy DA, Mc Enaney B, Misono PA, et al. Tailored porous materials. Chem Mater 1999;11:2633–56.
[2] Selvam P, Sakthivel A. In: Kholdeeva Oxana A, Clerici M, editors. Liquid phase oxidation via heterogeneous catalysis is organic synthesis and industrial applications. Wiley Interscience; 2013, ISBN 978-0-470-91552-3. p. 95–125.
[3] Xu R, Pang W, Yu J, Huo Q, Chen J. Chemistry of zeolites and related porous materials: synthesis and structure. Wiley; 2008.
[4] Sing KSW, Everett DH, Haul RAW, Moscou L, Pierotti RA, Pierotti J, et al. Reporting physisorption data for gas/solid systems with special reference to the determination of surface area and porosity. Pure Appl Chem 1985;57:603–19.
[5] Szostak R. Molecular sieves: principles of synthesis and identification. Springer Science and Business Media; 1998.
[6] Vermeiren W, Gilson JP. Impact of zeolites on the petroleum and petrochemical industry. Top Catal 2009;52(9):1131–61.

[7] Akhmedov VM, Al-Khowaiter SH. Recent advances and future aspects in the selective isomerization of high n-Alkanes. Catal Rev 2007;49(1):33–139.
[8] Climent MJ, Corma A, Iborra S. Heterogeneous catalysts for the one-pot synthesis of chemicals and fine chemicals. Chem Rev 2010;111(2):1072–133.
[9] Yadav R, Sakthivel A. Silicoaluminophosphate molecular sieves as potential catalysts for hydroisomerization of alkanes and alkenes. Appl Catal A: Gen 2014;481:143–60.
[10] Dyer A. An introduction to zeolite molecular sieves. Avon, GB: John Wiley and Sons, Ltd; 1987.
[11] Payra P, Dutta PK. Zeolites: a primer. USA: Handbook of zeolite science and technology; 2003.
[12] Breck DW. Zeolite molecular sieves: structure. Chemistry and use. New York: Wiley; 1974.
[13] Artioli G, Pluth JJ, Smith JV. Synthetic phosphorus-substituted analcime, $Na_{13}Al_{24}Si_{13}P_{11}O_{96}\cdot 16H_2O$ with ordered Al and Si/P. Acta Cryst, Sect C 1984;40:214–7.
[14] Van Bekkum H, Flanigen EM, Jansen JC. Introduction to zeolite science and practice. Stud Surf Sci Catal 1991;58:xiii–xiv.
[15] Wilson ST, Lok BM, Messina CA, Cannan TR, Flanigen EM. Aluminophosphate molecular sieves: a new class of microporous crystalline inorganic solids. J Am Chem Soc 1982;104(4):1146–7.
[16] Wilson ST, Lok BM, Flanigen EM. U.S. Patent No. 4,310,440. Washington, DC: U.S. Patent and Trademark Office; 1982.
[17] Weitkamp J. Zeolites and catalysis. Solid State Ionics 2000;131(1):175–88.
[18] Kärger J, Ruthven DM. Diffusion in zeolites and other microporous solids. New York: Wiley; 1992 (p. 472–7).
[19] Beck JS, Vartuli JC, Roth WJ, Leonowicz ME, Kresge CT, Schmitt KD, et al. A new family of mesoporous molecular sieves prepared with liquid crystal templates. J Am Chem Soc 1992;114(27):10834–43.
[20] Misaelides P. Application of natural zeolites in environmental remediation: a short review. Microporous Mesoporous Mater 2011;144:15–8.
[21] Tanev PT, Pinnavaia TJ. A neutral templating route to mesoporous molecular sieves. Science 1995;267:865–7.
[22] Huo QS, Margolese DI, Ciesla U, Feng PY, Gier TE, Sieger P, et al. Generalized synthesis of periodic surfactant/inorganic composite materials. Nature 1994;368:317–21.
[23] Deng Y, Wei J, Sun Z, Zhao D. Large-pore ordered mesoporous materials template from non-Pluronic amphiphilic block copolymers. Chem Soc Rev 2013;42:4054–70.
[24] Kresge CT, Roth WJ. The discovery of mesoporous molecular sieves from the twenty year perspective. Chem Soc Rev 2013;42:3663–70.
[25] Sakthivel A, Badamali SK, Selvam P. *para*-Selective *t*-butylation of phenol over mesoporous H-AlMCM-41. Microporous Mesoporous Mater 2000;39:457–63.
[26] Sakthivel A, Selvam P. Vapour phase tertiary butylation of phenol over mesoporous gallosilicate molecular sieves. Catal Lett 2002;84:37–43.
[27] Badamali SK, Sakthivel A, Selvam P. Tertiary butylation of phenol over mesoporous H–FeMCM-41. Catal Lett 2000;65:153–7.
[28] Sakthivel A, Dapurkar SE, Selvam P. Allylic oxidation of cyclohexene over chromium containing mesoporous molecular sieves. Appl Catal A, Gen 2003;246:283–93.
[29] Sakthivel A, Selvam P. Mesoporous (Cr) MCM-41: a mild and efficient heterogeneous catalyst for selective oxidation of cyclohexane. J Catal 2002;211:134–43.
[30] Sakthivel A, Zhao J, Kuhn FE. Grafting of the η^5-CpMo(CO)$_3$ moiety on pure and surface modified SBA-15 molecular sieves. Microporous Mesoporous Mater 2005;27:341–8.
[31] Linares N, Silvestre-Albero AM, Serrano E, Silvestre-Albero J, García-Martínez J. Mesoporous materials for clean energy technologies. Chem Soc Rev 2014;43:7681–717.
[32] Li K, Valla J, Garcia-Martinez J. Realizing the commercial potential of hierarchical zeolites: new opportunities in catalytic cracking. ChemCatChem 2014;6:46–66.
[33] On DT, Kaliaguine S. Ultrastable and highly acidic, zeolite-coated mesoporous aluminosilicates. Angew Chem Int Ed 2002;41:1036–40.

[34] Karlsson A, Stöcker M, Schmidt R. Composites of micro and mesoporous materials: simultaneous syntheses of MFI/MCM-41 like phases by a mixed template approach. Microporous Mesoporous Mater 1999;27:181–92.
[35] de Moor PEA, Beelen TPM, van Santen RA, Tsuji T, Davis ME. SAXS and USAXS investigation on nanometer-scaled precursors in organic-mediated zeolite crystallization from gelating systems. Chem Mater 1999;11:36–43.
[36] Huang L, Guo W, Deng P, Xue Z, Li Q. Investigation of synthesizing MCM-41/ZSM-5 composites. J Phys Chem B 2000;104:2817–23.
[37] Meng X, Nawaz F, Xiao F-S. Templating route for synthesizing mesoporous zeolites with improved catalytic properties. Nano Today 2009;4:292–301.
[38] Kirschhock CEA, Buschmann V, Kremer S, Ravishankar R, Houssin CJY, Mojet BL, et al. Zeosil nanoslabs: building blocks in $n\mathrm{Pr_4N^+}$-mediated synthesis of MFI zeolite. Angew Chem Int Ed 2001;40:2637–9.
[39] Zhang Z, Han Y, Zhu L, Wang R, Yu Y, Qiu S, et al. Strongly acidic and high-temperature hydrothermally stable mesoporous aluminosilicates with ordered hexagonal structure. Angew Chem Int Ed 2001;40:1258–62.
[40] Liu Y, Pinnavaia TJ. Aluminosilicate mesostructures with improved acidity and hydrothermal stability. J Mater Chem 2002;12(11):3179–90.
[41] Prokešová P, Mintova S, Čejka J, Bein T. Preparation of nanosized micro/mesoporous composites via simultaneous synthesis of Beta/MCM-48 phases. Microporous Mesoporous Mater 2003;64:165–74.
[42] Xia Y, Mokaya R. On the synthesis and characterization of ZSM-5/MCM-48 aluminosilicate composite materials. J Mater Chem 2004;14:863–70.
[43] Egeblad K, Christensen CH, Kustova M, Christensen CH. Templating mesoporous zeolites. Chem Mater 2008;20:946–60.
[44] Pérez-Ramírez J, Christensen CH, Egeblad K, Christensen CH, Groen JC. Hierarchical zeolites: enhanced utilization of microporous crystals in catalysis by advances in materials design. Chem Soc Rev 2008;37:2530–42.
[45] Liu Y, Pinnavaia TJ. Assembly of wormhole aluminosilicate mesostructures from zeolite seeds. J Mater Chem 2004;14:1099–103.
[46] Sakthivel A, Huang S-J, Chen W-H, Lan Z-H, Chen K-H, Kim T-W, et al. Replication of mesoporous aluminosilicates molecular sieves (RMMs) with zeolite framework from mesoporous carbons (CMKs). Chem Mater 2004;16:3168–75.
[47] Sakthivel A, Huang S-J, Chen W-H, Lan Z-H, Chen K-H, Lin H-P, et al. Direct synthesis of highly stable mesoporous molecular sieves containing zeolite building units. Adv Funct Mater 2005;15:253–8.
[48] Sakthivel A, Komura K, Sugi Y. Fe(III) containing MCM-48 type mesoporous molecular sieves with framework zeolite secondary building units. J Nanosci Nanotechnol 2008;8:400–4.
[49] Laha SC, Sakthivel A, Komura K, Huang SJ, Wu PH, Liu SB, et al. Highly stable aluminosilicates with a dual pore system: simultaneous formation of mesophase with zeolitic building units. Chem Lett 2009;38:548–9.
[50] Ivanova I, Knyazeva EE. Micro-mesoporous materials obtained by zeolite recrystallization: synthesis, characterization and catalytic applications. Chem Soc Rev 2013;42:3671–88.
[51] Koletstra KR, Zandbergen HW, Jansen JC, van Bekkum H. Overgrowth of mesoporous MCM-41 on faujasite. Microporous Mater 1996;6:287–93.
[52] Sakthivel A, Chen W-H, Liu S-H, Huang S-J, Lo A-Y, Hsu Y-H, et al. Acidity and catalytic behaviors of ordered mesoporous aluminosilicate materials containing zeolite building units. Catal Lett 2006;108:173–8.
[53] Qian XF, Li B, Hu YY, Niu GX, Zhang DYH, Che RC, et al. Exploring meso-/microporous composite molecular sieves with core-shell structures. Chem Eur J 2012;18:931–9.
[54] Ivanova II, Kuznetsov AS, Yuschenko VV, Knyazeva EE. Design of composite micro/mesoporous molecular sieve catalysts. Pure Appl Chem 2004;76:1647–58.

[55] Sakthivel A, Nakamura R, Komura K, Sugi Y. Esterification of glycerol by lauric acid over aluminium and zirconium containing mesoporous molecular sieves in supercritical carbon dioxide medium. J Supercrit Fluids 2007;42:219–25.
[56] Sakthivel A, Komura K, Huang S-J, Wu P-H, Liu S-B, Sasaki Y, et al. Calcium-incorporated mesoporous aluminosilicates: synthesis, characterization, and applications to the condensation of long-chain fatty acid with long-chain amine and alcohol. Ind Eng Chem Res 2010;49:65–71.
[57] Ganjala VSP, Neeli CKP, Pramod CV, Khagga M, Rao KSR, Burri DR. Eco-friendly nitration of benzenes over zeolite-β-SBA-15 composite catalyst. Catal Commun 2014;49:82–6.
[58] Laha SC, Venkatesan C, Sakthivel A, Komura K, Kim TH, Cho SJ, et al. Highly stable aluminosilicates with a dual pore system: simultaneous formation of meso- and microporosities with zeolitic BEA building units. Microporous Mesoporous Mater 2010;133:82–90.
[59] Li T, Duan A, Zhao Z, Liu B, Jiang G, Liu J, et al. Synthesis of ordered hierarchically porous L-SBA-15 material and its hydro-upgrading performance for FCC gasoline. Fuel 2014;117:974–80.
[60] Mohanty S, Mouli KC, Soni K, Adjaye J, Dalai AK. Catalytic hydrotreatment using NiMo/MAS catalysts synthesized from ZSM-5 nano-clusters. Appl Catal A: Gen 2012;419–420:1–12.
[61] Serrano DP, García RA, Vicente G, Linares M, Procházková D, Čejka J. Acidic and catalytic properties of hierarchical zeolites and hybrid ordered mesoporous materials assembled from MFI protozeolitic units. J Catal 2011;279:366–80.
[62] Poladi RHPR, Landry CC. Oxidation of octane and cyclohexane using a new porous substrate, Ti-MMM-1. Microporous Mesoporous Mater 2002;52:11–8.
[63] Reichinger M, Schmidt W, van den Berg MWE, Aerts A, Martens JA, Kirshhock CEA, et al. Alkene epoxidation with mesoporous materials assembled from TS-1seeds – is there a hierarchical pore system? J Catal 2010;269:367–75.
[64] Jin C, Li G, Wang X, Wang Y, Zhao L, Sun D. A titanium containing micro/mesoporous composite and its catalytic performance in oxidative desulfurization. Microporous Mesoporous Mater 2008;111:236–42.
[65] Wilson ST, Lok BM, Messina CA, Cannan TR, Flanigen EM. Aluminophosphate molecular sieves: a new class of microporous crystalline inorganic solids. J Am Chem Soc 1982;104:1146–51.
[66] Martens JA, Jacobs PA. Crystalline microporous phosphates: a family of versatile catalysts and adsorbents. Stud Surf Sci Catal 1994;85:653–85.
[67] Singh AK, Yadav R, Sakthivel A. Synthesis, characterization, and catalytic application of mesoporous SAPO-34 (MESO-SAPO-34) molecular sieves. Microporous Mesoporous Mater 2013;181:166–74.
[68] Yadav R, Singh AK, Sakthivel A. Mesoporous silico-aluminophosphates derived from microporous precursors as promising catalyst for hydroisomerization. Catal Today 2015;245:155–62.
[69] Singh AK, Kondamudi K, Yadav R, Upadhyayula S, Sakthivel A. Uniform mesoporous silicoaluminophosphate derived by vapor phase treatment: Its catalytic and kinetic studies in hydroisomerization of 1-octene. J Phys Chem C 2014;118:27961–72.

CHAPTER 2

Functionalized Mesoporous Materials as Sustainable Catalyst in Liquid Phase Catalytic Transformations

Nabanita Pal[1,2], **Asim Bhaumik**[1]

[1]Department of Materials Science, Indian Association for the Cultivation of Science, Kolkata, West Bengal, India; [2]Surface Physics and Materials Science Division, Saha Institute of Nuclear Physics, Kolkata, West Bengal, India

1. Introduction

The panoramic journey towards exploration of mesoporous materials starts with the Mobil scientists in the United States who found out a high surface area ordered inorganic porous silica material having a pore dimension of >2 nm, named MCM-41 (Mobil Composition of Matter-41). This was synthesized based on a self-assembled supramolecular templating mechanism [1]. Before this invention, the largest pore dimensions reported for a templated porous material was mostly based on microporous zeolites or silicalites with a pore diameter well below the 2.0-nm regime and these materials are conventionally synthesized hydrothermally by using a single molecule template [2,3]. In a limited time, mesoporous materials have attracted widespread attention and emerged as the most popular category of nanoporous solids particularly to scientists belonging to materials chemistry. Having an exceptionally high surface area, well-defined tuneable pore diameter and a good possibility of surface functionalization, mesoporous materials open various new windows while experimenting on their potential applications. Liquid phase catalytic reactions are one of the most challenging applications of functionalized mesoporous solids.

The perception of catalysis formulated by Wilhelm Ostwald was that 'there is probably no chemical reaction which cannot be influenced catalytically' [4]. The economic development as well as advancement of a country largely depends on the industrial catalytic processes and >20% of the world's gross national products relies on these catalytic technologies [5]. In the present world, about 60% of the industrially important chemical products are produced in different chemical processes among which 90% are based on catalytic reactions [6]. The term 'catalysis' was coined by the Swedish chemist

Sustainable Catalytic Processes. http://dx.doi.org/10.1016/B978-0-444-59567-6.00002-9

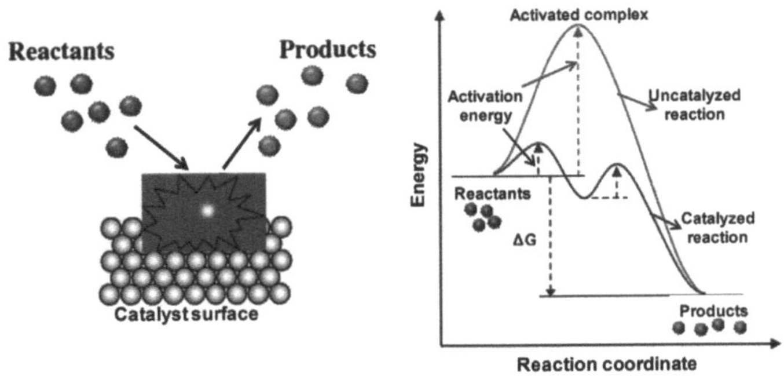

Figure 1
A simple pictorial scheme for catalytic reaction and the corresponding energy diagram.

Jöns Jacob Berzelius in 1835, and it refers to the change in the rate of a chemical reaction mediated by a substance named 'catalyst' [4]. Actually, a catalyst fulfils the criteria (1) of offering an alternative pathway of a lower activation energy than the respective un-catalysed reaction, resulting in a faster reaction rate but (2) does not hamper the overall thermodynamics of the reaction (Figure 1). By facilitating a chemical reaction, a catalyst can be said to be a *chemical marriage broker*. Later, a well-accepted mechanism and kinetics of the catalytic reaction was proposed by Cyril Norman Hinshelwood in 1927 based on this proposition. Catalysis is of two types: (1) homogeneous where the catalyst remains in the same phase as that of the reaction media (all are generally present in the liquid phase) and (2) heterogeneous, where the catalyst is generally solid and remains as a distinct phase from the liquid or gaseous reaction media. The major problem of homogeneous catalysis is the difficulty of separation and recovery, which makes it unsuitable for application in industrial purpose. The deficiency has been overcome by replacing homogeneous with heterogeneous solids [6]. The birth of industrial catalysis took place in Europe with the production of sulphuric acid in the 'contact' process discovered by Knietsch in 1898 [7]. Industrial processes which depend extremely on homogeneous catalysts are modified by 'heterogenization' of those catalysts, which possess all the properties of their homogeneous counterparts but have reusability like heterogeneous [5]. Organic functionalization or immobilization of soluble numerous metal complexes over solid inorganic mesoporous silica surface can build suitable organic–inorganic hybrid functionalized silicas, which can act as true heterogeneous catalysts in various important chemical reactions [8].

Catalytic reactions consist of steps like diffusion and adsorption of the reactants on the catalyst surface, successful reaction on the active site located at the surface and then desorption and finally diffusion of the products to the bulk reaction media. Since the catalyst surface plays a crucial role in the process, mesoporous materials having a high surface area compared to the nonporous material can thus contribute significantly in catalysis. Moreover, functionalization of mesoporous silicas with different organic moieties largely fulfils the requirement of generating a wide range of active sites for different organic catalytic transformations. These reactions result in a high turnover frequency, TOF (TOF = the number of moles of the substrate converted per mole of the active site of the catalyst in unit time). Using examples from recent literature, this review illustrates and refines the synthetic outlines and applications of those sustainable heterogeneous hybrid catalysts in many traditional as well as new organic transformations [9].

2. Synthesis and Types of Functionalized Mesoporous Materials

To introduce organic functionalities or metal complexes at the surface of an inorganic porous support, mesoporous silica is used extensively owing to its high surface area and versatility to condense with other active organic groups [8]. Recently, purely organic porous polymers containing various functional groups have been designed, which can accommodate different metal complexes and result in the formation of organic–inorganic hybrid materials [10]. There are also some reports on hybrid metallophosphate-based materials and metal-organic frameworks, which are synthesized generally through a non-templating pathway. They contain small pores within their network and are also very effective in catalytic transformations [11]. This review mainly highlights mesoporous hybrid materials, and thus, we will give special emphasis to silica- and polymer-based organic–inorganic hybrid catalysts, which have shown good potentials in sustainable organic transformations.

In this context, it is pertinent to mention that the concept of the introduction of bridging organic groups into the pore walls of mesoporous silica was first invented by Inagaki et al. who reported successful immobilization of the ethane/benzene ring within the mesoporous silica pore wall [12]. General chemical pathways and the mechanism to design such a type of organic–inorganic hybrid materials have been illustrated and reviewed in a huge number of journal articles [13,14]. Syntheses of these hybrid solids to form a chemical bond between an organic molecule and the pore wall of the inorganic support are usually carried out through several methods (Figure 2): (1) 'post-grafting' technique where the inorganic silica pore wall is 'grafted' by subsequent modification with other organic groups, (2) 'co-condensation' route leading to simultaneous condensation of inorganic silica and organosilica precursors to form hybrid silica, (3) sol–gel route for the formation of 'periodic mesoporous organosilica' (PMO), that is, bridging organic units are directly incorporated in the three-dimensional (3D) network structure of the silica matrix through two covalent bonds [14].

In the case of 'post-grafting' or post-functionalization, the –Si-OH groups of mesoporous pure silica surface are condensed with another silica precursor containing various organic groups via covalent bond formation and organo-grafted silica is formed. The organic group

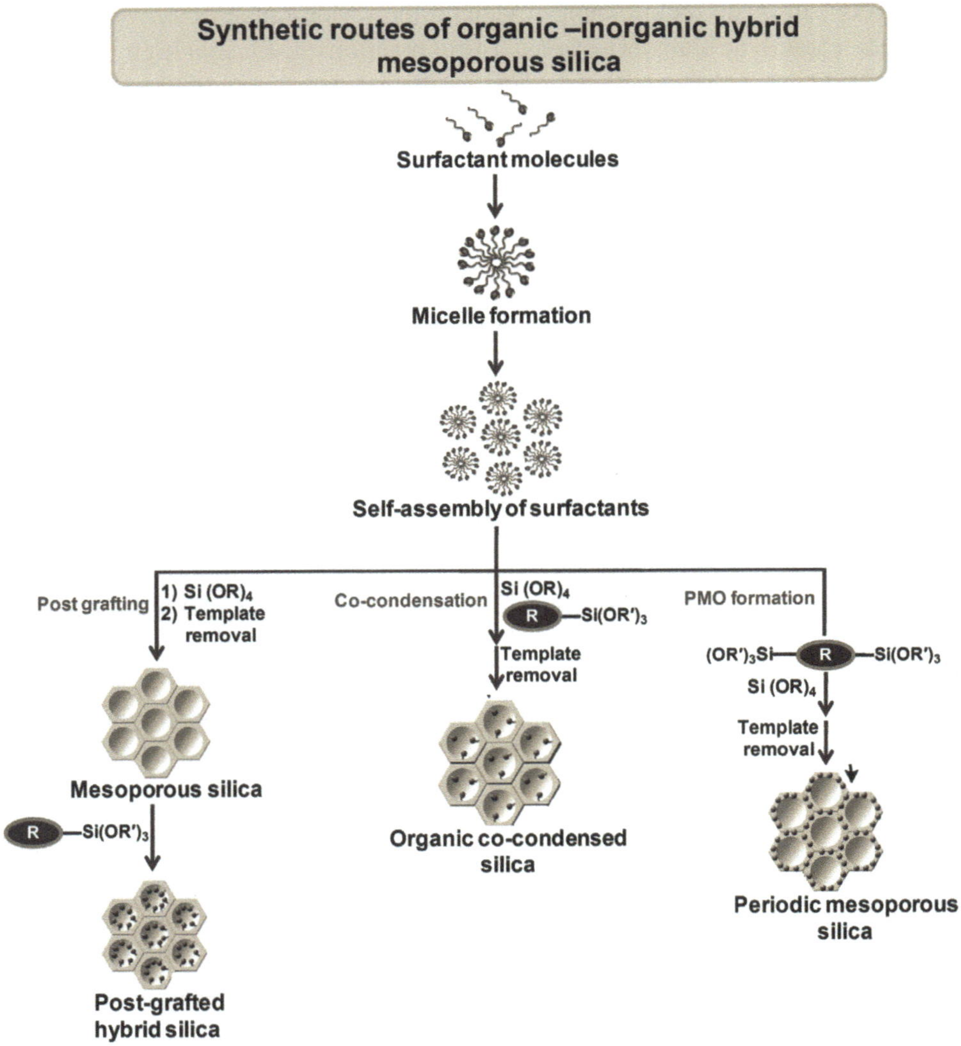

Figure 2

Schematic representation of the synthetic route of organic–inorganic hybrid mesoporous silica.

attached to silica can be further functionalized depending upon the application purpose (Figure 3(a)) [15]. The 'Co-condensation' route is a one-pot method for the formation of organic–inorganic hybrid silica and is preferable to 'post-grafting' if uniform surface modification with organic groups is required. However, the latter provides a hydrolytically more stable and well-defined structure than in the one-pot condensation approach. Here, simultaneous covalent bond formation occurs between inorganic silica and organosilica precursors in the presence or absence of a template to form hybrid silica with mesopores (Figure 3(b)) [16]. PMOs are another class of materials, which involves the incorporation

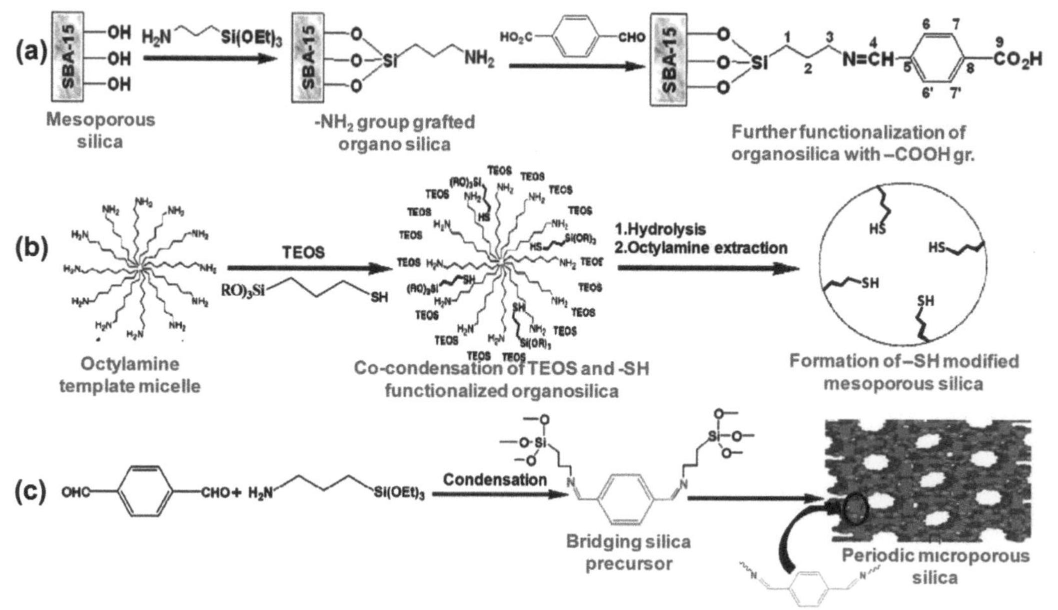

Figure 3

Schematic representation of the formation of organic–inorganic hybrid silica (a) via the 'post-grafting' method [14], (b) via the 'co-condensation' route [15] and (c) periodic microporous silica formation.

of a bridging organic group and its homogeneous distribution into the 3D network structure of pure silica generally via a self-condensation pathway. The process may be templated or non-templated, though the non-templated route results in microporosity in the newly formed hybrid silica (Figure 3(c)) [17]. High affinity of different transition metals and metal oxide nanoparticles or metal complexes towards these organically functionalized silicas is able to generate metal-grafted organic–inorganic hybrid silicas, which have great potential and flexibility for industrial catalytic processes [18]. This strategy of grafting a metal complex on the surface of hybrid silica is a process of heterogenization of homogeneous catalysts and this causes a huge enhancement in the sustainability of solid catalysts [19,20].

Mesoporous organic–inorganic hybrid polymers are less explored than hybrid silica materials due to some disadvantages to generate open pores in solution as the formation energy of the carbon–carbon bond is very high [21]. Recently, the surfactant templating pathway has been applied by Modak et al. to synthesize a purely organic highly ordered 2D-hexagonal mesoporous framework of triazine functionalized polymer, which effectively binds Pd metal at the pore surface to act as a good heterogeneous catalyst in green organic syntheses (Figure 4) [10]. Similarly, another interesting material is 3D

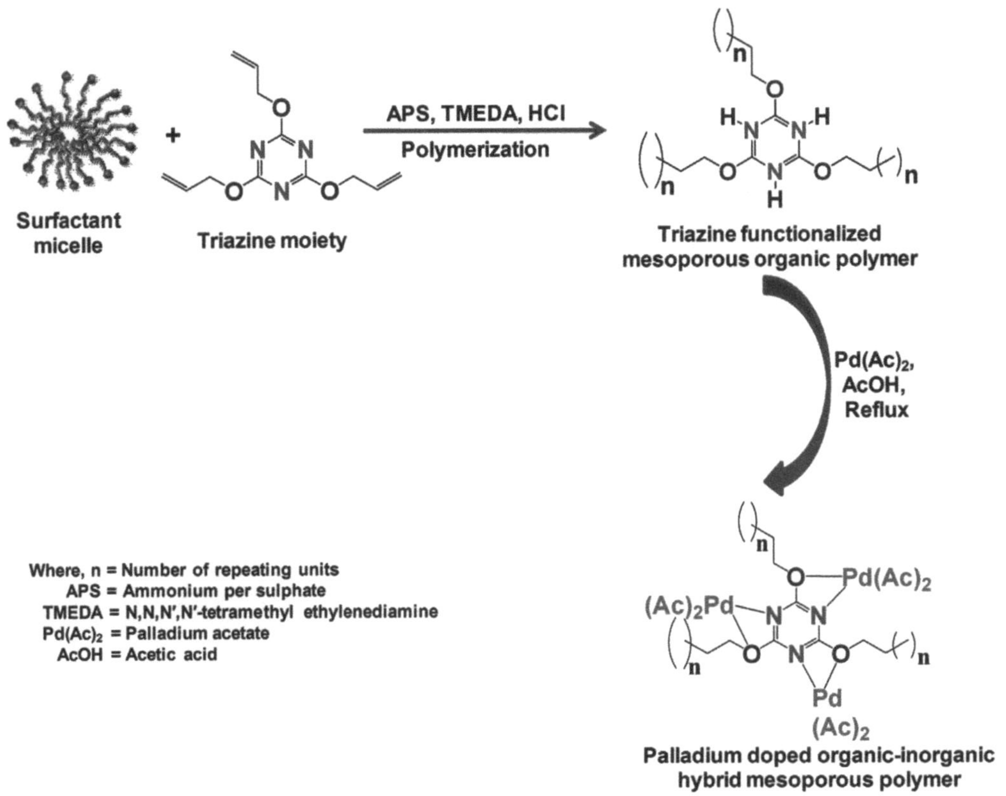

Figure 4
Schematic illustration of the synthesis of organic–inorganic hybrid mesoporous polymer [10].

porous polymer networks containing various transition metal loaded porphyrin rings which can be designed experimentally and show a huge potential as sustainable catalysts. Grafting of active organic groups at the surface of silica-based mesoporous materials can be carried out under different synthetic conditions like conventional sol–gel process [23], hydrothermal condition [17] or applying a particular refluxing temperature in the presence of organic solvents [10,22].

3. *Different Organic Transformations over Functionalized Mesoporous Catalysts*

Pure mesoporous silica devoid of any active functional group is not effective in catalytic reactions. Proper loading of organic functionalities and metal complexes on porous silica surface may lead to simple to excellent catalytic activity observed under suitable reaction conditions. The reason behind the utilization of this catalytically inactive silica matrix as a good support for organic incorporation is the appreciably high Brunauer–Emmett–Teller

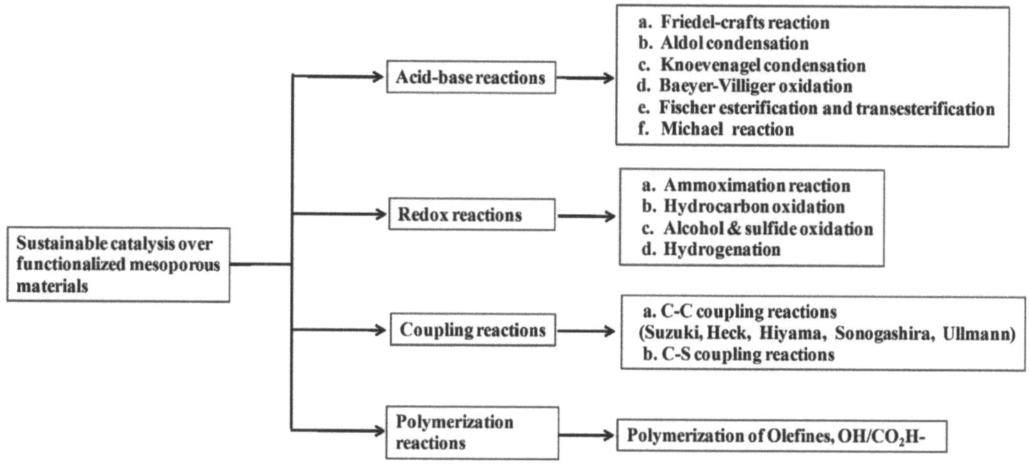

Scheme 1

Schematic representation of the catalytic reactions over functionalized mesoporous materials discussed herein.

(BET) surface area (c. 200–1400 m^2/g) and tuneable pore diameter (2–50 nm). A wide pore width helps in easy accommodation and transfer of mass. Also, surface area provides an extended space for catalytic reaction and higher accumulation of active site per unit mass of the material [24]. Herein, an elaborated idea about the role and function of the organic active site of hybrid silica in numerous heterogeneous organic transformations with proper examples is presented systematically. Functionalized mesoporous catalyst-mediated sustainable liquid phase reactions are divided into different categories (Scheme 1). These are acid–base reactions, redox reactions, coupling reactions, polymerization reactions, etc.

3.1 Acid–Base Catalysed Reactions on Functionalized Mesoporous Materials

Acidic or basic property in a solid catalyst is generated due to the presence of a Lewis or Brønsted acid or base sites in the material. Lewis acidic property is mainly generated through the metal site of the metal complex linked with the mesoporous support through a coordination bond and Brønsted acidity is observed in $-COOH$, $-SO_3H$, $-NR_3^+$, etc. groups attached to the surface of these materials [15]. Similarly, $-NH_2$, $-NMe_2$, etc. and $-COO^-$ are responsible for Lewis and Brønsted basicity, respectively. The catalytic activity of these groups and the selectivity for any organic reactions depend on the number and strength of those sites along with the morphology, that is, the surface area and pore width of the mesoporous support [5]. The following reactions are few examples of functionalized mesoporous silica-mediated acid- or base-catalysed sustainable organic transformations.

3.1.1 Friedel–Crafts Reaction

Alkylation and acylation of aromatic rings through Friedel–Crafts (FC) reaction is one of the most important chemical processes for laboratory synthesis and industrial production purposes [25]. These liquid phase reactions are habitually catalysed by Lewis acidic sites present in the catalyst. The use of liquid acids like H_2SO_4 or solid acids like $AlCl_3$, $FeCl_3$ and BF_3 in the homogeneous phase suffers from a lot of environmental hazards and inconvenience due to metal leaching, non-recyclability, etc. Thus, the replacement of these homogeneous catalysts by a green, completely nontoxic, recyclable heterogeneous catalyst is a significant approach towards organic synthesis [26]. Many organically functionalized silicas have been reported in recent times; these are applied for successful FC alkylation (or benzylation) or acylation (or benzoylation) (Figure 5). Alkylation reactions using alcohol instead of alkyl chloride require the involvement of Brønsted acidic sites present in the catalyst.

In 2005, Alvaro et al. synthesized perfluoroalkylsulphonic acid (analogue of 'Nafion') modified MCM-41 – an organic–inorganic hybrid mesoporous silica by a single step post-functionalization of 1,2,2-trifluoro-2-hydroxy-1-trifluoromethyl-ethane sulphonic acid beta-sultone on the silica pore wall in toluene refluxing conditions. This Nafion-silica organocatalyst has shown good efficiency in acylation of anisole using acetic anhydride as the acylating agent [27]. Similarly, chloroferrate ionic liquid-grafted silica synthesized via post-grafting of 1-trimethoxysilylpropyl-3-methylimidazolium chloride–$FeCl_3$ onto siliceous MCM-41 has been reported as an efficient and recyclable heterogeneous support for FC benzylation of substituted benzene to form derivatives of diphenylmethane at a moderate temperature. Although this catalyst contains $FeCl_3$ on its active site, it shows

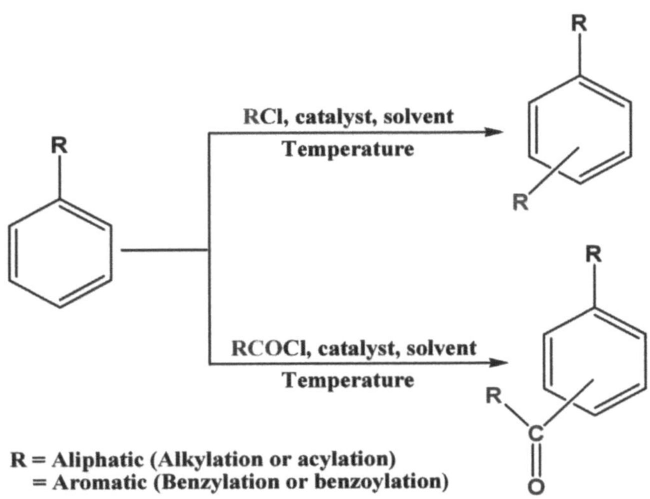

Figure 5
General scheme for the FC alkylation and acylation reactions.

minimum leaching of the metal after 10 times of recycling [28]. Zhao et al. reported another type of ionic liquid (IL)-modified organic–inorganic hybrid silica for preparation of diphenylmethane via FC benzylation of benzene with a 100% selectivity. Their catalyst has been synthesized by the co-condensation of N-(3-triethoxysilylpropyl), N(3)-(3-trimethoxysilylpropyl-4,5-dihydroimidazolium chloride with tetraethyl orthosilicate in the presence of P123 surfactant. The resulting mesoporous IL-PMO obtained after template removal was incorporated with $InCl_3$ salt under refluxing conditions in toluene to form In-IL-PMO; this has higher recycling activity towards the FC reaction than the corresponding In-doped SBA-15 [29]. Pure organosilicas such as sulphonic acid-functionalized phenylene-bridged cubic or 3D-hexagonal mesoporous organosilica deficient in any metal is also proved to be very efficient in acylation reactions under mild conditions with a >80% yield [30]. Thus, FC reactions using organically functionalized silica has got an imperative role in the fine chemical industry.

3.1.2 Aldol Condensation Reactions

This is a well-known organic reaction of the enol or enolate ion with a carbonyl compound to form β-hydroxyketone, which on dehydration produces an α,β-unsaturated carbonyl compound (Figure 6(b)). This important C–C bond formation reaction involved

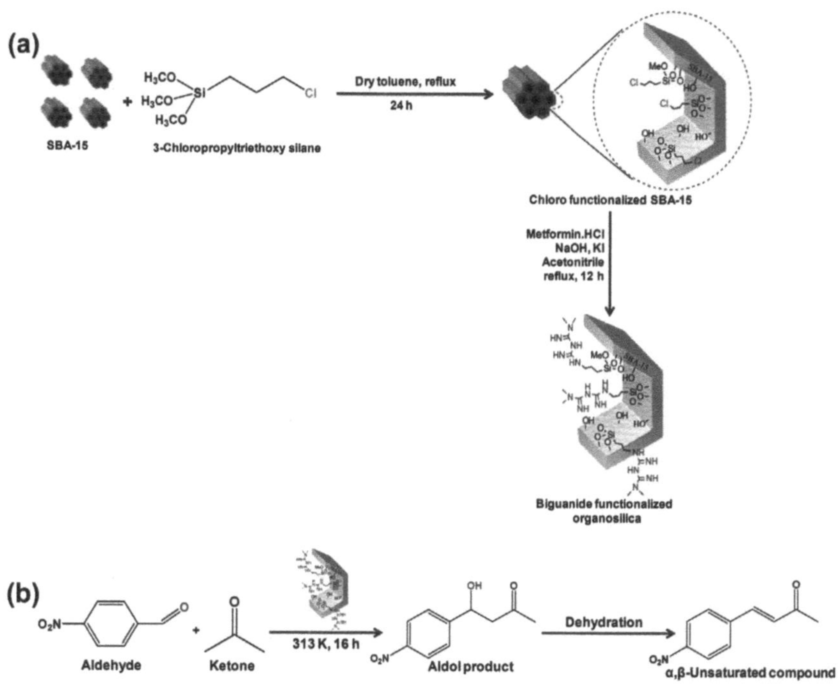

Figure 6

Schematic model (a) for the synthesis of biguanide functionalized organosilica [37] and (b) aldol condensation reaction of 4-nitrobenzaldehyde and acetone over this catalyst.

in various large biomolecules and industrial chemical synthesis follows both acid- and base-catalysed pathways to form the enol and enolate ions, respectively, in liquid media. Homogeneous bases like NaOH, $Ca(OH)_2$, $Ba(OH)_2$; Brønsted acids such as acetic acid; Lewis acid like $M(OTf)_n$ (M = Zn, Cu, Pb, Sc; OTf – trifluoro methane sulphonate) normally used for this reaction produce hazardous, toxic by-products. Thus, heterogeneous catalysts bearing Lewis acid/bases or Brønsted acid/bases centres are very much desirable for such reactions [31].

Different mesoporous metal-grafted silica, aluminophosphates, metal oxides, mixed oxides, etc. have been used as heterogeneous catalysts for this purpose, though totally organically modified mesoporous inorganic silica are the most convenient, environment friendly, green catalyst for this condensation process [31,32]. Recently, Shanmuganathan et al. have published an article on silica-immobilized piperazine-catalysed aldol condensation of few aldehydes with acetone at room temperature [33]. The aldol product was obtained exclusively over this Lewis basic group-modified recyclable organocatalyst, though the Mannich product is produced as a by-product when pure homogeneous base pyrrolidine was used. This type of efficiency and higher activity of heterogeneous organocatalyst over the homogeneous one to catalyse asymmetric and non-asymmetric aldol reaction has also been observed in various Lewis acidic or basic or bifunctional mesoporous organocatalysts [32,34,35]. By heterogenization, homogeneous catalysts like Sc(III) complex anchored on nanostructured polymer acts as Lewis acid for Mukaiyama aldol reaction in water at ambient temperature [36]. Mesoporous SBA-15 post-grafted or pure silica co-condensed with biguanide base has been found to be an excellent organocatalyst for such reactions with good catalyst recyclability (Figure 6) [37]. Nitroaldol condensation (Henry reaction) of nitroalkane with aldehyde or ketone to produce nitroalkene is also feasible over Lewis basic functional groups such as propylamine-supported MCM-41 as a highly efficient, recyclable organocatalyst [38].

3.1.3 Knoevenagel Condensation Reaction

This is a modified version of aldol reaction, which involves the condensation of the carbonyl carbon with a compound containing an active methylene group followed by dehydration to form a C–C bond resulting in an α,β-unsaturated compound. This reaction occurs generally in basic medium and it claims to be one of the most important environmentally benign C–C bond formation reactions since water is the only by-product of this condensation (Figure 7) [6]. Though soluble bases like primary, secondary or tertiary amines are habitually used for getting highest yield in this transformation, but primary amine loaded mesoporous organosilica has been found to be a suitable reusable, stable, non-air-sensitive heterogeneous support for this reaction [23]. The reaction proceeds via an enol intermediate formation from an active methyl compound after one of

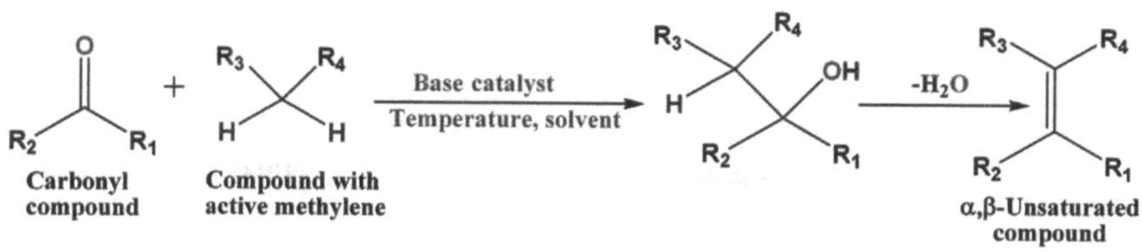

R_1, R_2 = H, alkyl, phenyl, sustituted phenyl etc.
R_3, R_4 = CHO, COOH, $COCH_3$, CO_2R, CN, NO_2, SO_2R, SOR, COOR etc.

Figure 7
Schematic representation for the Knoevenagel reaction.

its hydrogen capture by a base. The enol attack on the carbonyl carbon results in an aldol type of product, and base-induced elimination of one water molecule gives the final product. In 2009, Trilla et al. have also reported the application of imidazolium and dihydroimidazolium salt-grafted organic–inorganic hybrid silica for the Knoevenagel reaction with significant reusability [39].

3.1.4 Baeyer–Villiger Oxidation

This oxidation is an important reaction for industrial and bioactive chemical production, involving an oxidative cleavage of the C–C bond adjacent to the carbonyl carbon to form an ester from acyclic ketones and lactones from cyclic ketones usually in the presence of an acid or sometimes a base as catalyst [40]. Previously, the reaction was carried out using different Brönsted acids like $C_6H_5CO_3H$ and CF_3CO_3H homogeneously and even in a heterogeneous medium also by grafting these organic peroxyacids onto hexagonal mesoporous silica as reported by Lambert et al. [41]. Later, these hazardous oxidizing agents are replaced by less hazardous and low cost hydrogen peroxide along with Lewis acid. Recently, reactions are being carried out by using molecular oxygen over Lewis acidic catalyst. Metalloporphyrin-based periodic mesoporous silicas have effectively converted cyclohexanone to ε-caprolactone in the presence of aerial oxygen [42]. Various metals such as iron-, copper-, magnesium-, tin-doped silica and zeolites are also very efficient in this reaction (Figure 8) [43]. Either homogeneous or heterogeneous, the mechanism in Baeyer–Villiger oxidation proceeds via a Criegee intermediate, which is followed by intramolecular rearrangement to give the final product as explained beautifully by Paul et al. [44]. The reaction is sometimes facilitated with a higher yield in the presence of benzonitrile as a cosolvent.

3.1.5 Fischer Esterification and Transesterification Reactions

Biodiesel, which comprises monoalkyl esters of long chain fatty acids, is widely used as an alternative and renewable source of fuel in most of the developed countries. Esterification and transesterification are common chemical processes for biofuel

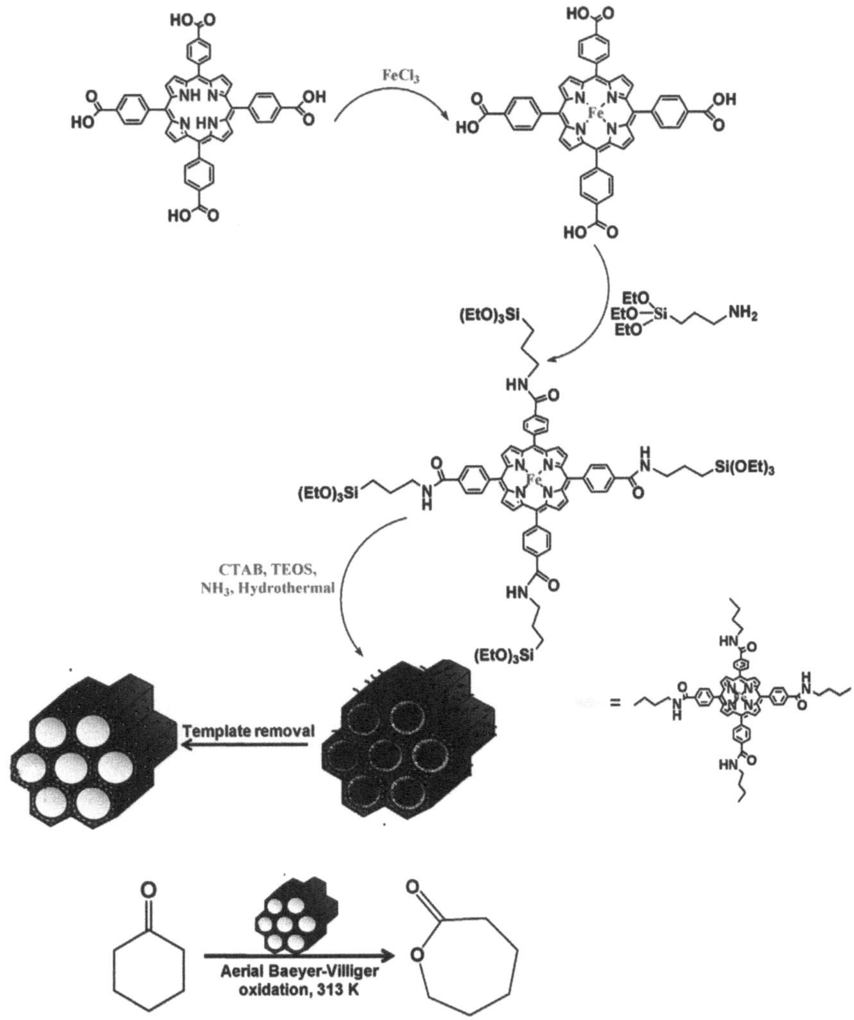

Figure 8

Schematic diagram for the synthesis of Fe porphyrin-based PMO and its activity towards the Baeyer–Villiger oxidation reaction.

production (Figure 9). The most simple and important pathway to form esters was developed by Emil Fischer (1852–1919); it involves the condensation of a carboxylic acid with an alcohol in the presence of any acid catalyst [45]. Transesterification is the process of exchanging an organic group of an ester with the 'R' group of an alcohol, and the reaction can be acid- or base-catalysed. Biodiesel production by this method involves conversion of all types of vegetable or animal fats and oils composed of triglyceride esters containing three free fatty acids being reacted with short-chain alcohols (typically methanol or ethanol) to form a monoalkyl ester [46].

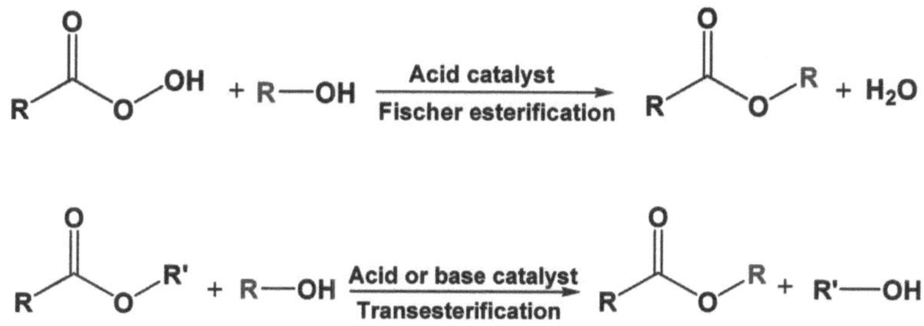

Figure 9
Fischer esterification and transesterification reactions over an acid and base catalyst.

Fischer esterification can be catalysed by both Lewis and Brönsted acids. The equilibrium reaction may be shifted towards the ester conversion by removing water from the reaction medium through azeotropic distillation or using molecular sieves or taking an excess of alcohol (Figure 9). Previously, strong Brönsted acids like H_2SO_4 and *p*-TsOH were used for this purpose. But recently, various heterogeneous, environmentally nonhazardous, nontoxic recyclable catalysts are extensively used for this biodiesel production. Karimi et al. reported an easy solvent-free esterification using Brönsted acid group $-SO_3H$ functionalized SBA-15 [47]. We found that organic–inorganic hybrid porous sulphonated Zn-phosphonate is also a very efficient and recyclable catalyst for the biodiesel synthesis from long chain fatty acids at room temperature [11].

Biodiesel production is performed mainly by the transesterification method that involves the mediation of acid or base catalysts. Acid catalyst promotes the reaction increasing the electrophilicity of the carbonyl carbon of carboxylic acid group while the base catalyst enhances the reaction rate by improving the nucleophilicity of the attacking alcohol [46,48]. The $-SO_3H$ group-functionalized periodic mesoporous silica has been used successfully for biodiesel production from different vegetable oils, whereas silica-supported guanidine is a true base catalyst for continuous flow biodiesel production [49]. The $-NH_2$-functionalized ordered mesoporous silica has also been utilized as a good reusable catalyst for transesterification of triglycerides in the presence of excess methanol [50].

Besides these, there are many reactions like the Michael reaction which involves a base-catalysed C–C bond formation between an α,β-unsaturated carbonyl compound and an active methylenic compound [51], base-induced ring opening of epoxy alcohol by piperidine-immobilized MCM-41 [52], acid-catalysed condensation reactions [53], etc. These are some well-known examples of acid–base catalysed reactions mediated by organic–inorganic hybrid mesoporous materials. There are some other organic reactions like hydroformylation of olefins, aldehydes, Biginelli condensation to form 3,4-dihydropyrimidin-2(1H)-ones, which are also found to be catalysed by some transition metals like Pd-, Fe-supported functionalized silica materials under mild conditions [18].

3.2 Redox Reactions Catalysed by Functionalized Mesoporous Materials

There are several reports in the literature on redox reactions catalysed by framework-modified mesoporous materials. In the past decades, Ti-, V-, Mn-doped mesoporous silica and Ni-, Cu-based other mesoporous solids are broadly utilized for oxidation as well as reduction reactions [54,55]. In recent times, these redox metal-loaded organic complex-incorporated hybrid mesoporous solids are widely used for such redox reactions. Metal complexes can be bound to the mesoporous support via coordination through the metal centre, bonding through one of the ligands or by encapsulation into the framework. Replacing homogeneous hazardous oxidants like chromate, permanganate or periodide, these metal-grafted organically functionalized materials are attracting great attention as nontoxic, environment friendly and mild sustainable catalysts. Some examples of redox reactions mediated by these solids are as follows:

3.2.1 Ammoximation Reactions

Liquid phase oxidation of ketones with ammonia and oxidants like H_2O_2 to synthesize oximes is a very crucial reaction in modern organic chemistry because these oximes act as intermediates for the synthesis of a series of fine chemicals like lactam and amides via different chemical reactions [56]. Transition metals like Ti- and V-doped mesoporous silicas are the most effective catalysts used for this purpose (Figure 10). We first reported the ammoximation of various cycloalkanes, namely, cyclohexanone and cyclodecanone, over Ti-containing organic–inorganic hybrid mesoporous silica, which resulted in an exclusive yield of oxime derivatives with high selectivity [57].

3.2.2 Oxidation of Hydrocarbons

Homogeneous complexes of transition metals like Mn, Ti, Cr and V have been heterogenized into a variety of inorganic silica matrices, and these materials are extensively studied for the liquid phase partial oxidation of various hydrocarbons like alkanes, alkenes, cycloalkenes and arenes [58]. Among these, olefin epoxidation is very important in organic transformation as a variety of highly demanding chemicals like polyurethanes, unsaturated resins and glycols are manufactured from epoxy compounds. Synthesis and isolation of epoxides from alkene sources are difficult also, since in many

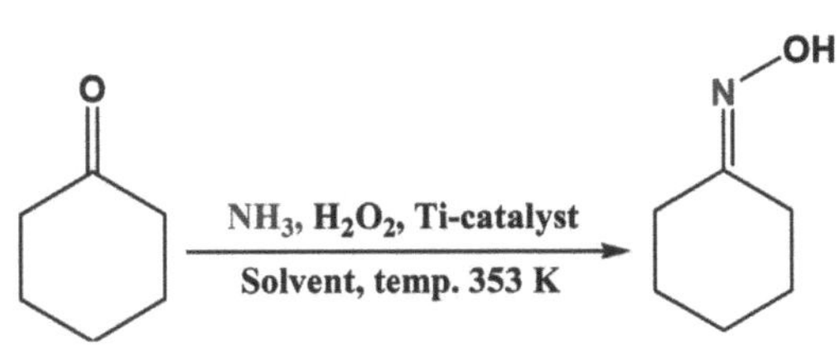

Figure 10
Ammoximation reaction over mesoporous Ti-based silica.

cases, further oxidation to other products minimizes the desired yield [54]. In 2003, Jia et al. reported an Mo complex-based MCM-41 catalyst, which shows a high efficiency and recyclability towards the highly selective conversion of cyclooctene oxide from cyclooctene in the presence of tertiary butyl hydroperoxide (TBHP) [58]. Cu- and Ni-functionalized periodic mesoporous organosilica is also an excellent oxidizing catalyst for partial oxidation of various olefins like cyclohexene, styrene, stilbene and norbornene under mild conditions, as is shown in Figure 11 [59]. The reaction proceeds

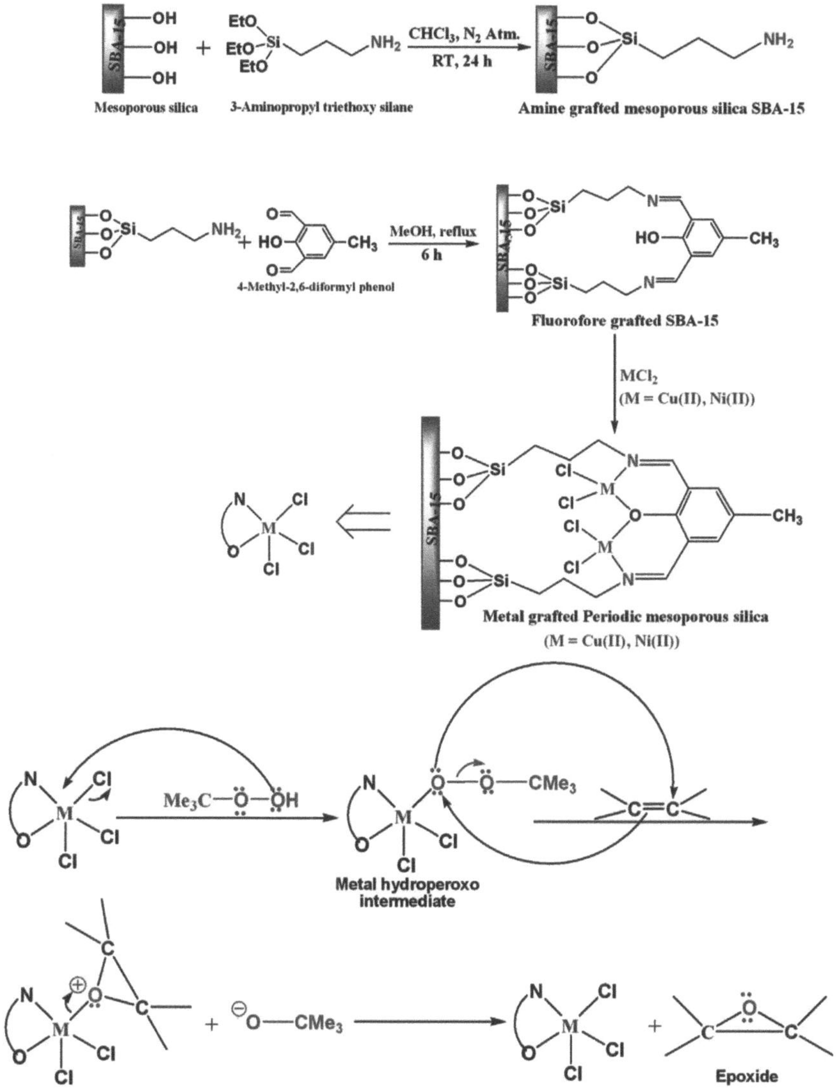

Figure 11

Epoxidation of various olefins over Cu- and Ni-doped mesoporous organosilica.

through the formation of metal a hydroperoxo active intermediate, which on interaction with the olefinic double bond generates the corresponding epoxide with an excellent yield.

In 2009, Singh et al. has described Cr-based silica, which is proved to play an excellent catalytic role in the aerial oxidation of arenes like ethylbenzene and common alkanes such as cyclohexane. The former produces acetophenone as major along with some benzaldehyde while later yields cyclohexanol, cyclohexanone and a small amount of benzoic acid [60]. Oxidation of hydrocarbons like methane at high temperatures has also been performed in a fixed-bed down-flow reactor over an Mo-SBA-15 catalyst and molecular oxygen gas as clean oxidant produced a significant amount of formaldehyde [61]. However, hydrogen peroxide has been used extensively as mild and nontoxic oxidant for liquid phase partial oxidation of different unsaturated alkenes and alkynes. For example, Vasylyev et al. launched a new mesoporous material composed of Zn-based polyoxometalate which successfully catalysed the epoxidation reactions of allylic alcohols utilizing H_2O_2 [62]. Different metal (Cu, Co, Ni)-based polyoxometalate-supported hexagonal mesoporous silicas are very promising also in the oxidation of propylene to acetone in the presence of molecular oxygen [63]. Different types of products generated due to hydrocarbon oxidation are comprehensively summarized in Table 1.

Table 1: Oxidation Products of Different Hydrocarbons.

		Different Oxidation Products		
Oxidant Types	**Hydrocarbons**	**Examples of Reactants**	**Possible Products Formed**	**References**
Hydrogen peroxide (H_2O_2), TBHP, aerial oxygen, ozone, air, etc.	Saturated (alkanes)	Methane, butane, isobutane	Formaldehyde, maleic anhydride, methacrolein	[61]
	Unsaturated (alkenes, allyl and alkynes)	Propylene	Acetone, propylene oxide, allyl alcohol	[63]
		Acetylene	Ethanol, glyoxal	[64]
		Allyl alcohol	Glycidol, glycerine	[62]
	Cycloalkanes and cycloalkenes	Cyclohexane	Cyclohexanol, cyclohexanone, adipic acid	[60]
		Cyclohexene, cyclooctene	Cyclohexene oxide, cyclohexene-1-one, cyclohexene-1-ol, adipic acid, cyclooctene oxide, etc.	[58,59]
	Aromatic hydrocarbons or arenes	Toluene, ethylbenzene, styrene, etc.	Acetophenone, benzaldehyde, benzoic acid, styrene oxide	[54,59, 60]

3.2.3 Oxidation of Alcohols and Sulphides

Transition metals containing organic–inorganic hybrid solids have also shown their capability in the selective oxidation of primary, secondary alcohols, sulphides, etc. The aldehydes and ketones obtained on oxidation of primary and secondary alcohols are valuable intermediates in organic synthesis, pharmaceuticals, agricultural chemicals, etc. Various conventional methods based on stoichiometric amounts of transition metals (Cr, Ru, Mn, etc.) result in the production of copious amounts of toxic waste and thus are not environment friendly as well as reusable [65]. Hence, metal-based heterogeneous supports are highly desirable for the green and sustainable development in the selective oxidation of alcohols using a non-pollutant, harmless oxidants like molecular oxygen and H_2O_2. A high yield of ketones from secondary alcohols, diols and α-hydroxy ketones under gentle conditions have been reported by Verma et al. who have employed a silica-supported oxo-vanadium Schiff base complex as a successful recyclable catalyst for this purpose [65]. These scientists used TBHP for the oxidation process, which yields some toxic by-products. But in the presence of molecular oxygen, Co Schiff base complex immobilized on SBA-15 via the 'click reaction' has played an outstanding catalytic role in the conversion of secondary alcohol to ketone at relatively lower temperatures [66]. Very recently, the transformation of a primary alcohol to an aldehyde has been carried out over Au@PMO at room temperature using molecular O_2 [67]. PMO acts as a good support for Au nanoparticles (Figure 12), and has been proved to be

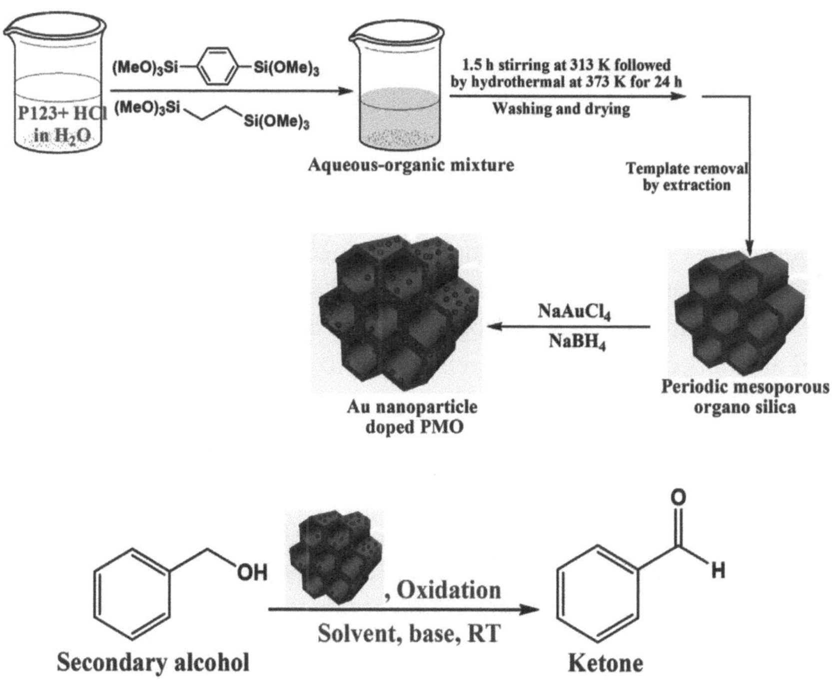

Figure 12
Schematic representation for the oxidation of alcohol over the Au-supported PMO catalyst.

highly reusable with minimum leaching of metal and is also environment friendly for the selective oxidation of benzyl alcohol to benzaldehyde under liquid phase conditions [67].

Oxidation of sulphides to obtain sulphoxide, sulphones, etc. are another good example of recyclable catalytic reactions, since the resulting products are a major part in anti-hypertension, cardiotonic agents, etc. and also they play an imperative role in organic and biological reactions [68]. Replacing homogeneous catalysts like halogenated compounds, nitrates and transition metal oxides industrially important heterogeneous solids has won the position to catalyse sulphide oxidation in order to produce highly selective products using mild oxidants like H_2O_2 and TBHP. [68]. Jain et al. have prepared a silica-based heterogeneous support immobilized with a vanadium Schiff base complex via the azide—alkyne [3 + 2] cycloaddition 'click reaction' and used the material as catalyst for the oxidation of sulphides at 333 K temperature using TBHP oxidant with an almost 100% yield of sulphoxide [69]. Another highly recoverable silica-based tungsten interphase catalyst has been developed, which exhibited good efficiency towards the selective conversion of aliphatic and aromatic sulphides to the corresponding sulphoxides even at room temperature in the presence of hydrogen peroxide as the oxidant [70]. Cr-grafted organic—inorganic hybrid mesoporous polymer reported by Salam et al. acts as an efficient vehicle for the liquid phase partial oxidative conversion of sulphides to the corresponding oxides with good recyclability along with no significant loss in activity (Figure 13) [68].

3.2.4 Hydrogenation Reactions

The process of hydrogenation, that is, treatment with hydrogen, invented by scientist Paul Sabatier in 1897 is a chemical reduction method applied widely in the field of industrial sectors like food, petrochemical, pharmaceutical and agricultural industries. The process is universally employed to reduce or saturate organic compounds [71]. Hydrogenation typically involves the addition of a pair of hydrogen atoms to unsaturated substrates containing a double or triple bond resulting in saturated hydrocarbons, its derivatives or sometimes addition to saturated compounds also for its ring opening [72]. Owing to some inconvenience on handling gaseous hydrogen, an alternative source of hydrogen such as hydrazine, and alcohols can act as hydrogen donors to transfer hydrogen (transfer hydrogenation). This is a very helpful technique in organic synthesis for the asymmetric reduction of polar unsaturated substrates, such as ketones, aldehydes and imines [55].

Since the non-catalytic pathway for the reduction reaction requires a very high temperature and pressure, catalyst-mediated procedures (homogeneous or heterogeneous) are relatively more desirable and cost effective [72]. Catalysts are also responsible for the enantioselective hydrogenation of a prochiral compound, resulting in optically active products. A very popular homogeneous catalyst for this reaction is Rh-based Wilkinson's catalyst and Ir-based Crabtree's catalyst [73]. Regarding industrial purpose, various heterogeneous catalysts are developed for such reactions and asymmetric hydrogenation

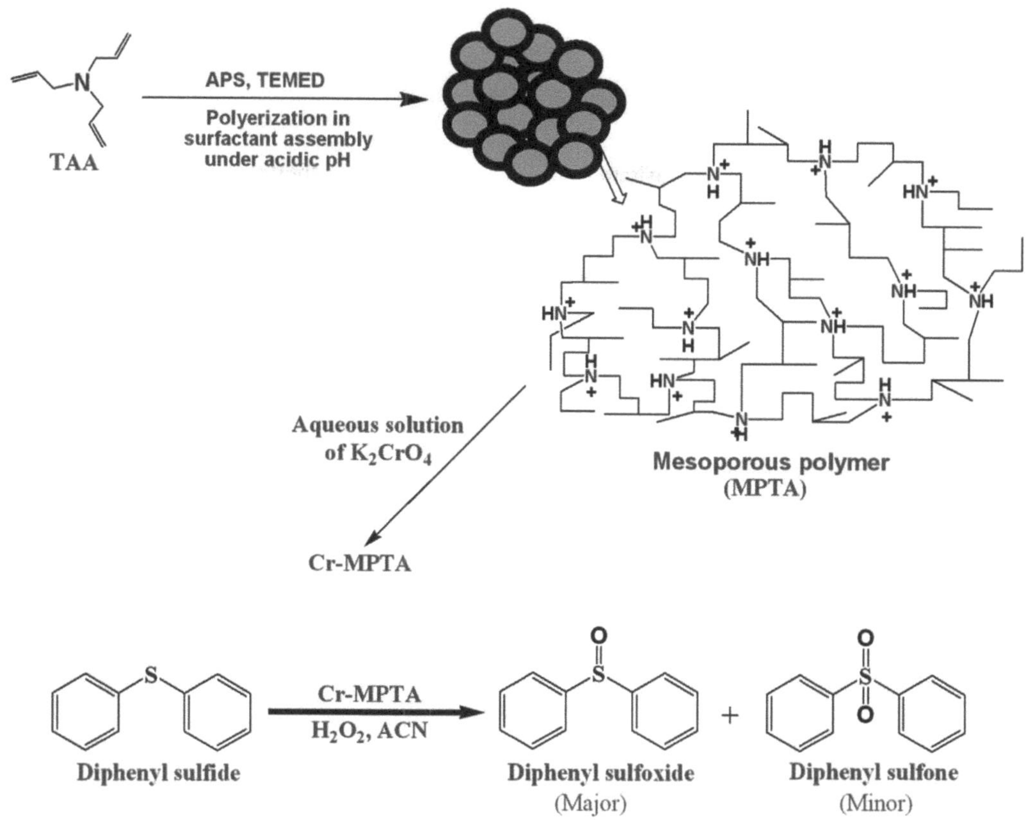

APS = Ammonium per sulphate
TEMED = N,N,N',N'-tetramethylenediamine
TAA = Triallyl amine
ACN = Acetonitrile

Figure 13

General schematic representation for the oxidation of sulphides to sulphoxides over a novel Cr-functionalized mesoporous polymer support.

has become possible over these by modifying the metal support with a chiral centre. Precious heavy metals, for example, Pd, Pt, Rh and non-precious Ni (Raney nickel) also act as active catalysts for hydrogenation [74]. Hydrogenation products of different substrates are given in Table 2.

A well-known heterogeneous optically active catalyst for hydrogenation of α-(acetyl-amino)-cinnamic acid and its methyl ester was an Rh complex-anchored mesoporous silica organic–inorganic hybrid matrix [75]. An interesting one-step hydrogenation/esterification reaction which has an immense demand in the upgradation of bio-oil derived from biomass fast pyrolysis has been carried out successfully with acetaldehyde and acetic acid

Table 2: Hydrogenation Products of Different Substrates.

Hydrogenating Agents	Substrates	Possible Products Formed	References
Gaseous H_2, hydrogen donor (cyclohexadiene, 2-propanol, hydrazine, etc.)	Alkene, $R_2C{=}CR'_2$	Alkane, $R_2CH{-}CHR'_2$	[77,81]
	Alkyne, $RC{\equiv}CR'$	Alkene, cis-$RHC{=}CHR'$	[72]
	Aldehyde, RCHO	Primary alcohol, RCH_2OH	[76]
	Ketone, $R_2C{=}O$	Secondary alcohol, $R_2CH\text{-}OH$	[79,80]
	Ester, RCO_2R'	Two alcohols, $RCH_2OH + R'OH$	[82]
	Imine, $RR'C{\equiv}NR''$	Secondary amine, RR'CHNHR	[81]
	Amide, $RC(O)NR'_2$	Tertiary amine, $RCH_2NR'_2$	[72]
	Nitrile, RCN	Primary amine, RCH_2NH_2	[71]
	Nitro, RNO_2	Amine, RNH_2	[55,78]

on metallic Pt modified mesoporous organosilica [76]. A good catalytic hydrogenation of tetraline followed by ring opening is observed on Pd−Pt-supported zirconium-doped mesoporous silica at temperatures above 500 K [77]. But a similar type of bimetal Pd−Ni nanoparticles deposited on Ti-doped hexagonal mesoporous silica by a facile photo-assisted deposition method function as a true heterogeneous catalyst for the hydrogenation of nitrobenzene to aminobenzene at 323 K under a 20 cm^3/min H_2 flow [78]. On the other hand, organometallic Ir complex loaded over an SBA-15-containing chiral ligand has become a promising catalyst for asymmetric hydrogenation of ketones at room temperature of $<$10−40 atm H_2 pressure with high yield and enantioselectivity (up to 93% ee) [79]. Another enantioselective transfer hydrogenation of prochiral ketones has been investigated by Parambadath et al. over Ru(II)-Chiral (1R,2S)-(+)-cis-1-amino-2-indanol immobilized over SBA-15 catalyst under mild conditions and 2-propanol was the hydrogen donor. Nowadays, this type of asymmetric transfer hydrogenation is considered an efficient method for the synthesis of chiral molecules with low fractional yields of side products and high product yields with a high enantiomeric purity [80]. Approximately 1 nmRh and Pt nanoparticles synthesized within a fourth-generation polyaminoamide (PAMAM) dendrimer and loaded on an SBA-15 support are reported by Huang et al. for the hydrogenation of pyrrole and ethylene to produce ethane and n-butylamine, respectively [81]. The schematic representation of the preparation of the catalyst and hydrogenation over it is shown in Figure 14. Mesoporous polymers are also a good support for catalytic hydrogenation. Very recently, mesoporous cross-linked polymer attached to a chiral ligand coordinated to ruthenium species exhibits a high enantioselective conversion as well as extraordinary reusability towards asymmetric hydrogenation of a β-keto ester in MeOH at 323 K [82]. Most of the hydrogenation reactions are based on simple mechanistic steps: firstly, binding of hydrogen and the substrate over a solid catalyst followed by the dissociation of H_2 into atomic hydrogen, addition of one atom of hydrogen to the substrate and then another atom to it, finally desorption of the hydrogenated product from the catalyst surface.

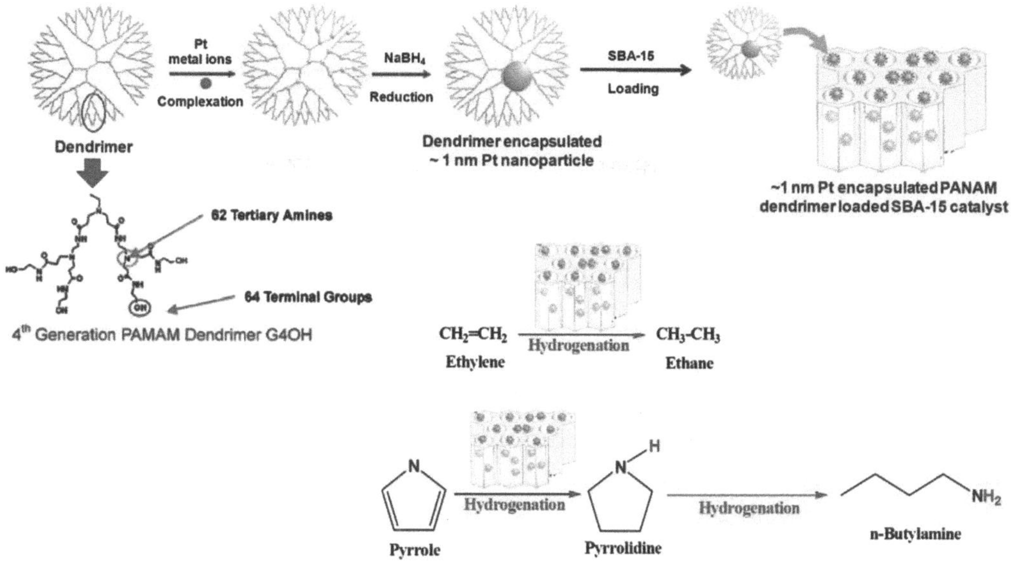

Figure 14
Schematic representation of the hydrogenation reactions of ethylene and pyrrole over PANAM dendrimer encapsulated Pt nanoparticle loaded SBA-15.

3.3 Different Coupling Reactions over Functionalized Mesoporous Materials

Application of organic–inorganic hybrid mesoporous solids for many hydrosilylation, cyanosilylation, Diels–Alder reactions along with C–C, C–N, C–H, C–S, C–O coupling and cross-coupling reactions in order to synthesize various organic fine chemicals is an active area of research. Organic modified or transition metals like Cu- and Ni- or noble metals like Pt, Pd etc. supported silicas, polymers etc. play an imperative role for these types of organic transformations [10]. Reactions that are not possible without a metal can be carried out by heterogenization of those metal ions on a mesoporous support so that the problem of using toxic metal ions and its leaching can be avoided. Some of these reactions that have a great impact on modern industrial technology are discussed here.

3.3.1 C–C Coupling Reactions

A series of metal-catalyzed C–C cross-coupling reactions for the synthesis of value-added biaryl derivatives are an interesting topic of research for scientists worldwide. Although Pd- and Fe-based homogeneous catalysts have been widely employed for these reactions, the reactions suffer from the drawback of complicated catalyst recovery and difficulty in regenerating the original activity [83]. Thus, metals containing a heterogeneous support with an exceptionally high surface area, wide and controllable pore opening in the nanoscale dimension are highly popular as catalysts for different C–C cross-coupling reactions [84] (Table 3).

Table 3: Different C—C Homo- and Cross-Coupling Reactions.

Reaction Name	Reactant A	Reaction B	Homo/Cross	References
Ullmann reaction	Ar—X	Ar—X	Homo	[84]
Kumada coupling	Ar-MgBr	Ar—X	Cross	[83]
Heck coupling	Alkene (aliphatic or aromatic)	R—X	Cross	[10]
Sonogashira coupling	RC≡CH	R—X	Cross	[87—89]
Suzuki reaction	$R-B(OH)_2$	R—X	Cross	[85,86]
Stille coupling	$R-SnR_3$	R—X	Cross	[89]
Hiyama coupling	$R-SiR_3$	R—X	Cross	[87]
Negishi coupling	R—Zn—X	R—X	Cross	[94]

The **Suzuki or Suzuki—Miyuara** cross-coupling reaction has been investigated and performed over Pd catalyst supported on numerous organic—inorganic hybrid mesoporous materials. This was first published by Akira Suzuki in 1979. This involves the organic coupling of an aryl- or vinyl-boronic acid with an aryl or vinyl halide catalysed by a palladium(0) complex with or without a base [10]. The reaction proceeds through the oxidative addition of a halide to Pd(0) to form organopalladium(II). This species on reaction with an organic base is transmetallated followed by the nucleophilic addition of hydroxyl borate. Finally, reductive elimination of the desired biaryl or divinyl restores the original palladium catalyst (Figure 15) [10]. Pd-doped 2D-hexagonal mesoporous organofunctionalized MCM-41 can be a good example for the Suzuki—Miyaura coupling of organoboron compounds with organic halides at 373 K in dimethyl formamide (DMF) solvent [85]. A base-free Suzuki reaction has been performed on Pd-loaded triazine-functionalized highly ordered mesoporous polymer (Figure 4), which allows the reaction in a milder condition [10]. Very recently, Mondal et al. have prepared in situ organofunctionalized mesoporous silica nanosphere that on successful Pd loading exhibits excellent activity for Suzuki—Miyaura cross-coupling reactions of aryl halides (iodides, bromides and chlorides) with phenylboronic acids in totally nontoxic green solvent water with high yields and without any loss of activity up to the eighth consecutive cycle [86].

Hiyama coupling is another palladium-catalysed cross-coupling reaction of organosilanes with organic halides used in organic natural product synthesis chemistry to form a carbon—carbon bond (C—C bond) with chemoselectivity and regioselectivity. This method generally also uses a base and a fluoride source, such as tetrabutylammonium fluoride for the synthesis of unsymmetrical biphenyl derivatives, which have widescale applications in materials science, drug discovery and the production of several bioactive natural products. The pathway mediated by an expensive fluoride source that is required for the activation of a silica precursor in hazardous organic solvents like DMF and dimethyl sulphoxide (DMSO) is recently modified by Modak et al. who used a newly developed Pd-grafted PMO for this Hiyama coupling reaction in water at an alkaline pH and in the absence of

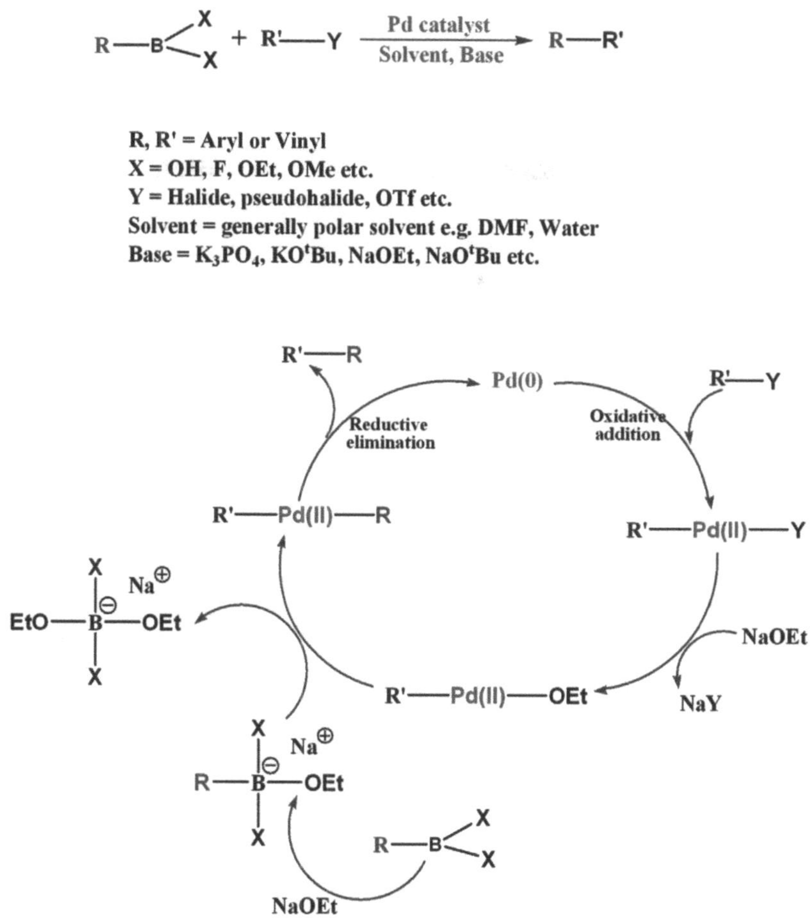

Figure 15

General scheme and mechanism for Suzuki–Miyuara cross-coupling reaction over Pd catalyst.

any fluoride source at a moderate temperature [87]. In Figure 16, the synthetic scheme of this Pd-grafted PMO and its catalytic activity towards fluoride-free Hiyama coupling of several organic halides along with a suggested mechanism (oxidative addition and reductive elimination) is given.

Sonogashira reaction is a cross-coupling reaction used in organic synthesis to form carbon–carbon bonds. It makes use of a palladium catalyst to form a carbon–carbon bond between a terminal alkyne and an aryl or vinyl halide (Figure 17). The reaction can be carried out under mild conditions, such as without heating or strong base or in water medium for the synthesis of complex molecules related to pharmaceuticals, natural products, organic materials and nanomaterials [88]. The reaction is initiated by the attack of a copper ion catalyst on the terminal alkyne to form a copper acetylide complex in the presence of a base and simultaneously oxidative addition of organic halides to Pd(0) to

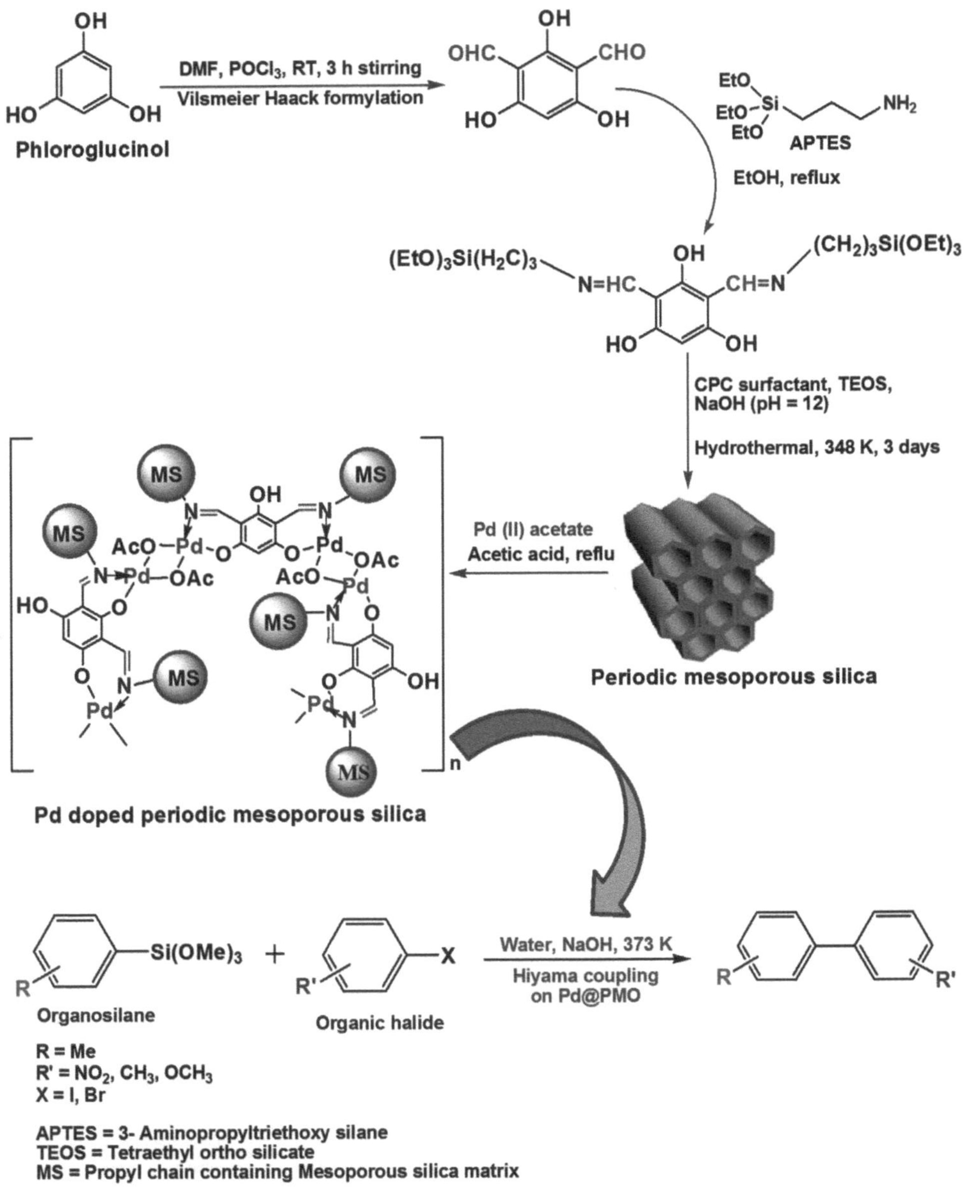

Figure 16

Schematic representation of the Hiyama coupling reaction over Pd-loaded periodic mesoporous silica.

form a Pd(II) intermediate complex. Finally, this intermediate on reaction with copper acetylide followed by reductive elimination gives the final C–C coupled product with regeneration of Pd(0) catalyst. Bhunia et al. have synthesized a catalyst by loading $PdCl_2$ to imine bond-functionalized MCM-41 in order to carry out the Sonogashira coupling in DMF solvent [89]. However, the same reaction can be performed in an aqueous solvent,

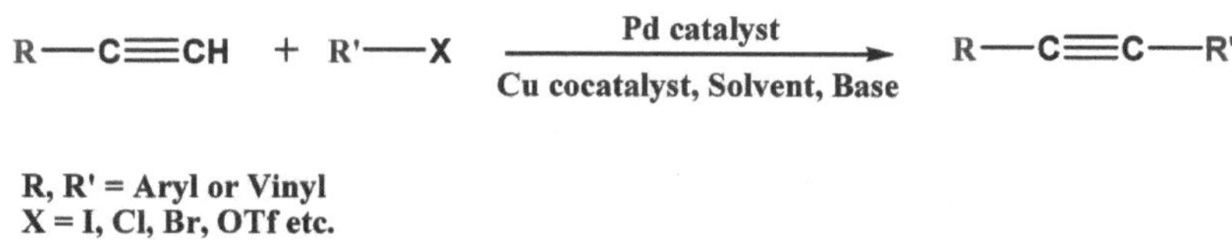

Figure 17
A general schematic outline for the Sonogashira cross-coupling reaction.

which is completely pollution free, and this is done with an ordered mesoporous polymer mentioned previously (Figure 4) [10]. Another approach has been made successfully by the same group of scientists carrying out the same reaction in copper-free condition using Pd-loaded organosilica (Figure 16) in the presence of an organic hexamine base [87].

Another C—C cross-coupling is the **Heck or Mizoroki—Heck** reaction, which involves the interaction of an unsaturated halide (or triflate) with an alkene in the presence of a base and palladium catalyst or palladium nanomaterial-based catalyst to form a substituted alkene. This coupling is very much important in organic synthesis ranging from natural products to bioactive compounds as well as to the development of organic electronics since the reaction allows one to do substitution reactions on planar centres [90]. The bases used in this reactions are generally triethylamine, potassium carbonate, sodium acetate, etc. Previously, the reactions were catalysed by mostly a Pd-based homogeneous support (metal complexes containing organic ligands), and due to the hydrophobic nature of these catalysts, dipolar aprotic organic solvents such as DMF, DMSO, CH_3CN, N-methylpyrrolidone (NMP), tetrahydrofuran, triethanolamine or dioxane were used instead of a nontoxic solvent water [91]. Pd-grafted heterogeneous catalysts provide ease of product separation, recovery of the catalyst by simple filtration and excellent recycling efficiency compared to their homogeneous counterparts [10]. In 2005, Crudden et al. reported an excellent catalyst containing Pd-loaded mercaptopropyl-modified mesoporous silica, which has been proved to be highly reusable with minimum Pd leaching and active for Heck coupling of various organics in DMF at 393 K in the presence of sodium acetate base [92]. Pd nanoparticles loaded on starch or polysaccharide-derived mesoporous solid are also a very useful recyclable catalyst for the microwave-assisted Heck coupling of iodobenzene and methyl acrylate in the presence of triethylamine [93]. Mesoporous organic polymers also play an essential role as a heterogeneous support of the Pd ion. Triazine functionalized polymer (schematically represented in Figure 4) or triallylamine-derived porous polymer (schematically represented in Figure 13) are very well-known examples of this type. They help the reaction proceed in aqueous media under mild conditions (Figure 18) [10,90]. Besides these, there are many other C—C cross-coupling reactions. Stille or Negishi reactions are common among those. In **Stille reaction** coupling used extensively in organic fine chemical synthesis, an organotin compound with an sp^2-hybridized organic halide

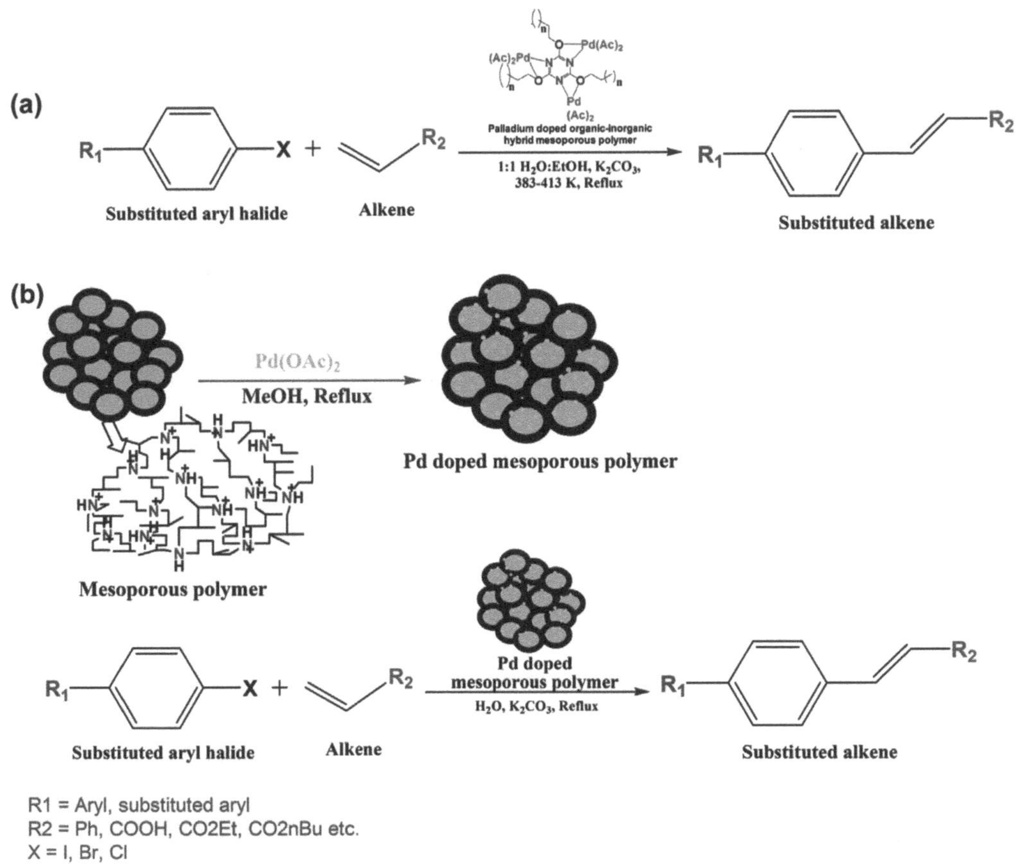

Figure 18
Mizoroki–Heck reaction over (a) Pd-doped triazine-functionalized polymer and (b) Pd-doped triallylamine-based mesoporous polymer.

catalysed by palladium is usually performed under an inert atmosphere. Pd-grafted functionalized MCM-41 works as an efficient catalyst for this reaction in air [89]. On the contrary, the **Negishi coupling** is the cross-coupling involving an organozinc compound, an organic halide in the presence of a nickel or palladium catalyst creating a new carbon–carbon covalent bond [94].

A well-known example of the C–C homo coupling reaction is the **Ullmann reaction**, which is the coupling of two molecules of similar aryl halides in the presence of copper metal. However, the reaction can be done using other metal catalysts like Pd and Au [84]. Pd-doped mesoporous silica-mediated Ullmann reaction of iodobenzene is a good example of heterogeneous catalysis for the clean organic synthesis of biphenyls by using water instead of organic solvents as the reaction medium [95]. The reaction is initiated by the oxidative addition of Ar–I to Pd(0) to form Ar–Pd(II)–I two

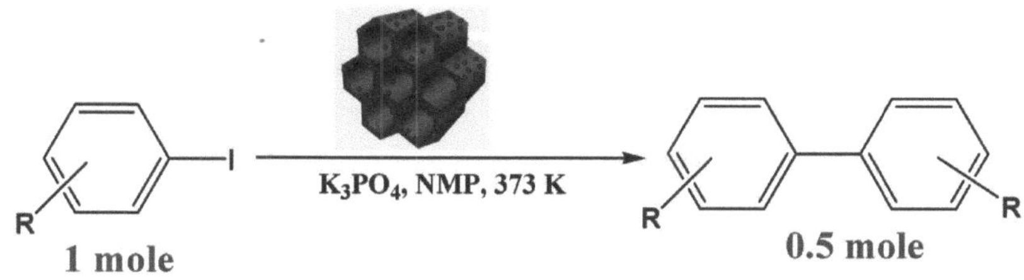

Figure 19
A general scheme for the Ullmann reaction of substituted iodobenzene over Au@PMO catalyst.

molecules of which form Ar—Pd—Ar releasing PdI_2 and a reductive elimination step forms Ar—Ar with the regeneration of Pd(0). HCOONa was used in the reaction media to reduce Pd(II) for the catalyst regeneration, whereas KOH helps to neutralize the HI generated due to the reduction of PdI_2. In a consecutive cycle, Ar—H is formed as the by-product [95]. Heterogeneous and reusable Pd on a mesoporous carbon catalyst has also been designed by Wang et al. for application in the Ullmann reaction of chlorobenzene in an aqueous medium, though selectivity towards biphenyl is not good owing to the formation of benzene as a by-product [96]. Unexpectedly, very highly selective yields of biphenyls from various aryl substituents are observed in NMP solvent using Au nanoparticle-doped bifunctional PMO Au@PMO (Figure 12), which is shown schematically in Figure 19 [84].

3.3.2 C—S Coupling Reactions

Formation of carbon—sulphur bond is an indispensable tool for designing new target molecules in synthetic organic, biological and pharmaceutical research. Reaction of aryl or substituted aryl halides with sulphur-containing sources like thiol [97], thiophenol [98], KSCN [99], thiourea [100] and potassium ethyl xanthogenate [101] used produces the particular diaryl sulphides (Figure 20). Transition metals like Fe-, Ni-, Pd-, Cu-, Co-based complexes are very effective catalysts for C—S coupling reactions in various polar organic solvents and an inert atmosphere [102]. But those catalytic reactions being homogeneous suffer from the disadvantage of catalyst separation, metal leaching, environmental hazards, etc. So a good heterogeneous catalyst based on these metals will be highly active and in

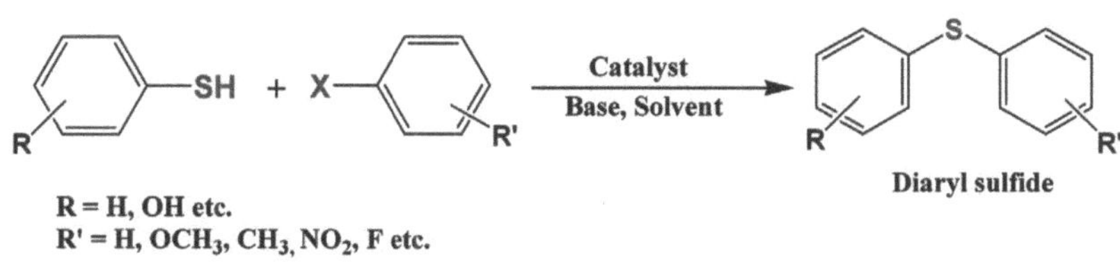

Figure 20
Simple scheme for the C—S coupling reaction of substituted aryl halides with substituted aryl sulphur to form disulphide.

demand for these aryl-sulphur coupling reactions. Various metal and mixed metal oxide nanoparticles are reported till date suggesting an easy, reusable, simple mechanistic heterogeneous pathway for the synthesis of diaryl sulphides [98,103]. In heterogeneous media, aerial catalytic reaction in water may also be helpful sometimes for successful reaction [98]. Another type of heterogeneous method for C–S coupling is the grafting of a homogeneous metal complex on the surface of an insoluble high surface area solid support so that the activity of the catalyst is enhanced as well as the criteria of recovery of the catalyst, reuse efficiency, minimum metal leaching can be fulfilled. Cu grafted on organic-inorganic hybrid silica or zirconia may be a suitable catalyst mentioned in this regard [97,104]. Cu-grafted furfural functionalized mesoporous organosilica has been prepared by a simple method and applied for the aryl-sulphur coupling of thiophenol source with aryl iodide in the presence of K_2CO_3 base and DMF solvent [97]. In another case, amine-functionalized mesoporous zirconia acts as true support for Cu ions and is thus employed for diaryl sulphide synthesis using the same sulphur source [104]. In both cases, the catalysts are recovered from the reaction mixture and recycled for another batch of coupling reaction.

These sulphur sources like volatile and toxic thiols or foul-smelling thiophenols used for C–S coupling reactions lead to environmental and safety problems and thus limits the use of this method for large-scale operation. Another source of sulphur that has attracted scientists' interest recently is thiourea. This is completely free from the above-mentioned problems [100]. The one-pot thioetherification reaction of numerous aryl halides and benzyl bromides with thiourea in aqueous media is a significant approach towards green organic synthesis made by the group of Mondal et al. (Figure 21). They employed a similar type of Cu-anchored mesoporous organosilica catalyst (synthesized via the post-grafting method) for this reaction carried out at 373 K in water with an excellent yield. There are also other coupling reactions like C–N and C–O, which are not observed till date to proceed over functionalized mesoporous materials. Although there are very few reports of the Buchwald–Hartwig amination, a carbon–nitrogen bond formation reaction over a Cu and Au metal homogeneous support [105], application over a heterogeneous support is very rare in this case.

Another type of coupling reaction reported to be successful over metal functionalized mesoporous organosilica is A^3 coupling, that is, the three-component coupling of aldehyde, alkyne and amine to form propargylamine, which is a major intermediate in the synthesis of many biological compounds and a structural component of many natural products (Figure 22). Numerous metal ions like Au, Ag, Au and In can catalyse this coupling reaction under mild conditions in homogeneous and heterogeneous media [106,107]. Very recently, Karimi et al. group has suggested a reaction pathway of the propargylamine synthesis via A^3 coupling over Au nanoparticle-supported PMO containing an imidazolium IL, a highly effective and recyclable catalyst [108].

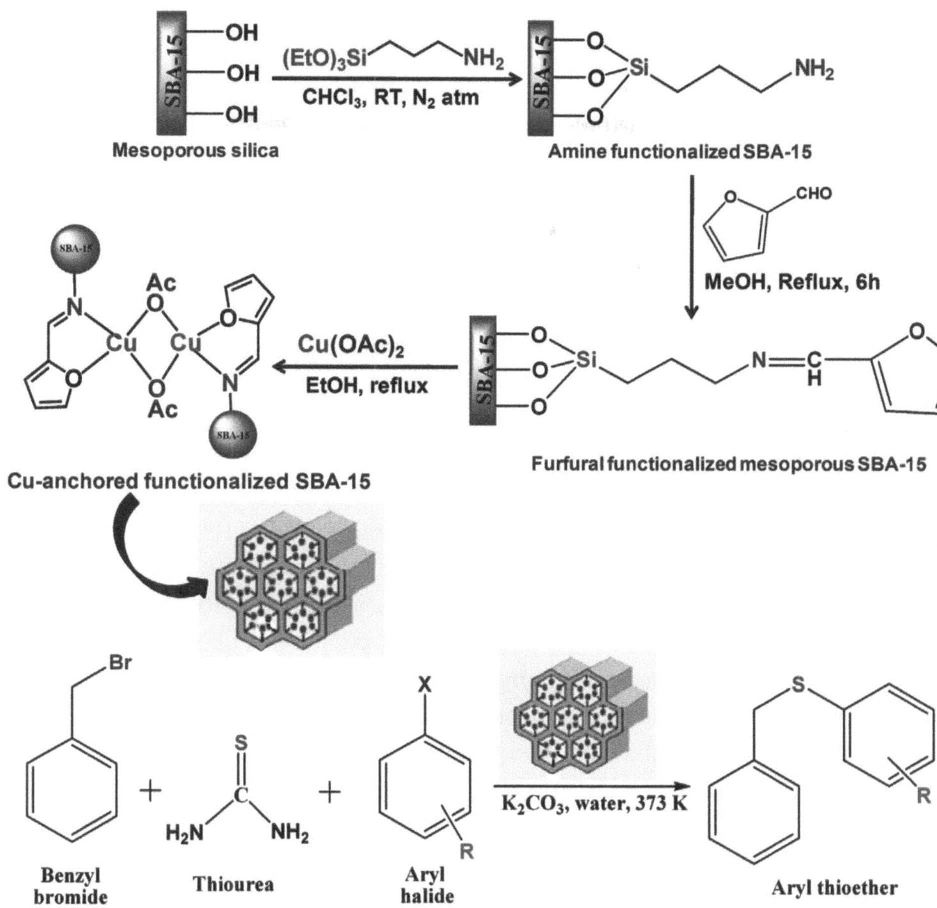

Figure 21
One-pot thioetherification using thiourea over Cu-doped organo-functionalized SBA-15 catalyst [100].

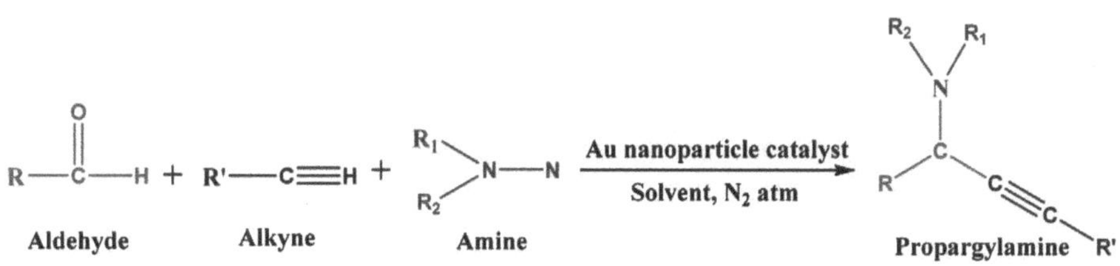

Figure 22
Au nanoparticle catalysed A^3 coupling reaction of aldehyde, amine and alkyne [107].

3.4 Polymerization Reactions

Polymerization or the process of combining monomer molecules together in a chemical reaction to form polymer chains or three-dimensional networks is a very important organic reaction for the manufacture of a wide range of automotive and electrical components to chemicals and versatile commercial equipments. Polymerized products of styrene (polystyrene), ethylene (polyethylene), propene, etc., are used widely in consumer products (e.g. trays, toys, kitchen wares, packing materials), fibres, films, etc. The Ziegler–Natta catalyst, a Ti-based compound-containing organoaluminium cocatalyst, is used widely for polyolefin production in a heterogeneous medium. Metallocene is another type that has a higher activity for olefin polymerization, but works in homogeneous media and thus faces the problem of catalyst separation, metal leaching, etc. [6]. So it will be advantageous to use these catalysts on a solid support for doing the polymerization more conveniently. Such an attempt has been made by some Korean scientists who confined Zr-based metallocene onto MCM-41 and utilized this successfully for propene polymerization [109]. Another Zr-based metallocene $(nBuCp)_2ZrCl_2$ or bis(butylcyclopentadienyl zirconium dichloride supported over mesoporous $-SO_3H$-functionalized SBA-15 has been employed to carry out ethylene polymerization in *n*-heptane [110]. A very convenient approach has been made very recently by Pramanik et al., where a tin-containing organic–inorganic hybrid porous phosphonate material has been employed as a heterogeneous catalyst for the polymerization of styrene under solvent-free conditions and at room temperature [111]. The rate of the radical polymerization of styrene has been largely enhanced over this tin-containing hybrid nanocatalyst in the presence of H_2O_2, which acts as a radical initiator (Figure 23). Thus, this method exhibits a new gateway for the synthesis of value-added polymers in a simple and convenient pathway. Recently, Staub et al. have demonstrated that the nature of interactions between the grafted organometallic ruthenium complex and surface silanols in Ru-containing functionalized mesoporous SBA-15 silica materials plays a crucial role in the olefin ring-closing metathesis catalysis [112]. The presence of surface silanols significantly decreases the longevity of the metathesis catalyst, whereas the passivation of the functionalized mesoporous surface by trimethylsilyl groups slows down the catalysis rate, but does not significantly affect the lifetime of the catalyst. A sustainable catalytic role of these metal-grafted functionalized mesoporous materials in various olefin polymerization reactions may motivate researchers to explore their potential in other polymerization reactions for the synthesis of value-added commercial products.

3.5 Role of Functionalization over Ordered Mesoporous Silicas

From the above discussion on the catalytic performances of organically functionalized mesoporous silica surface in reactions like redox, acid–base, coupling, polymerizations, it is quite evident that with the variation of the type of functional group, concentration of this surface functionality and periodicity of the mesopores, the catalytic activity and the yield of

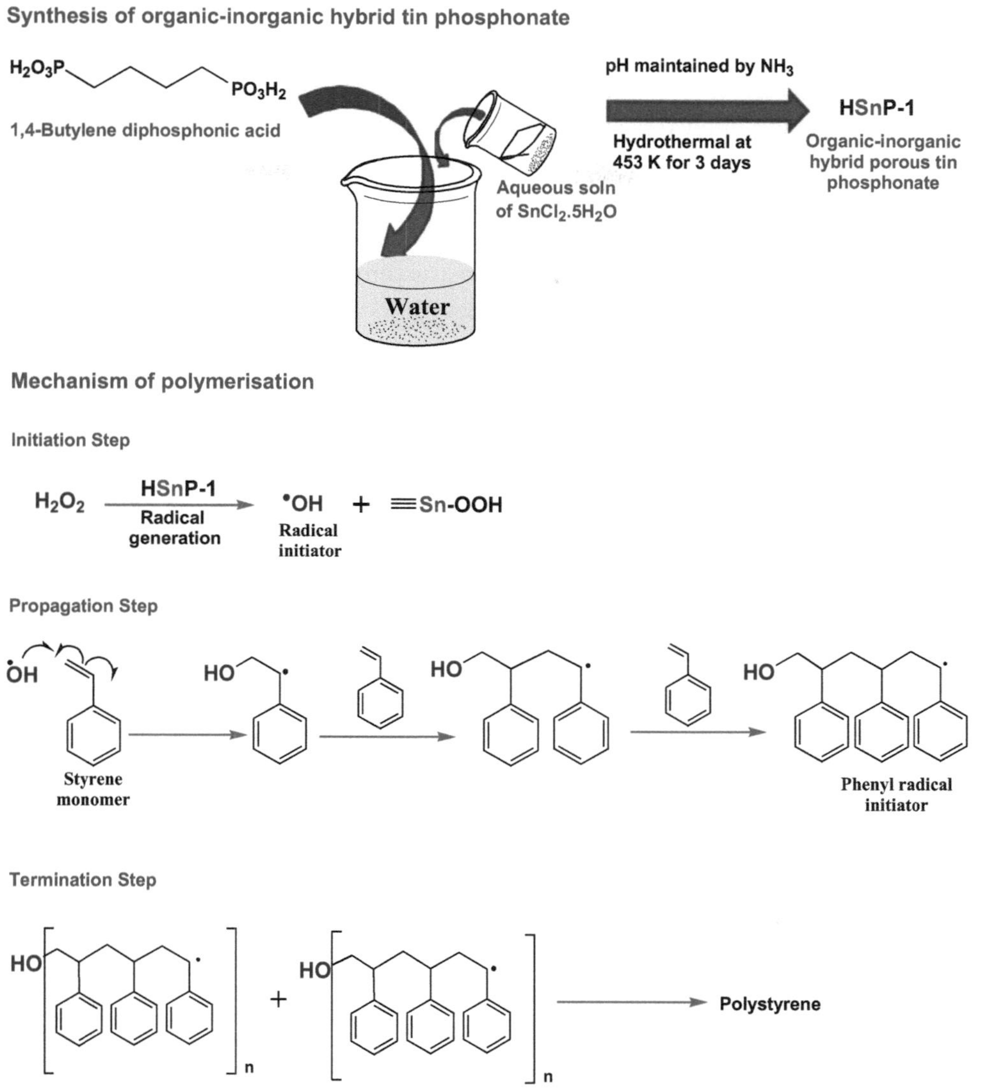

Figure 23

Polymerization of styrene over organic–inorganic hybrid porous tin phosphonate material.

products change considerably. The 2D- and 3D-hexagonal and cubic mesophases of the periodic mesostructures possess a much larger BET surface areas than do the disordered mesopores with a similar loading of functional groups. Moreover, the well-organized functional groups in these functionalized mesoporous materials could allow much facile diffusion of reactant and product molecules during reactions and the tuneable mesopore diameter could promote good selectivity for the desired products. A high selectivity provides the added advantages for the wide range of versatile catalytic responses observed over this functionalized mesoporous surface.

4. Summary

This book chapter describes how mesoporous materials are functionalized with a variety of organic ligands, metal complexes, etc. in order to get an advantage in a variety of industrially important liquid phase heterogeneous catalyses. Today, considering the high potential of these functionalized mesoporous materials for the development of a wide range of essential industrial fine chemical and pharmaceutical products, budding scientists are giving more attention to modifying the mesostructure with versatile organic groups or organometallic complexes. As an outcome of these in-depth studies, a large number of multitalented ordered organofunctionalized mesoporous architectures have been invented and these materials found a significant role in some of the frontline research areas like biomass conversion and biological sensing. Heterogeneous organocatalysts, namely, metal-free organic–inorganic hybrid mesoporous silica and purely organic porous polymers, fabricated recently can provide an alternative, green and convenient route for catalysis, if their potential could be explored properly. Thus, it can be concluded that a detailed investigation on these porous functionalized porous materials may earn a good opportunity of sustainable research for future scientists.

List of Abbreviations

2D Two dimensional
3D Three dimensional
BET Brunauer–Emmett–Teller
DMF Dimethyl formamide
DMSO Dimethyl sulphoxide
FC Friedel–Crafts
GNP Gross National Products
MOF Metal-organic framework
OTf Trifluoro methane sulphonate
pH $-[\log H^+]$
PMO Periodic mesoporous silica
SBA Santa Barbara
TBHP Tertiary butyl hydroperoxide
TEOS Tetraethyl orthosilicate
THF Tetrahydrofuran
TOF Turnover frequency

References

[1] Beck JS, Vartuli JC, Roth WJ, Leonowicz ME, Kresge CT, Scmitt KD, et al. A new family of mesoporous molecular sieves prepared with liquid crystal templates. J Am Chem Soc 1992;114:10834–43.

[2] Beck JS, Vartuli JC, Kennedy GJ, Kresge CT, Roth WJ, Schramm SE. Molecular or supramolecular templating: defining the role of surfactant chemistry in the formation of microporous and mesoporous molecular sieves. Chem Mater 1994;6:1816–21.

[3] Wilson ST, Lok BM, Messina CA, Cannan TR, Flanigen EM. Aluminophosphate molecular sieves: a new class of microporous crystalline inorganic solids. J Am Chem Soc 1982;104:1146–7.
[4] Lindströma B, Pettersson LJ. A brief history of catalysis. Cattech 2003;7:130–8.
[5] Wilson K, Clark JH. Solid acids and their use as environmentally friendly catalysts in organic synthesis. Pure Appl Chem 2000;72:1313–9.
[6] Valkenberg MH, Hölderich WF. Preparation and use of hybrid organic-inorganic catalysts. Catal Rev 2002;44:321–74.
[7] Lewis WK, Ries ED. Influence of reaction rate on operating conditions in contact sulfuric acid manufacture. Ind Eng Chem 1925;17:593–8.
[8] Wight AP, Davis ME. Design and preparation of organic-inorganic hybrid catalysts. Chem Rev 2002;102:3589–614.
[9] Yang X-Y, Li Y, Lemaire A, Yu J-G, Su B-L. Hierarchically structured functional materials: synthesis strategies for multimodal porous networks. Pure Appl Chem 2009;81:2265–307.
[10] Modak A, Mondal J, Sasidharan M, Bhaumik A. Triazine functionalized ordered mesoporous polymer: a novel solid support for Pd-mediated C–C cross-coupling reactions in water. Green Chem 2011;13:1317–31.
[11] Pramanik M, Nandi M, Uyama H, Bhaumik A. Organic-inorganic hybrid porous sulfonated zinc phosphonate material: efficient catalyst for biodiesel synthesis at room temperature. Green Chem 2012;14:2273–81.
[12] Inagaki S, Guan S, Ohsuna T, Terasaki O. An ordered mesoporous organosilica hybrid material with a crystal-like wall structure. Nature 2002;416:304–7.
[13] Sanchez C, Soler-Illia GJ de AA, Ribot F, Lalot T, Mayer CR, Cabuil V. Designed hybrid organic-inorganic nanocomposites from functional nanobuilding blocks. Chem Mater 2001;13:3061–83.
[14] Hoffmann F, Cornelius M, Morell J, Fröba M. Silica-based mesoporous organic-inorganic hybrid materials. Angew Chem Int Ed 2006;45:3216–51.
[15] Nandi M, Mondal J, Sarkar K, Yamauchi Y, Bhaumik A. Highly ordered acid functionalized SBA-15: a novel organocatalyst for the preparation of xanthenes. Chem Commun 2011;47:6677–9.
[16] Mercier L, Pinnavaia TJ. Direct synthesis of hybrid organic–inorganic nanoporous silica by a neutral amine assembly route: structure-function control by stoichiometric incorporation of organosiloxane molecules. Chem Mater 2000;12:188–96.
[17] Paul M, Pal N, Rajamohanan PR, Rana BS, Sinha AK, Bhaumik A. New organic-inorganic hybrid microporous organosilica having high metal ion adsorption capacity. Phys Chem Chem Phys 2010;12:9389–94.
[18] Mondal J, Sen T, Bhaumik A. Fe_3O_4@mesoporous SBA-15: a robust and magnetically recoverable catalyst for one-pot synthesis of 3,4-dihydropyrimidin-2(1H)-ones via the Biginelli reaction. Dalton Trans 2012;41:6173–81.
[19] Saikia L, Srinivas D, Ratnasamy P. Comparative catalytic activity of Mn(Salen) complexes grafted on SBA-15 functionalized with amine, thiol and sulfonic acid groups for selective aerial oxidation of limonene. Micropor Mesopor Mater 2007;104:225–35.
[20] Zhang D, Wang X, Qiao Z-A, Tang D, Liu Y, Huo Q. Synthesis and characterization of novel lanthanide(III) complexes-functionalized mesoporous silica nanoparticles as fluorescent nanomaterials. J Phys Chem C 2010;114:12505–10.
[21] Meng Y, Gu D, Zhang F, Shi Y, Cheng L, Feng D, et al. A Family of highly ordered mesoporous polymer resin and carbon structures from organic–organic self-assembly. Chem Mater 2006;18:4447–64.
[22] Modak A, Nandi M, Mondal J, Bhaumik A. Porphyrin based porous organic polymers: novel synthetic strategy and exceptionally high CO_2 adsorption capacity. Chem Commun 2012;48:248–50.
[23] Mondal J, Modak A, Bhaumik A. One-pot efficient Heck coupling in water catalyzed by palladium nanoparticles tethered into mesoporous organic polymer. J Mol Catal A Chem 2011;335:236–41.

[24] Taguchi A, Schüth F. Ordered mesoporous materials in catalysis. Micropor Mesopor Mater 2005;77:1–45.
[25] Tsuchimoto T, Tobita K, Hiyama T, Fukuzawa S-i. Scandium(III) triflate-catalyzed Friedel-Crafts alkylation reactions. J Org Chem 1997;62:6997–7005.
[26] Tahir N, Abdelssadek Z, Halliche D, Saadi S, Chebout R, Cherifi O, et al. Mg-Fe-hydrotalcite as catalyst for the benzylation of benzene and other aromatics by benzyl chloride reactions. Surf Interface Anal 2008;40:254–8.
[27] Alvaro M, Corma A, Das D, Fornés V, García H. "Nafion"-functionalized mesoporous MCM-41 silica shows high activity and selectivity for carboxylic acid esterifications and Friedel-Crafts acylation reactions. J Catal 2005;231:48–55.
[28] Wang G, Yu N, Peng L, Tan R, Zhao H, Yin D, et al. Immobilized chloroferrate ionic liquid: an efficient and reusable catalyst for synthesis of diphenylmethane and its derivatives. Catal Lett 2008;123:252–8.
[29] Zhao H, Yu N, Wang J, Zhuang D, Ding Y, Tan R, et al. Preparation and catalytic activity of periodic mesoporous organosilica incorporating Lewis acidic chloroindate(III) ionic liquid moieties. Micropor Mesopor Mater 2009;122:240–6.
[30] Kapoor MP, Kasama Y, Yanagi M, Yokoyama T, Inagaki S, Shimada T, et al. Cubic phenylene bridged mesoporous hybrids from allylorganosilane precursors and their applications in Friedel-Crafts acylation reaction. Micropor Mesopor Mater 2007;101:231–9.
[31] Noda Pérez C, Pérez CA, Henriques CA, Monteiro JLF. Hydrotalcites as precursors for Mg,Al-mixed oxides used as catalysts on the aldol condensation of citral with acetone. Appl Catal A Gen 2004;272:229–40.
[32] Zeidan RK, Hwang SJ, Davis ME. Multifunctional heterogeneous catalysts: SBA-15-containing primary amines and sulfonic acids. Angew Chem Int Ed 2006;45:6332–5.
[33] Shanmuganathan S, Greiner L, Domínguez de María P. Silica-immobilized piperazine: a sustainable organocatalyst for aldol and Knoevenagel reactions. Tetrahedron Lett 2010;51:6670–2.
[34] Zamboulis A, Rahier NJ, Gehringer M, Cattoën X, Niel G, Bied C, et al. Silica-supported L-proline organocatalysts for asymmetric aldolisation. Tetrahedron Asymm 2009;20:2880–5.
[35] Corma A, Garcia H. Silica-bound homogenous catalysts as recoverable and reusable catalysts in organic synthesis. Adv Synth Catal 2006;348:1391–412.
[36] Gu W, Zhou W-J, Gin DL. A nanostructured, scandium-containing polymer for heterogeneous Lewis acid catalysis in water. Chem Mater 2001;13:1949–51.
[37] Alizadeh A, Khodaei MM, Kordestani D, Fallah AH, Beygzadeh M. The successful synthesis of biguanide-functionalized mesoporous silica catalysts: excellent reactivity combined with facile catalyst recyclability. Micropor Mesopor Mater 2012;159:9–16.
[38] Demicheli G, Maggi R, Mazzacani A, Righi P, Sartori G, Bigi F. Supported organic catalysts: synthesis of (E)-nitrostyrenes from nitroalkanes and aromatic aldehydes over propylamine supported on MCM-41 silica as a reusable catalyst. Tetrahedron Lett 2001;42:2401–3.
[39] Trilla M, Pleixats R, Man MWC, Bied C. Organic–inorganic hybrid silica materials containing imidazolium and dihydroimidazolium salts as recyclable organocatalysts for Knoevenagel condensations. Green Chem 2009;11:1815–20.
[40] Baeyer A, Villiger V. Einwirkung des caro'schen reagens auf ketone. Chem Ber 1899;32:3625–33.
[41] Lambert A, Elings JA, Macquarrie DJ, Carr G, Clark JH. The Baeyer-Villiger oxidation of ketones using HMS supported peroxycarboxylic acids. Synlett 2000;7:1052–4.
[42] Jeong E-Y, Ansari MB, Park S-E. Aerobic Baeyer Villiger oxidation of cyclic ketones over metalloporphyrins bridged periodic mesoporous organosilica. ACS Catal 2011;1:855–63.
[43] Sasidharan M, Kiyozumi Y, Mal NK, Paul M, Rajamohanan PR, Bhaumik A. Incorporation of tin in different types of pores in SBA-15: synthesis, characterization and catalytic activity. Micropor Mesopor Mater 2009;126:234–44.
[44] Paul M, Pal N, Mondal J, Sasidharan M, Bhaumik A. New mesoporous magnesium-aluminum mixed oxide and its catalytic activity in liquid phase Baeyer-Villiger oxidation reaction. Chem Eng Sci 2012;71:564–72.

[45] Kiss AA, Dimian A, Rothenberg CG. Solid acid catalysts for biodiesel production-towards sustainable energy. Adv Synth Catal 2006;348:75–81.
[46] Singh AK, Fernando SD, Hernandez R. Base-catalyzed fast transesterification of soybean oil using ultrasonication. Energy Fuel 2007;21:1161–4.
[47] Karimi B, Vafaeezadeh M. SBA-15-functionalized sulfonic acid confined acidic ionic liquid: a powerful and water-tolerant catalyst for solvent-free esterifications. Chem Commun 2012;48:3327–9.
[48] Pal N, Paul M, Bhaumik A. Highly ordered Zn-doped mesoporous silica: an efficient catalyst for transesterification reaction. J Solid State Chem 2011;184:1805–12.
[49] Balbino JM, de Menezes EW, Benvenutti EV, Cataluña R, Ebeling G, Dupont J. Silica-supported guanidine catalyst for continuous flow biodiesel production. Green Chem 2011;13:3111–6.
[50] Guerrero VV, Shantz DF. Amine-functionalized ordered mesoporous silica transesterification catalysts. Ind Eng Chem Res 2009;48:10375–80.
[51] Jain SL, Modak A, Bhaumik A. Novel mesoporous silica grafted organocatalyst via click method for Michael addition reaction. Green Chem 2011;13:586–90.
[52] Jaenicke S, Chuah GK, Lin XH, Hu XC. Organic-inorganic hybrid catalysts for acid- and base-catalyzed reactions. Micropor Mesopor Mater 2000;35-36:143–53.
[53] Juvaste H, Iiskola EI, Pakkanen TT. Aminosilane as a coupling agent for cyclopentadienyl ligands on Silica. J Organomet Chem 1999;587:38–45.
[54] Chandra D, Bhaumik A. Highly active 2D hexagonal mesoporous titanium silicate synthesized using a cationic-anionic mixed surfactant assembly. Ind Eng Chem Res 2006;45:4879–83.
[55] Paul M, Pal N, Bhaumik A. Mesoporous nickel-aluminum mixed oxide: a promising catalyst in hydride-transfer reactions. Eur J Inorg Chem 2010:5129–34.
[56] Bordoloia A, Halligudi SB. Catalytic properties of WOx/SBA-15 for vapor-phase Beckmann rearrangement of cyclohexanone oxime. App Catal A Gen 2010;379:141–7.
[57] Bhaumik A, Kapoor MP, Inagaki S. Ammoximation of ketones catalyzed by titanium-containing ethane bridged hybrid mesoporous silsesquioxane. Chem Commun 2003:470–1.
[58] Jia M, Seifert A, Thiel WR. Mesoporous MCM-41 materials modified with oxodiperoxo molybdenum complexes: efficient catalysts for the epoxidation of cyclooctene. Chem Mater 2003;15:2174–80.
[59] Nandi M, Roy P, Uyama H, Bhaumik A. Functionalized mesoporous silica supported copper(II) and nickel(II) catalysts for liquid phase oxidation of olefins. Dalton Trans 2011;40:12510–8.
[60] Singh AP, Torita N, Shylesh S, Iwasa N, Arai M. Catalytic aerobic oxidation of cyclohexane and ethyl benzene over chromium-containing mesoporous organosilicas. Catal Lett 2009;132:492–9.
[61] Lou Y, Tang Q, Wang H, Chia B, Wang Y, Yang Y. Selective oxidation of methane to formaldehyde by oxygen over SBA-15-supported molybdenum oxides. Appl Catal A Gen 2008;350:118–25.
[62] Vasylyev MV, Neumann R. New heterogeneous polyoxometalate based mesoporous catalysts for hydrogen peroxide mediated oxidation reactions. J Am Chem Soc 2004;126:884–90.
[63] Liu Y, Murata K, Inaba M, Mimura N. Selective oxidation of propylene to acetone by molecular oxygen over $M_{x/2}H_{5-x}[PMo_{10}V_2O_{40}]$/HMS ($M=Cu^{2+}$, Co^{2+}, Ni^{2+}). Catal Commun 2003;4:281–5.
[64] Alzueta MU, Borruey M, Callejas A, Millera A, Bilbao R. An experimental and modeling study of the oxidation of acetylene in a flow reactor. Combust Flame 2008;152:377–86.
[65] Verma S, Nandi M, Modak A, Jain SL, Bhaumik A. Novel organic-inorganic hybrid mesoporous silica supported oxo-vanadium Schiff base for selective oxidation of alcohols. Adv Synth Catal 2011;353:1897–902.
[66] Rana BS, Jain SL, Singh B, Bhaumik A, Sain B, Sinha AK. Click on silica: systematic immobilization of cobalt (II) Schiff base complexes to the mesoporous silica *via* click reaction and their catalytic activity for aerobic oxidation of alcohols. Dalton Trans 2010;39:7760–7.
[67] Karimi B, Esfahani FK. Gold nanoparticles supported on the periodic mesoporous organosilicas as efficient and reusable catalyst for room temperature aerobic oxidation of alcohols. Adv Synth Catal 2012;354:1319–26.
[68] Salam N, Mondal P, Mondal J, Roy AS, Bhaumik AS, Islam M. Chromium(VI) grafted mesoporous polyaniline as a reusable heterogeneous catalyst for oxidation reactions in aqueous medium. RSC Adv 2012;2:6464–77.

[69] Jain SL, Rana BS, Singh B, Sinha AK, Bhaumik A, Nandi M, et al. Click chemistry: an improved approach to high yielding immobilization of vanadium Schiff base on mesoporous silica for oxidation of sulfides. Green Chem 2010;12:374–7.
[70] Karimi B, Ghoreishi-Nezhad M, Clark JH. Selective oxidation of sulfides to sulfoxides using 30% hydrogen peroxide catalyzed with a recoverable silica-based tungstate interphase catalyst. Org Lett 2005;7:625–8.
[71] Braos-Garćıa P, Maireles-Torres P, Rodríguez-Castellón E, Jiménez-López A. Gas-phase hydrogenation of acetonitrile on zirconium-doped mesoporous silica-supported nickel catalysts. J Mol Catal A Chem 2003;193:185–96.
[72] Kuhn JN, Huang W, Tsung C-K, Zhang Y, Somorjai GA. Structure sensitivity of carbon-nitrogen ring opening: impact of platinum particle size from below 1 to 5 nm upon pyrrole hydrogenation product selectivity over monodisperse platinum nanoparticles loaded onto mesoporous silica. J Am Chem Soc 2008;130:14026–7.
[73] Harmon RE, Gupta SK, Brown DJ. Hydrogenation of organic compounds using homogeneous catalysts. Chem Rev 1973;73:21–52.
[74] Adkins H, Billica HR. The preparation of Raney nickel catalysts and their use under conditions comparable with those for platinum and palladium catalyst. J Am Chem Soc 1948;70:695–8.
[75] Nagel U, Kinzel E. The first stereospecific catalytic hydrogenation with a polymer supported optically active rhodium complex. Chem Commun 1986:1098–9.
[76] Tang Y, Miao S, Shanks BH, Zheng X. Bifunctional mesoporous organic-inorganic hybrid silica for combined one-step hydrogenation/esterifications. Appl Catal A Gen 2010;375:310–7.
[77] Carrión MC, Manzano BR, Jalón FA, Eliche-Quesada D, Maireles-Torres P, Rodríguez-Castellón E, et al. Influence of the metallic precursor in the hydrogenation of tetralin over Pd–Pt supported zirconium doped mesoporous silica. Green Chem 2005;7:793–9.
[78] Fuku K, Sakano T, Kamegawa T, Mori K, Yamashita H. Enhanced hydrogenation activity of nano-sized Pd–Ni bimetal particles on Ti-containing mesoporous silica prepared by a photo-assisted deposition method. J Mater Chem 2012;22:16243–7.
[79] Liu G, Wang J, Huang T, Liang X, Zhang Y, Li H. Mesoporous silica-supported iridium catalysts for asymmetric hydrogenation reactions. J Mater Chem 2010;20:1970–5.
[80] Parambadath S, Singh AP. Ru(II)-Chiral (1R,2S)-(+)-cis-1-amino-2-indanol immobilized over SBA-15 for asymmetric transfer hydrogenation reaction of prochiral ketones. Catal Today 2009;141:161–7.
[81] Huang W, Kuhn JN, Tsung C-K, Zhang Y, Habas SE, Yang P, et al. Dendrimer templated synthesis of one nanometer Rh and Pt particles supported on mesoporous silica: catalytic activity for ethylene and pyrrole hydrogenation. Nano Lett 2008;8:2027–34.
[82] Sun Q, Meng X, Liu X, Zhang X, Yang Y, Yang Q, et al. Mesoporous cross-linked polymer copolymerized with chiral BINAP ligand coordinated to a ruthenium species as an efficient heterogeneous catalyst for asymmetric hydrogenation. Chem Commun 2012;48:10505–7.
[83] Fürstner A, Leitner A, Méndez M, Krause H. Iron-catalyzed cross-coupling reactions. J Am Chem Soc 2002;124:13856–63.
[84] Karimi B, Esfahani FK. Unexpected golden Ullmann reaction catalyzed by Au nanoparticles supported on periodic mesoporous organosilica (PMO). Chem Commun 2011;47:10452–4.
[85] Dhara K, Sarkar K, Srimani D, Saha SK, Chattopadhyay P, Bhaumik A. A new functionalized mesoporous matrix supported Pd(II)-Schiff base complex: an efficient catalyst for the Suzuki-Miyaura coupling reaction. Dalton Trans 2010;39:6395–402.
[86] Mondal P, Banerjee S, Roy AS, Mandal TK, Islam SM. In situ prepared mesoporous silica nanospheres supported palladium(II) 2-aminopyridine complex catalyst for Suzuki-Miyaura cross-coupling reaction in water. J Mater Chem 2012;22:20434–42.
[87] Modak A, Mondal J, Bhaumik A. Pd-grafted periodic mesoporous organosilica: an efficient heterogeneous catalyst for Hiyama and Sonogashira couplings, and cyanation reactions. Green Chem 2012;14:2840–55.

[88] Sonogashira K. Development of Pd-Cu catalyzed cross-coupling of terminal acetylenes with sp^2-carbon halides. J Organomet Chem 2002;653:46–9.
[89] Bhunia S, Koner S. Heterogeneous Stille and Sonogashira cross-coupling reactions over palladium anchored mesoporous silica catalyst. Ind J Chem 2011;50A:1380–7.
[90] Mondal J, Modak A, Bhaumik A. One-pot efficient Heck coupling in water catalyzed by palladium nanoparticles tethered into mesoporous organic polymer. J Mol Catal A Chem 2011;350:40–8.
[91] Ozawa F, Kubo A, Hayashi T. Generation of tertiary phosphine-coordinated Pd(0) species from $Pd(OAc)_2$ in the catalytic heck reaction. Chem Lett 1992;21:2177–80.
[92] Crudden CM, Sateesh M, Lewis R. Mercaptopropyl-modified mesoporous silica: a remarkable support for the preparation of a reusable, heterogeneous palladium catalyst for coupling reactions. J Am Chem Soc 2005;127:10045–50.
[93] Budarin VL, Clark JH, Luque R, Macquarrie DJ, White RJ. Palladium nanoparticles on polysaccharide-derived mesoporous materials and their catalytic performance in C—C coupling reactions. Green Chem 2008;10:382–7.
[94] King AO, Okukado N, Negishi E-i. Highly general stereo-, regio-, and chemo-selective synthesis of terminal and internal conjugated enynes by the Pd-catalysed reaction of alkynylzinc reagents with alkenyl halides. J Chem Soc Chem Commun 1977:683–4.
[95] Li H, Chen J, Wan Y, Chai W, Zhang F, Lu Y. Aqueous medium Ullmann reaction over a novel Pd/Ph-Al-MCM-41 as a new route of clean organic synthesis. Green Chem 2007;9:273–80.
[96] Wang H, Wan Y. Synthesis of ordered mesoporous Pd/carbon catalyst with bimodal pores and its application in water-mediated Ullmann coupling reaction of chlorobenzene. J Mater Sci 2009;44:6553–62.
[97] Mondal J, Modak A, Dutta A, Bhaumik A. Facile C—S coupling reaction of aryl iodide and thiophenol catalyzed by Cu-grafted furfural functionalized mesoporous organosilica. Dalton Trans 2011;40:5228–35.
[98] Pal N, Bhaumik A. Self-assembled $NiO-ZrO_2$ nanocrystals with mesoscopic void space: an efficient and green catalyst for C—S cross-coupling reaction in water. Dalton Trans 2012;41:9161–9.
[99] Reddy KHV, Reddy VP, Kumar AA, Kranthi G, Nageswar YVD. Nano copper oxide catalyzed synthesis of symmetrical diaryl sulfides under ligand free conditions. Beilstein J Org Chem 2011;7:886–91.
[100] Mondal J, Modak A, Dutta A, Basu S, Jha SN, Bhattacharyya D, et al. One-pot thioetherification of aryl halides using thiourea and benzyl bromide in water catalyzed by Cu-grafted furfural functionalized mesoporous SBA-15. Chem Commun 2012;48:8000–2.
[101] Prasad DJC, Sekar G. Cu-catalyzed one-pot synthesis of unsymmetrical diaryl thioethers by coupling of aryl halides using a thiol precursor. Org Lett 2011;13:1008–11.
[102] Zhang J, Medley CM, Krause JA, Guan H. Mechanistic insights into C—S cross-coupling reactions catalyzed by nickel bis(phosphinite) pincer complexes. Organometallics 2010;29:6393–401.
[103] Bhadra S, Sreedhar B, Ranu BC. Recyclable heterogeneous supported copper-catalyzed coupling of thiols with aryl halides: base-controlled differential arylthiolation of bromoiodobenzenes. Adv Synth Catal 2009;351:2369–78.
[104] Mallick S, Rana S, Parida K. A facile method for the synthesis of copper modified amine-functionalized mesoporous zirconia and its catalytic evaluation in C—S coupling reaction. Dalton Trans 2011;40:9169–75.
[105] Shafir A, Buchwald SL. Highly selective room-temperature copper-catalyzed C—N coupling reactions. J Am Chem Soc 2006;128:8742–3.
[106] Zhang Y, Li P, Wang M, Wang L. Indium-catalyzed highly efficient three-component coupling of aldehyde, alkyne, and amine via C—H bond activation. J Org Chem 2009;74:4364–7.
[107] Kidwai M, Bansal V, Kumar A, Mozumdar S. The first Au-nanoparticles catalyzed green synthesis of propargylamines *via* a three-component coupling reaction of aldehyde, alkyne and amine. Green Chem 2007;9:742–5.

[108] Karimi B, Gholinejad M, Khorasani M. Highly efficient three-component coupling reaction catalyzed by gold nanoparticles supported on periodic mesoporous organosilica with ionic liquid framework. Chem Commun 2012;48:8961–3.
[109] Ko YS, Han TK, Park JW, Woo SI. Propene polymerization catalyzed over MCM-41 and VPI-5-supported Et(ind)$_2$ZrC1$_2$ catalysts. Macromol Rapid Commun 1996;17:749–58.
[110] Casas E, van Grieken R, Escola JM. Polymerization of ethylene with (nBuCp)$_2$ZrCl$_2$ supported over mesoporous SBA-15 functionalized with sulfonic acid groups. Appl Catal A Gen 2012;437-438:44–52.
[111] Pramanik M, Nandi M, Uyama H, Bhaumik A. Organic-inorganic hybrid tinphosphonate material with mesoscopic void spaces: an excellent catalyst for the radical polymerization of styrene. Catal Sci Technol 2012;2:613–20.
[112] Staub H, Guillet-Nicolas R, Even N, Kayser L, Kleitz F, Fontaine FG. Substantiating the influence of pore surface functionalities on the stability of grubbs catalyst in mesoporous SBA-15 silica. Chem Eur J 2011;17:4254–65.

CHAPTER 3

Sustainable Catalysis Systems Based on Ionic Liquids

Jianji Wang, Anlian Zhu, Lingjun Li
School of Chemistry and Chemical Engineering, Key Laboratory of Green Chemical Media and Reactions, Ministry of Education, Henan Normal University, Xinxiang, Henan, PR China

1. Introduction

Ionic liquids are organic salts that have low melting points as compared to traditional inorganic salts and benefit from negligible vapour pressure compared with traditional organic solvents. They also have a wide liquid range, high conductivity, wide electrochemical window and excellent solubility for organic, inorganic and organometallic compounds. The first ionic liquid described in 1914 by Walden is ethyl ammonium nitrate ($[EtNH_3][NO_3]$) with a melting point of 12.5 °C [1]. However, its function and potential in the industry were underestimated until the 1960s; the truly first generation of ionic liquids composed of dialkylimidazolium and alkylpyridinium cations was synthesized. The introduction of this type of ionic liquids opened up many novel aspects of application, although counter-anions such as chloroaluminate and other metal halides were found to be sensitive to water and air [2]. After the 1990s, the second generation of ionic liquids composed of weakly coordination anions such as $[BF_4]^-$ or $[PF_6]^-$ has attracted increasing attention due to their high air and water stability and then broadened their applicable fields [3]. More recently, the third generation of ionic liquids coming into the picture of scientists are the low-transition-temperature mixtures, which can be directly prepared by mixing natural high-melting-point starting materials to form a liquid through the formation of hydrogen bonds. The third generation of ionic liquids was believed to overcome some limitations of the traditional ionic liquids such as high preparation cost, limited biocompatibility and renewability [4].

The superior characteristics of ionic liquids including their convenient designability have made ionic liquids to have great potential in various fields such as energy storage and conversion [5], biomass transformation [6], organic synthesis and catalysis [7] and

Sustainable Catalytic Processes. http://dx.doi.org/10.1016/B978-0-444-59567-6.00003-0

extraction and separation [8]. In organic synthesis and catalysis fields, ionic liquids were initially used as a replacement for organic solvents due to their negligible vapour pressure and excellent solubility for organic compounds or metal catalysts, but now with the development of functional ionic liquids and the insight into the structures of ionic liquid-based systems, they can act as reaction media, catalysts, co-catalysts, structure modulation regents, separation media and even multifunctional materials in improving the sustainability of catalysis procedures. In this chapter, we will show the development and prospect of the utilization of ionic liquids in some typical sustainable catalysis procedures.

2. Physical Properties of Ionic Liquids and Their Potentials in Sustainable Catalysis

Physical properties of ionic liquids such as liquid range, vapour pressure, polarity, solubility, viscosity and density are the essential factors in controlling their performance in catalysis procedures. Therefore, it is necessary to depict the physical properties of ionic liquids and the relationship with their structures in order to obtain useful information on the rational design of ionic liquids for particular processes.

2.1 Liquid Range

Ionic liquids are often composed of bulky asymmetric cations and weakly coordinating anions, which makes them have a low packing efficiency in the lattice structure and thus a low melting point and low glass transition temperature [9]. The increased symmetry of cations and the size difference in cations and anions will increase the melting point. For example, melting points of pyridinium-based ionic liquids are obviously higher than those of imidazolium-based ionic liquids. For the imidazolium-based ionic liquids, the length of the side chain can significantly affect their melting point. Holbery et al. [10] determined the melting point of a series of 1-alkyl-3-methylimidazolium ([$C_n$mim]) ionic liquids (Figure 1). It was found that when the length of the alkyl chain was between 5 and 9, the ionic liquids showed the lowest melting point or glass transition temperature, which can be as low as $-90\ ^\circ$C. Further increase of the chain length increased their melting point. They believed that the rapidly decreased glass transition temperatures with increasing alkyl chain length was due to the elongation of the molecular chain, which disrupted the crystal packing and reduced the lattice energy; however, the further increase of the alkyl chain length enhanced the attractive van der Waals interactions and the amphiphilic character, and then led to a significant increase of melting points. In addition, it was also found that the substitution of hydrogen at C-2 by the methyl group would also increase the melting point of [$C_n$mim] ionic liquids because of the interruption of the hydrogen bonds between the cations and the anions [11].

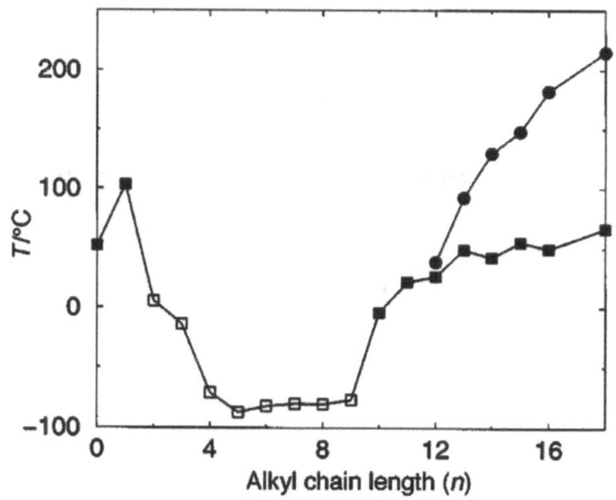

Figure 1
Phase diagram for 1-alkyl-3-methylimidazolium tetrafluoroborate ionic liquids and liquid crystals showing the melting (closed square), glass (open square) and clearing (circle) transitions measured by differential scanning calorimetry [10].

The high ionic characteristic of ionic liquids makes them difficult to vaporize and this leads to negligible vapour pressure and very high boiling points. This results in the high upper temperature limits of ionic liquids, which are often bounded by their decomposition rather than by their vaporization. Therefore, the understanding of thermal stability of ionic liquids is important for their safe utilization and storage. It is found that protic ionic liquids are less thermally stable than their aprotic homologues because the hydrogen bound to the nitrogen on the cations can be easily neutralized via proton transfer to the anions to form a neutral species that is rather volatile [12] (but this is not always the case [13]). The stability of aprotic ionic liquids such as 1,3-dialkylimidazolium ionic liquids increases with the increase of the side chain length, the anion size and the substituent group type. For example, introduction of vinyl and phenyl groups increases the stability of ionic liquids [14], while the increase of nucleophilicity of the anion decreases their stability. An Ab initio calculation study combined with thermogravimetric analysis and mass spectrometry suggests that the decomposition of aprotic ionic liquids follows an E2 type mechanism, which may explain the experimental observation that the increase in the basicity of the anion and acidity of the cation can both unstabilize ionic liquids [15]. Therefore, the weakly coordinating anions and the substitution of the C-2 hydrogen of imidazolium can increase the thermostability of the ionic liquids.

It is clear that the low melting point and high decomposition temperature make it possible for ionic liquids to have a wide liquid range, which allows one to use ionic liquids in a wide temperature range without the risk of exposure to a toxic organic atmosphere.

Meanwhile, the use of ionic liquids avoids volatilization of the solvent, reduces loss of the reactant and then increases sustainability of the related catalytic systems.

2.2 Density and Viscosity

The densities of ionic liquids are mainly determined by the type of cation and anion and their interaction, and the typical values are between 1.1 and 1.6 g/cm^3. For the ionic liquids with the same anion, their densities are affected by the alkyl chain length on the cations. For example, densities of 1,3-dialkylimidazolium hexafluorophosphate decrease with increasing alkyl chain length [16], and the introduction of hydroxyl or ether group at alkyl chains increases the densities [17]. For the ionic liquids with the same cation, their densities are mainly affected by the volume and coordination ability of the anions; the ionic liquids with large size and weak coordination ability have higher density values. For instance, the density for the imidazolium-based ionic liquids with different anions follows the sequence: $(CF_3SO_2)_2N^- > C_3F_7CO_2^- > CF_3CO_2^- > PF_6^- > BF_4^- > Cl$ [16a].

The viscosities of ionic liquids are usually much higher than those of molecular solvents mainly due to the strong electrostatic interactions between the cation and the anions although van der Waals and hydrogen bond interactions can also increase their viscosities. Dzyuba et al. [18] systematically studied the change tendency in the viscosities of $[C_n\text{mim}][(CF_3SO_2)_2N]$ and $[C_n\text{mim}][PF_6]$, and they found that viscosities of the both series increased with the increase of their alkyl length. On the other hand, the viscosities of ionic liquids increase with the increase of the symmetry of anions [19]. For $[C_n\text{mim}]$-based ionic liquids, the viscosities decrease in the order: $PF_6 > SbF_6 > BF_4 > (CF_3SO_2)_2N$, $CF_3COO > CF_3SO_3$. The high viscosities of ionic liquids limit their utilization due to the difficulty in mass and thermal transfers; thus, the development of ionic liquids with low viscosity is important for broadening their industrial application. Related results showed that the functionalization of the cation with siloxyl chains [20] or alkoxy chains [21] can significantly reduce the viscosity of ionic liquids.

2.3 Polarity

The qualitative idea about polarity is based on the observation of 'like dissolves like', which has been widely accepted and understood, but the salvation ability of ionic liquids is significantly different from that of traditional organic solvents and often does not comply with the traditional solubility principles. The currently accepted definition of polarity [22] is described by the sum of all possible intermolecular interactions between the solvent and any potential solutes, excluding those interactions leading to a definite chemical change of the solutes. This includes Columbic interactions, dipole interactions, hydrogen bond interactions, and electron pair acceptor and donor interactions. There are many empirical solvent polarity scales attempting to give quantitative estimates of solvent polarity, although none of them is perfect because they are sensitive to different aspects of polarity and often lead to contradictive results. However, it is still possible to draw some tentative conclusions, for example, the polarity of most of the ionic liquids falls in the range of that of short and medium alkyl chain length alcohols.

The most commonly used polarity scale is the dielectric constant. The static dielectric constants of ionic liquids have been derived from microwave dielectric spectroscopy. It was found that most of the dielectric constant values for ionic liquids ranged from 9 to 15 as moderate polar liquids [23], but the values for some protic ionic liquids were found to be >30 [24], suggesting that the dielectric constant of ionic liquids is affected by hydrogen bond networks. Other methods using single molecule spectroscopic probes and Kamlet–Taft parameters have also been applied to estimate the polarity of ionic liquids [7,25–27]. Measurements of the hydrogen bond donor ability of ionic liquids using the Kamlet–Taft approach indicated that the strong hydrogen bond donor ability of some ionic liquids correlated well with their performance in catalytic reactions [25]. However, the polarity scales from other approaches such as fluorescence spectra of coumarin 153 [26] and ultraviolet visible (UV-vis spectrum of $Fe(phen)_2(CN)_2$ [27] suggested that these ionic liquids were poor hydrogen bond donors. Welton et al. [7] indicated that when using empirical polarity scales to gain an insight into ionic liquid effects on solvent-dependent phenomena, it is important to consider the nature of the solute species being investigated. This means that the phenomena arising from neutral solutes will be better modelled using polarity scales based upon neutral probes, whereas those arising from charged solutes are probably modelled better using polarity scales based upon charged probes.

3. The Utilization of Ionic Liquids in Sustainable Catalysis Procedures

As has been mentioned above, ionic liquids are initially used as solvents in synthesis due to their tuneable solubility for organic compounds as well as ionic type catalysts. However, with the exploitation of functional ionic liquids, it is realized that besides solvents, ionic liquids can act as special ligands for metal complexes and can provide a unique stabilization effect for some reaction intermediates, such as free radicals and carboions, and then increase the catalytic activity and selectivity of related catalytic systems. Ionic liquids can also provide an ingenious environment for enzymes and increase their stability and activity, and thus optimize their catalytic performance. It was also found that ionic liquids are good solvent candidates for biomass, and ionic liquid-based systems have been developed for sustainable biomass transformation. In addition, acidic and basic functional ionic liquids have been widely used as acidic and basic catalysts in catalytic transformations. Here, we will discuss the utilization of ionic liquids in such aspects and the increased sustainability due to the introduction of ionic liquids.

3.1 The Utilization of Ionic Liquids in Transition Metal Catalysis Reactions

Obviously, the utilization of ionic liquids instead of volatile organic solvents can avoid atmospheric pollution, and reduce the cost of the process. The use of ionic liquids can also simplify the work-up procedures by distillation of the volatile products. In addition, the safety of related processes can be significantly improved. For example, in the oxidation reactions using air or oxygen as oxidant, traditional organic solvents can easily form explosive mixtures with oxygen in the gas phase, while the replacement of organic solvents with ionic liquids does overcome this problem. Thus, the use of ionic liquids in these transformations makes the catalysis processes more economic, environmental friendly and safe.

In transition metal catalysis, the design and synthesis of special ligands are often needed in order to guarantee the rational solubility of catalysts in the reaction system. The introduction of ionic liquids to this area can overcome this limitation because ionic liquids have a strong salvation ability for ionic type catalysts and their salvation ability can be tuned through the feasible combination of different cations and anions. Due to the high cost, recovery and reutilization of transition metal complexes are crucial in industrial applications. However, it is difficult to separate homogeneous catalysts from the products and by-products after the reaction was completed in traditional organic solvents. The high solubility of catalysts in ionic liquids also makes it easy to separate the catalysts from the products by solvent extraction. In addition, the special solubility of ionic liquid towards organic reactants or reaction intermediates and the coordination effect for transition metals can significantly improve the catalytic reactivity and selectivity. Therefore, ionic liquids have been found to be excellent reaction systems for transition metal catalysis processes [28–30]. In order to save space, here we focus our attention only on the utilization of ionic liquids in hydrogenation and oxidation reactions and their contributions towards sustainability of the catalytic procedures (Scheme 1).

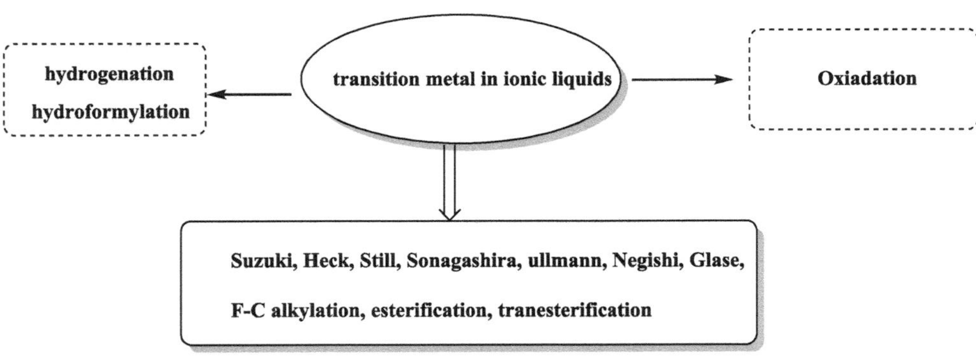

Scheme 1
Transition metal catalysis reaction types in ionic liquids.

3.1.1 Hydrogenation in Ionic Liquids

Catalytic hydrogenation, especially selective hydrogenation reaction, has important applications in the pharmaceutical industry. However, the traditional homogeneous catalytic systems suffer from low selectivity and the difficulty in the separation of the homogeneous catalysts from the products, and thus, the reutilization of catalyst is restricted. Therefore, it is necessary to develop facile catalytic systems that can combine the superiority of both homogeneous and heterogeneous catalyses. Because of the unique characteristics, ionic liquids have been regarded as intriguing solvent systems for hydrogenation. However, the hydrogenation using ionic liquids as solvents has not attracted sufficient attention due to the limited solubility of hydrogen gas in ionic liquids and the high viscosity of ionic liquids, which results in difficulty in mass transfer. For example, Dyson et al. [30] studied the catalytic activity of ionic cluster $[H_3Os_4(CO)_{12}]^-$ on the hydrogenation of styrene in a number of ionic liquids, and they found that the ionic liquids with a higher viscosity led to a lower turnover frequency, suggesting that mass transfer restricts catalytic performance. Therefore, development of low viscosity ionic liquids is necessary in this aspect.

Due to the high viscosity and high polarity of ionic liquids, the ionic type of metal complex often has a reasonable solubility in ionic liquids, also metal nanoparticles can be well dispersed in ionic liquids. Thus, extensive studies have been reported on the utilization of ionic liquids in hydrogenation under biphasic reaction conditions. In 1995, Chauvin et al. [31] developed a biphasic hydrogenation reaction of 1-pentene and cyclohexene using 1-butyl-3-methylimidazolium ionic liquids paired with a weak coordinating anion such as tetrafluoroborate and hexafluorophosphate as the reaction medium. It was found that although the reactant has a low solubility in the ionic liquid phase containing an ionic-type Rh catalyst, the reaction rate of hydrogenation in the ionic liquids was almost improved five times compared with that performed in acetone under similar reaction conditions. This reaction system also benefits from increased selectivity and feasible separation. In addition, the ionic catalyst solubilized in ionic liquids can be easily recycled, and its leaching loss into the organic phase is below the detection limit of 0.02%.

It is worthy to note that due to the higher solubility of dienes compared with that of monoenes in ionic liquids, the selective hydrogenation of monoenes can be achieved successfully in biphasic systems. For example, Dupont et al. [32] synthesized a series of Pd, Fe, Co, Ni complexes and Pd nanoparticles. It was found that all of them can achieve high catalytic activity and high selectivity for diene hydrogenation, and the product and catalyst can be separated by simple extraction, the catalysts supported in ionic liquids can be easily recovered and reused. Han et al. [33] immobilized nanoparticle Pd onto molecular sieves using functional ionic liquid 1,1,3,3-tetramethylguanidium lactate as a

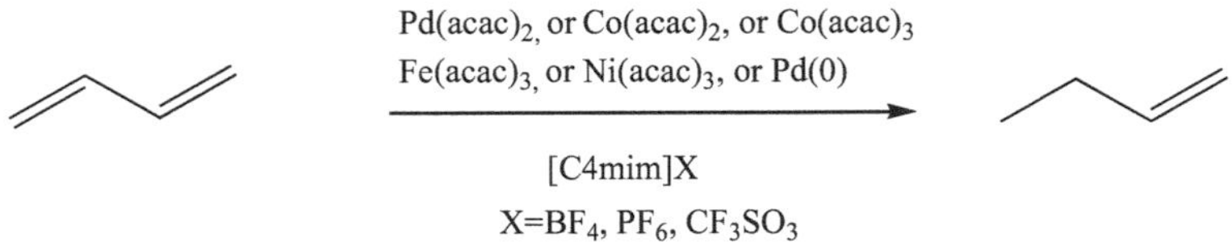

Scheme 2
Selective hydrogenation of diene in ionic liquids.

potential ligand, and the immobilized catalytic system was used for solvent-free hydrogenation reactions. The authors noted that this nanocatalyst showed unprecedented activity compared to the biphasic ionic liquid system or with a directly supported nanocatalyst. They believed that the combination of nanoparticles, ionic liquids and molecular sieves showed synergistic effects to enhance the activity and durability of the catalyst. The selectivity for the hydrogenation of cyclohexadiene to cyclohexene was very high. The authors attribute this to the properties of Pd metal and the nanoscale size of the particle (Scheme 2).

It is well known that many transition metal catalysts are water sensitive and very low concentration of water can quench their catalytic activity. However, most of the ionic liquids including hydrophobic ionic liquids can adsorb water from the atmosphere if no strict action is taken. This usually has a negative effect on the hydrogenation reactions in ionic liquids. Nevertheless, this is not always the case. For example, Rossi et al. [34] found that the catalytic activity of hydrated Ru nanoparticles in ionic liquids is about 50 times higher than in its anhydrous version, indicating that water in ionic liquids is not strictly forbidden and its existence may have an activation effect on the catalyst precursor. This discovery indicates another superiority for the transition metal-catalysed hydrogenation reactions in ionic liquids: no strict moisture control is necessary to guarantee the catalytic activity.

3.1.2 Oxidation in Ionic Liquids

As a kind of important functional group transformation reactions, oxidation has an important role in catalysis and organic synthesis. The main challenge in this field is selective oxidation and development of green catalysts and oxidants. Without doubt, O_2 or air is the ideal oxidant, but the related catalytic systems suffer from a low diffusion rate of O_2, and unstability of solvents and free radicals. In addition, as an efficient method to improve the selectivity such as the aldehyde selectivity over acid from the oxidation of alcohols, electrooxidation can avoid the use of conventional chemical oxidizing and reducing reagents, but it is limited by low electronic conductivity, small chemical window of traditional organic solvents and laborious separation of the products.

As has been mentioned above, ionic liquids have a high polarity and high ionic environment. They can act as an excellent reaction medium to provide a stabilization effect for the free radicals produced in the oxidation process as well as excellent and modulated solubility for ionic type catalysts. Yao [35] developed a strategy using ionic liquid $[C_4mim][PF_6]$ and 4-(dimethylamino)pyridine as efficient immobilizer for OsO_4-catalyzed olefin dihydroxylation reactions of various olefins. It was found that both the catalyst and the ionic liquid can be recycled and reused with only a slight activity decrease after six cycles. Using 1,3-dialkylimidazolium-based ionic liquids as immobilizers for the osmium complex and reaction solvents for the asymmetric Sharpless dihydroxylation, Afonso et al. [36] found that a series of alkenes including hex-1-ene, styrene, methyl-trans-styrene, methyl-cyclohexene and 1,2-diphenylethene can be oxidized selectively with high yields and high enantiomeric excess. Through the combination of the use of supercritical carbon dioxide in the separation process, the product diol can be isolated without any contamination with toxic osmium complex, which in a great extent decreases the risk coming from the metal contamination. This separation technology does not use organic solvents and makes the reutilization of the osmium complex possible even after 12 cycles.

Adipic acid is an important industrial material for the production of Nylon-66, million tons of which was needed per year in the world [37]. Its traditional industrial synthesis involves nitric acid oxidation, and this method inevitably produces nitrous oxide which is commonly thought to cause ozone depletion, acid rain and smog. Inspired by Noyori's work [38] on the oxidation using $[CH_3(C_8H_{17})_3N][HSO_4]$ as a phase transfer catalyst in the presence of a small amount of $Na_2WO_4 \cdot 2H_2O$, Sun et al. [39] synthesized a series of SO_3H-functionalized ionic liquids and studied their co–catalytic activity with $Na_2WO_4 \cdot 2H_2O$ for the oxidation of cyclohexene to adipic acid. It was found that the ionic liquids with the *p*-toluene-sulphonate (Tos) anion have a much higher catalytic activity than those with the HSO_4 anion, although the latter ones have a higher acidity. The authors believed that this is due to the more oleophilic character of the Tos anion which enhances the dissolution of cyclohexene with the ionic liquid-supported catalyst and increases its catalytic performance. After the completion of reaction, the catalyst and the products can be separated by filtration. The recovered ionic liquid can be reused with little decrease in catalytic selectivity. Hashemi et al. [40] developed a heterogeneous catalyst system, silica sulphamic acid-supported task-specific ionic liquid $[C_4mim][WO_4]$, for the oxidation of cyclohexene to adipic acid using H_2O_2 as the oxidant. They suggested that the combination of silica-supported acid catalyst with hydrophilic task-specific ionic liquids can generate an amphiphilic reaction environment and provide numerous nanoscale reactors containing all necessary components for cyclohexene oxidation in an aqueous–organic process. Overall, the designability of ionic liquids makes their diverse utilization possible in a single process.

The electrochemical oxidation of benzyl alcohol has been conducted in an ionic liquid/supercritical CO_2 biphasic system. Two ionic liquids $[C_4mim][BF_4]$ and $[C_4mim][PF_6]$ were used as solvents and electrolytes for the electro-oxidation [41]. It was found that benzyl alcohol could be selectively oxidized to benzaldehyde, and the selectivity is dependent on the pressure of CO_2. This was attributed to an optimization of the solubility difference between the reactants and products. This work indicated that the products could be easily recovered from the ionic liquids using supercritical CO_2 extraction after the electrolysis, and the ionic liquids could be easily reused. Wu et al. [42] developed a homogeneous supercritical CO_2/[C_4mim][PF_6]/MeCN solution for the electro-oxidation of benzyl alcohol. They found that the high selectivity and yield of benzaldehyde could be obtained by tuning electrolysis conditions such as electrolysis potential, electric quantity and system pressure, and both the selectivity and the yield of benzaldehyde could be >99.8% under optimized reaction conditions.

3.2 Enzymatic Reactions in Ionic Liquids

Enzymatic reactions have many benefits in the context of green chemistry and sustainable chemistry. Such reactions can be performed under mild conditions including physiological pH, ambient temperature and atmosphere pressure. Enzymatic reactions of multifunctional molecules can proceed with excellent chemoselectivity, regioselectivity, and stereoselectivity without the need for functional protection and activation, which is often unavoidable in traditional organic syntheses. Moreover, enzymatic cascade processes integrating multiple steps are more feasible than traditional chemical or chemo-catalytic processes because most biocatalytic processes are performed under rough conditions. Enzymatic procedures in ionic liquids have received increasing attention in recent years not only due to the wide liquid range, adjustable physical properties such as density, viscosity, polarity and solubility of ionic liquids but also due to their stabilization effect for the structure of enzymes. Various ionic liquids have been involved in enzymes systems such as lipase, proteases, esterase, lysozyme, epoxide hydrolase, dehydrogenases, oxidases and aldolases and some of the investigations have been well reviewed [43]. Therefore, we only like to discuss the influence of ionic liquids on the activity and stability of enzymes as well as the enzymatic reactions for the production of biofuels in the presence of ionic liquids (Scheme 3).

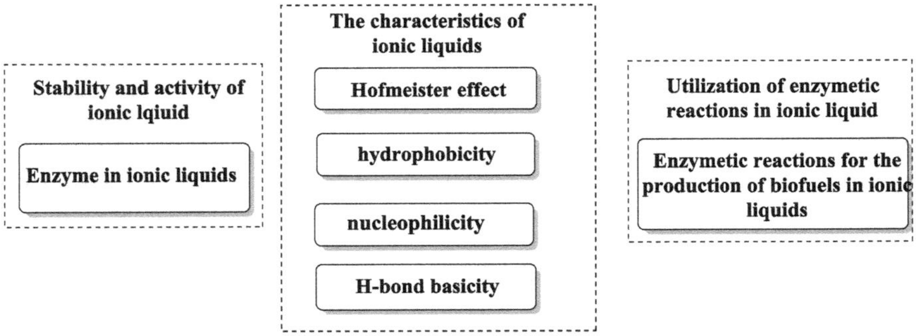

Scheme 3
Main contents discussed in the section of enzymatic reactions.

3.2.1 Hofmeister Effect of Ionic Liquids on the Stability and Activity of Enzymes

Salt effect in aqueous solutions is very important because biological processes often occur at very high salt concentrations. The stability and activity of proteins are greatly influenced by the type of ions in the way of Hofmeister order. In 2007, Weingartner et al. [44] used differential scanning calorimetry to systematically characterize the effect of different ionic liquids on the thermal denaturation of ribonuclease A. They found that the effects of cations and anions of ionic liquids were practically additives, and it is possible to establish mutually independent cation and anion series to interpret the stability of enzyme in ionic liquids:

$K^+ > Na^+ > [C1,1,1,1N]^+ > Li^- > [C2,2,2,2N]^+$,
$[C_2mim]^+ > [C_4mpyrr]^+ > [C_4mim]^+$, $[C3,3,3,3N]^+ > [C_6mim]^+ > [C4,4,4,4N]^+$ for cations, and
$[SO_4]^{2-} > [HPO_4]^{2-} > Cl^- > [EtOSO_3]^- > [BF_4]^-$,
$Br^- > [MeOSO_3]^- > [TfO]^- > [SCN]^-$, $[N(CN)_2]^- > [(CF_3SO_2)_2N]^-$ for anions.

In the traditional classification of ion effect on the enzyme, Kosmotropes are considered to promote water structure, while chaotropes are assumed to destroy water structure. Several research groups investigated the Hofmeister effects of the cation and anion of ILs on enzyme performance in aqueous solution [45]. It was shown that under most circumstances, enzymes followed the general Hofmeister effect, that is, kosmotropic anions and chaotropic cations stabilize enzymes, while chaotropic anions and kosmotropic cations destabilize enzymes. Yang [45a] summarized the Hofmeister effect of different ions on the activity and stability of some enzymes in aqueous solutions. His result can be used to direct the design of ionic liquids through the rational combination of chaotropic cation with a kosmotropic anion in order to enhance the stability and activity of a special enzyme. [Choline][H_2PO_4] is a successful example designed from this strategy [45b]. Cytochrome c dissolved in [choline][H_2PO_4] hydrated with 20 wt% water can maintain its activity and structure after 18 months of storage at room temperature. Comparably, if this protein is stored in an aqueous buffer solution such as Tris–HCl or phosphate buffer, its activity will be lost in only 1–2 weeks.

However, the complex anion does not always strictly follow the Hofmeister sequence. For example, the initial reaction rate of lipase-catalysed enantioselective hydrolysis of D,L-phenylglycine methyl ester in the ionic liquid-containing aqueous buffer varied in the order $[C_4mim][BF_4] > [C_4mim][Cl] > [C_4mim][Br] > [C_4mim][NO_3] > [C_4mim][HSO_4]$, which is different from the Hofmeister order of anions [45c]. These facts imply that the ionic kosmotropicity/chaotropicity of the ionic liquids is not the only factor in affecting enzyme performance.

3.2.2 Effect of Hydrophobicity of Ionic Liquids on the Stability and Activity of Enzymes

In imidazolium ionic liquids with longer alkyl chains on the cations, enzymes are usually found to have higher activity [46,47] and higher stability [45a], suggesting that hydropobicity of ionic liquids is also an important factor that influences the performance of enzymes. Russel and co-workers [48] observed that the free lipase showed promising activity in the hydrophobic ionic liquid, [C_4mim][PF_6], but was inactive in the hydrophilic ionic liquids such as [C_4mim][CH_3COO], [C_4mim][NO_3] and [C_4mim][CF_3COO]. In these cases, especially when there has been no water activity control, hydrophobicity may provide a better explanation for the experimental phenomenon.

LogP scale is used to quantify the hydrophobicity of solvents, which is the logarithm of its partition coefficient P between octanol and water. For the ILs with [C_4mim] as cation, the logP values decrease in the order $Tf_2N^- > PF_6^- > Cl^- > NO_3^- > Br^- > BF_4^-$, and for the ionic liquids with the most hydrophobic anion, Tf_2N^-, the logP values increase with the increase of the alkyl chain length on the imidazolium ring of the cation [49]. In the study of using ionic liquids as additives for sol–gel immobilization of *Candida rugosa* lipase, a linear relationship was observed between the residual activity of the immobilized lipase after 5 days' incubation and the logP values of the ILs [50]. It was also reported that the initial rates of the lipase-catalysed transesterification reactions showed an obvious trend of progressively rising with an increase in the logP values of the ionic liquids used [51]. These results imply that the hydrophobic ionic liquids are more favourable for stabilizing and activating enzymatic reactions.

However, the hydrophobicity of ionic liquids is not always positive in enzymatic reactions. For example, Rudolph et al. [52] investigated the influence of a series of imidazolium ionic liquids on the renaturation of two enzymes. It was found that more hydrophobic imidazolium cations with longer alkyl chain destabilized the enzyme, while the cations with hydrophilic terminal hydroxyl group stabilized the enzyme. Yuan and co-workers [53] demonstrated that the conformation stability and transesterification activity of *Candida antarctica* lipase B (CALB) dissolved in hydrophilic hydroxyl-functionalized ionic liquids were much higher than that dissolved in nonhydroxyl-functionalized companions.

3.2.3 Effects of Nucleophilicity and H-Bond Basicity of Ionic Liquids on the Stability and Activity of Enzymes

Based on their observation that *Candida rugosa* lipase was active only in [C_4mim][PF_6], but not in other ionic liquids with NO_3, CF_3CO_2, CH_3CO_2, CF_3SO_3 and CH_3SO_3 as anions, Kaar et al. [48] suggested the importance of nucleophilicity in enzymatic reactions. They indicated that these more nucleophilic anions may coordinate strongly with the positively charged sites in the enzyme structure, and cause changes in the enzyme conformation. Thus, the use of ILs with low anion nucleophilicity is essential for enzyme activity. The importance of H-bond basicity was brought to attention by Lau et al. [54]. They found that the low activity of CALB in [C_4mim][lactate] was accompanied by its denaturation upon dissolution, and suggested that the strong ability of lactate to form stable H-bonds with the polypeptide backbone of the enzyme should be responsible for this situation. Therefore, it was proposed that the ionic liquid with a strong H-bond basicity anion may cause dissociation of the hydrogen bonds that maintain the structural integrity of the α-helices and β-sheets, which in turn cause the protein to unfold.

3.2.4 Enzymatic Reactions for the Production of Biofuels in Ionic Liquids

Enzyme-catalysed transesterification is a very powerful method for the production of biodiesel fuels [55]. In this procedure, the by-product glycerol can be easily separated and the purification of fatty methyl esters is very simple. The main hurdle of this enzymatic reaction is that recovery of the enzymes is difficult, which increases the production cost to a great extent. Dupont and co-workers [56] developed an alternative green strategy with a lipase supported on ionic liquids for the production of biodiesel from the alcoholysis of soybean oil. Different lipases and imidazolium ionic liquids with different anions were screened to optimize this reaction. It was found that *Pseudomonas cepacia* lipase and hydrophobic ionic liquid [C_4mim][$(CF_3SO_2)_2N$] were the best combination. In this catalysis system, the transesterification reaction can be well performed at room temperature in the presence of water. In addition, the other intriguing aspects of this catalytic system include the wide alcohol compatibility and the product biodiesel can be separated by simple decantation. The recovered ionic liquid and enzyme can be reutilized at least four times without the loss of catalytic activity and selectivity (Scheme 4).

However, the utilization of ionic liquid [C_4mim][$(CF_3SO_2)_2N$] still suffers from the high cost of ionic liquid preparation and its poor biocompatibility. More recently, Zhao and co-workers [57] synthesized a series of eutectic ionic liquids derived from choline acetate and choline chloride coupled with biocompatible hydrogen bond donors. As has been realized, these eutectic solvents have favourable properties such as low viscosity, high biodegradability, and excellent compatibility with enzyme. It was also found that Novozyme 435, supported in choline acetate/glycerol (with molar ratio of 1:1.5), showed high catalytic activity for the enzymatic transesterification of Miglyol oil 812 with methanol, and the conversion was as high as 97%. In their later work, these authors [58]

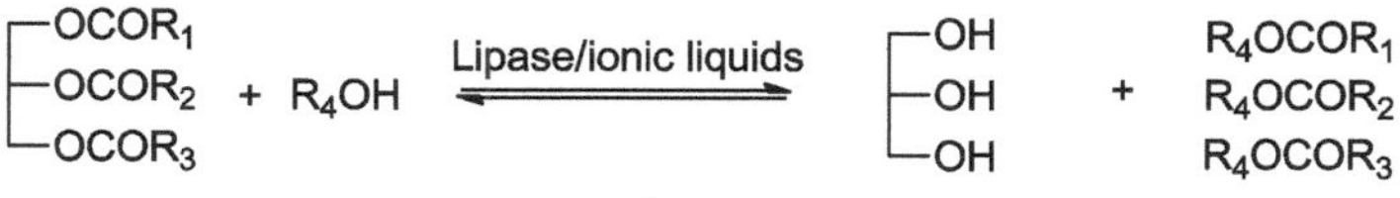

Scheme 4

Transesterification reaction of vegetable oil in ionic liquids.

also applied this catalytic system for the preparation of biodiesel from soybean oil. Under optimized reaction conditions, the triglyceride can achieve up to 88% conversions in 24 h, and the eutectic ionic liquids and enzymatic system can be reused for at least four times without a significant decrease in activity.

Overall, a number of physicochemical properties of ionic liquids contribute to affect the enzyme function in ionic liquids, although none of these seems to be universal for a certain enzymatic reaction because biocatalytic systems contain so many complicated interactions. Some factors specific to the enzyme and the components involved in the reaction system have to be considered as well. For example, the pH values of the medium [59], co-solvent nature [60] and halide impurity [61,62] may influence the physicochemical properties of ionic liquids, and then affect the activity and stability of enzymatic reactions in ionic liquids. Therefore, although there are several successful examples about the use of ionic liquids to increase the catalytic performance of enzyme, there is still a long way to rationally design a specific ionic liquid according to the specific enzymatic reaction.

3.3 The Dissolution and Conversion of Biomass in Ionic Liquids

Biomass conversion to fuels and chemicals has attracted significant interest in the past decade due to the necessity for a renewable source of carbon [63]. Many different reactions and processes to convert biomass into high-value products and fuels have been developed, among which, special attention is given to the conversion of lignocellulosic biomass [64–66]. Lignocellulosic biomass is composed of cellulosic, hemicellulosic and ligninic components. Herein, cellulose is used as an example to discuss the effects of ionic liquids on its dissolution, depolymerization and conversion to fuels and fine chemicals such as 5-hydroxymethylfurfural (HMF) (Figure 2).

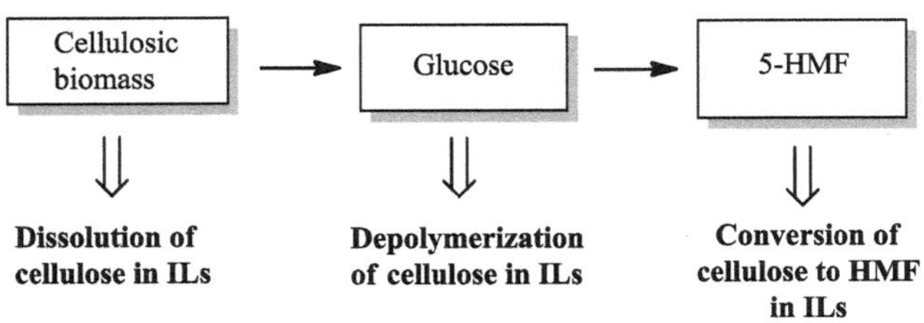

Figure 2
Cellulosic biomass conversion procedure in ionic liquids.

3.3.1 The Dissolution of Cellulose in Ionic Liquids

Cellulose is a polysaccharide consisting of a linear chain of several hundreds to over ten thousand β-(1→4)-linked glucose repeating units (Figure 3) [67], and there are numerous intermolecular and intramolecular hydrogen bonds in cellulose, which make cellulose hard to dissolve in water and traditional organic solvents. A key problem in the separation and conversion of cellulose is to find a proper medium to dissolve cellulose [68]. Heizne and co-workers [69,70] reviewed the solvents used in the functionalization of cellulose, but these traditional aqueous and nonaqueous solvents suffer from toxicity, volatility, high cost, generation of poisonous gas and difficulty in solvent recovery. In 2002, Rogers and coworkers [71] reported that cellulose could be dissolved in 1-butyl-3-methylimidazolium chloride ([C_4mim]Cl) with high concentrations, which led to the development of a new class of cellulose solvent systems. Additionally, it is suggested that during the dissolution of cellulose in ionic liquids, some physical barriers of cellulose have been broken down, such as crystallinity index, surface area, morphology and degree of polymerization (DP) [72], which make subsequent catalysed depolymerization and functionalization of cellulose in ILs more convenient and effective.

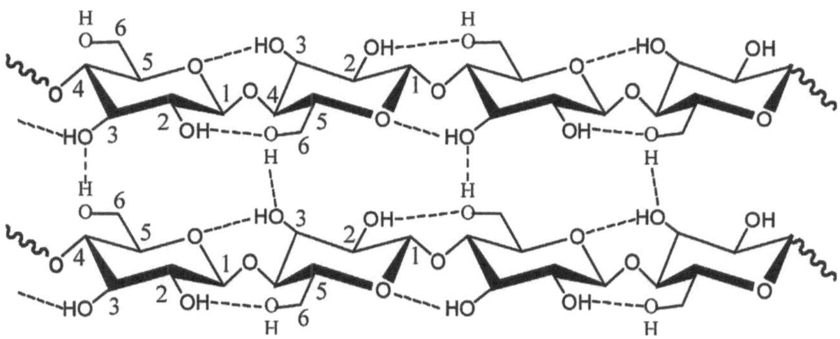

Figure 3
Schematic structure and numbering of cellulose.

Several research groups investigated the effects of physicochemical properties of ionic liquids on the solubility of cellulose. The current results suggest that the high hydrogen bond basicity and dipolarity of anion favour dissolution of cellulose in ionic liquids. Rogers and co-workers [73] studied the interactions between the anion of ionic liquids and cellulose using ^{13}C-NMR (nuclear magnetic resonance spectroscopy) and $^{35/37}$Cl-NMR spectroscopy. It was found that the solvation of cellulose by [C_4mim]Cl involved hydrogen bonding formation between the carbohydrate hydroxyl protons and chloride ions of the ionic liquids in a 1:1 stoichiometry. After that, Zhang and co-workers [74] investigated the dissolution mechanism of cellulose in ionic liquids by using cellobiose and 1-ethyl-3-methylimidazolium acetate ([C_2mim][Ac]) as a model system under various conditions by conventional and variable-temperature NMR spectroscopy. In dimethyl sulphoxide-d_6 solution, NMR data of the model system suggested that hydrogen bonding was formed between hydroxyls of cellobiose and both the anion and the cation of [C_2mim][Ac]. The CH_3COO^- anion favoured the formation of hydrogen bonds with hydrogen of hydroxyls. After acetylation of all hydroxyls in cellobiose, the interactions between cellobiose octaacetate and [C_2mim][Ac] became weaker, which implied that hydrogen bonding could be the crucial factor for the dissolution of cellobiose in this ionic liquid. Liu and co-workers [75] applied molecular dynamics simulations to study the interactions between cellulose and ionic liquids. An all-atom force field was developed for [C_2mim][Ac] and the behaviour of cellulose in this ionic liquid was examined using molecular dynamics simulations of a series of (1–4)-linked β-D-glucose oligomers with a DP $n = 5$, 6, 10 and 20. It was found that the interaction energy between the polysaccharide chain and the ionic liquid was stronger than that for either water or methanol. The anion acetate could form strong hydrogen bonds with hydroxyl groups of the cellulose. Their results support the dissolution mechanism of cellulose in ionic liquids by the formation of hydrogen bonds between the ionic liquid anions and the hydroxyl group protons of the glucose. Some other experiments were also designed to investigate the relationship between the hydrogen bond accepting ability of anions and the solubility of cellulose. For example, we prepared a series of ionic liquids by pairing the [C_4mim]$^+$ cation with different Bronsted basic anions [70], and determined hydrogen bond accepting ability of the anions by using solvatochromatic methods. It was found that the hydrogen bond accepting ability of the anions decreased in the order $Ac^- > HSCH_2COO^- > HCOO^- > (C_6H_5)COO^- > H_2NCH_2COO^- > HOCH_2COO^- > CH_3CHOHCOO^- > DCA^-$, and solubility of cellulose in these ILs increased almost linearly with increasing hydrogen bond accepting ability of the anions. It is noted that the trend of hydrogen bond accepting ability of the anions does not correlate with the established pKa values of the parent acids: HAc (4.75) > $(C_6H_5)COOH$ (4.18) > $CH_3CHOHCOOH$ (3.87) > $HOCH_2COOH$ (3.83) > HCOOH (3.72) > $HSCH_2COOH$ (3.68) > H_2NCH_2COOH (2.35). However, not all of the data support the conclusion that the higher the hydrogen bond accepting ability of the anions, the stronger their ability to dissolve cellulose. For example, Vitz et al. [76] reported that [C_4mim]Cl showed a higher ability to dissolve cellulose than [C_4mim][Ac].

Schubert and co-workers [77] systematically investigated the effect of alkyl chain length of imidazolium-based ionic liquids on the solubility of cellulose in ionic liquids, and a distinct odd-even effect was observed. It was shown that cellulose was more soluble in 1-alkyl-3-methylimidazolium-based ILs with even-numbered (carbon atom) alkyl chains compared to those with odd-numbered (carbon atom) alkyl chains. Marsh et al. [78] attributed this even-odd effect to the additional polarity in the heteroatomic substituent on the imidazolium ring. The high bulkiness of cations might reduce the number of hydrogen bonds between cellulose and cations. A hydroxyl group at the end of side chains would compete with cellulose in forming hydrogen bonds with anions, which was considered to play a key role during dissolution of cellulose in ionic liquids. In another NMR spectroscopy study, Rogers and co-workers [79] found that the interactions between the cations of ionic liquids and the solutes were nonspecific. In all the cases of [C_4mim], [C_2mim] and [C=C_2mim], the change in the relaxation time of carbon nuclei of the cations as a function of carbohydrate concentration was small and consistent with the variation in solution viscosities. Besides, Novoselov and co-workers [80] also indicated that the binding of cation of [C_4mim]Cl with the hydroxyl oxygen in cellulose was difficult due to steric hindrance. In this ionic liquid, the cation remained associated with the Cl^- anion, but the association became weaker with the increase of cellulose concentration. Based on these experimental results, Rogers et al. [68] suggested that the role of the cation in solvating cellulose might be in keeping the anionic hydrogen-bonded cellulose moieties dispersed, and they also indicated that if this consideration is correct, the ability of the cations to organize in solution to shield the negative charges and still pack well with other cations would be important. This can also explain the odd–even effect of the solubility of cellulose in ionic liquids.

3.3.2 The Depolymerization of Cellulose in Ionic Liquids

The direct hydrolysis of cellulose to glucose is considered to be one of the most promising processes due to its excellent industrial potentials [81,82]. H_2SO_4 [83] and heterogeneous solid acids [84,85] were used as active catalysts for the hydrolysis of cellulose in water, but only low glucose yields were obtained.

Zhao and co-workers [86] reported an efficient acid-catalysed hydrolysis of cellulose in ionic liquid. Under atmospheric pressure and 100 °C, 43% yield of glucose and 77% yield of total reductive sugar (TRS) could be obtained in the H_2SO_4/[C_4mim]Cl system. The hydrolysis of celluloses with different DPs ranging from 100 to 450 could also give good results (20% to 39% glucose and 62% to 73% TRS yields) in the same catalytic system. In their later study, Zhao et al. [87] also showed that the HCl/[C_4mim]Cl system could effectively catalyse the hydrolysis of different lignocellulosic materials to give TRS, and 66%, 74%, 81% and 68% TRS yields could be obtained from corn stalk, rice straw, pine wood and bagasse, respectively. Furthermore, kinetics of the acid-catalysed hydrolysis of cellobiose in [C_2mim]Cl was studied as a model system for general lignocellulosic biomass hydrolysis in ionic liquid systems [88]. The results showed that the rate of the two competing reactions, polysaccharide hydrolysis and sugar decomposition, varied with acid strength. For the acids with an aqueous p*K*a value below approximately zero, the hydrolysis reaction was significantly faster than the degradation of glucose, thus allowing hydrolysis to be performed with a high selectivity in glucose. Moreover, some Lewis acids showed catalytic activity for the hydrolysis of cellulose in ionic liquids. For example, Tao et al. [89] reported that catalytic amount of $FeCl_2$ in an acidic ionic liquid, 1-(4-sulphonic acid)-butyl-3-methylimidazolium hydrogen sulphate, could promote effective hydrolysis of cellulose with >84% of conversion yield.

Recently, solid acid catalysts have attracted interest in the conversion of biomass due to their intrinsic advantages such as heterogeneous catalysis, feasible reusability and easy work-up procedure [90–92]. In 2008, Rinaldi and co-workers [86] reported that solid acid Amberlyst 15DRY could act as a powerful catalyst for the hydrolysis of cellulose dissolved in ionic liquids. Cellulose underwent selective depolymerization to produce cellulose oligomers (with a DP around 30) with a 90% yield without any substantial formation of side products. They also found that even wood, a lignocellulosic material, could be hydrolysed using their methodology. Other solid acid catalyst like zeolite was also reported to promote the hydrolysis of cellulose under microwave heating at 240 W [91]. H-form zeolites with a lower Si/Al molar ratio and a larger surface area showed a relatively higher catalytic activity. Under the optimized conditions, a typical hydrolysis reaction with Avicel cellulose could produce glucose in around 37% yield within 8 min.

In addition, some new protocols including tandem hydrolysis/alkylation [93] and hydrolysis/hydrogenation [94] have also been developed to transfer the depolymerization of cellulose to sugar derivatives in ionic liquids.

3.3.3 The Conversion of Cellulose to HMF in Ionic Liquids

As the product of hydrolysis of cellulose, glucose can be converted to several polyalcohols and other fine chemicals by hydrogenation and dehydration [95–98]. Among these conversions, special attention is given to the conversion of glucose to HMF due to the perspective of HMF as versatile and key intermediate in biofuel chemistry and petrochemical industry [98]. Furthermore, the conversion of glucose to 5-HMF can also make the hydrolysis product of cellulose to be easily separated from ionic liquids by simple extraction.

In 2007, Zhang and co-workers [99] reported the catalytic conversion of sugars to HMF with high yield using metal halides in [$C_n$mim]Cl as the catalyst. A wide range of metal halides was found to catalyse the conversion of fructose to HMF, among which chromium(II) chloride was found to be uniquely effective, leading to the conversion of glucose to HMF with a yield of about 70%. Only a negligible amount of levulinic acid was formed in these reactions. The catalytic mechanism of this reaction was also proposed, in which the $[CrCl_3]^-$ intermediate formed between [C_2mim]Cl and $CrCl_2$ played a key role in proton transfer, which facilitated the isomerization formation of fructose, followed by a rapid dehydration to 5-HMF (Figure 4). After that, many chromium(II) salts with different ionic liquids systems have been tested to promote the conversion of sugars into 5-HMF [100,101]. For example, an N-heterocyclic carbene (NHC)-Cr/ionic liquid system has been developed for the conversion of sugars into 5-HMF with a yield up to 96% and 81%, respectively, for fructose and glucose [100]. Li et al. [101] reported that 90% isolated yield from glucose could be obtained under microwave irradiation in the presence of $CrCl_3$ and ionic liquid. Furthermore, exceptional effectiveness of $CrCl_3$ in the above processes also attracted the interest of theoretical chemists. Hensen and co-workers [102] proposed a possible mechanism to unravel the molecular-level details of the $CrCl_2$-catalysed glucose dehydration in [C_2mim]Cl. Combined with the kinetic experiments, in situ X-ray absorption spectroscopy and density functional theory (DFT) calculations, they proposed that the transient formation of Cr^{2+} dimers through coordination to a second molecule of the catalyst promoted the isomerization of glucose to fructose, and explained the unique ability of $CrCl_2$ to catalyse the dehydration of glucose to 5-HMF in ionic-liquid media.

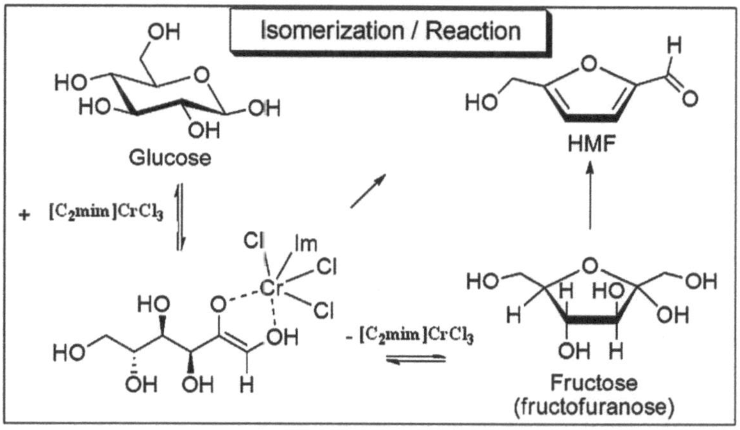

Figure 4
The proposed interaction of metal halide with glucose in [C_4mim]Cl/$CrCl_2$-catalysed transformation of glucose to HMF. *Adapted from Ref. [99].*

In addition to chromium(II) catalyst, Han and co-workers [103] found that the Lewis acid $SnCl_4$ can effectively promote the conversion of glucose into 5-HMF in [C_2mim][BF_4]. Similarly, they suggested that the formation of five-membered ring chelate complex between the Sn atom and glucose might be responsible for the formation of HMF in this conversion. Moreover, heteropoly acids/ionic liquid systems [104] were also reported recently as highly active and selective catalysts for the dehydration of glucose.

Compared to the conversion of monosaccharide to HMF, however, the direct catalytic conversion of cellulose or lignocellulosic biomass to HMF in ionic liquids is a more difficult procedure. Zhang and co-workers [105] reported a successful single-step catalytic protocol for hydrolysis of cellulose and dehydration of glucose to HMF in ionic liquids under mild conditions. A pair of metal chlorides ($CuCl_2$ and $CrCl_2$) dissolved in [C_2mim] Cl at the temperatures of 80–120 °C collectively catalyzed this single-step process of converting cellulose to HMF with an unrefined 96% purity among recoverable products (at $55.4 \pm 4.0\%$ HMF yield). After extractive separation of HMF from the solvent, the catalytic performance of recovered [C_2mim]Cl was investigated. It was shown that the catalyst can be reused three times without obvious catalyst deactivation. They also found that single metal chlorides at the same total loading showed considerably less activity under similar conditions. Zhao et al. [101] reported another efficient strategy for the direct conversion of cellulose into HMF with $CrCl_3$ as catalyst in ionic liquid, and 60% isolated yield was obtained under microwave irradiation. For the reaction mechanism, the authors suggested that the coordination chemistry involving $CrCl_3$ played an important role in this process. Moreover, a direct conversion of lignocellulosic biomass (e.g. corn stover and pine sawdust) to HMF was reported by Raines [106] (Figure 5). The reaction was performed in N,N-dimethylacetamide (DMA)/LiCl solvent by using $[C_2mim]Cl$ as an additive in the presence of chromium(II) chloride and hydrochloric acid, and 54% HMF yield was reported from cellulose at 140 °C within 2 h. Mechanistic analyses revealed that loosely ion-paired halide ions in DMA/LiCl are critical for the remarkable rapidity (1−5 h) and yield (up to 92%) in this low-temperature (≤140 °C) process. In this catalysis system, the lignocellulosic biomass could also be readily converted to HMF without any pretreatment. In addition to lignocellulosic biomass, the macroalgae-derived carbohydrates can also act as the feedstock for the HMF production.

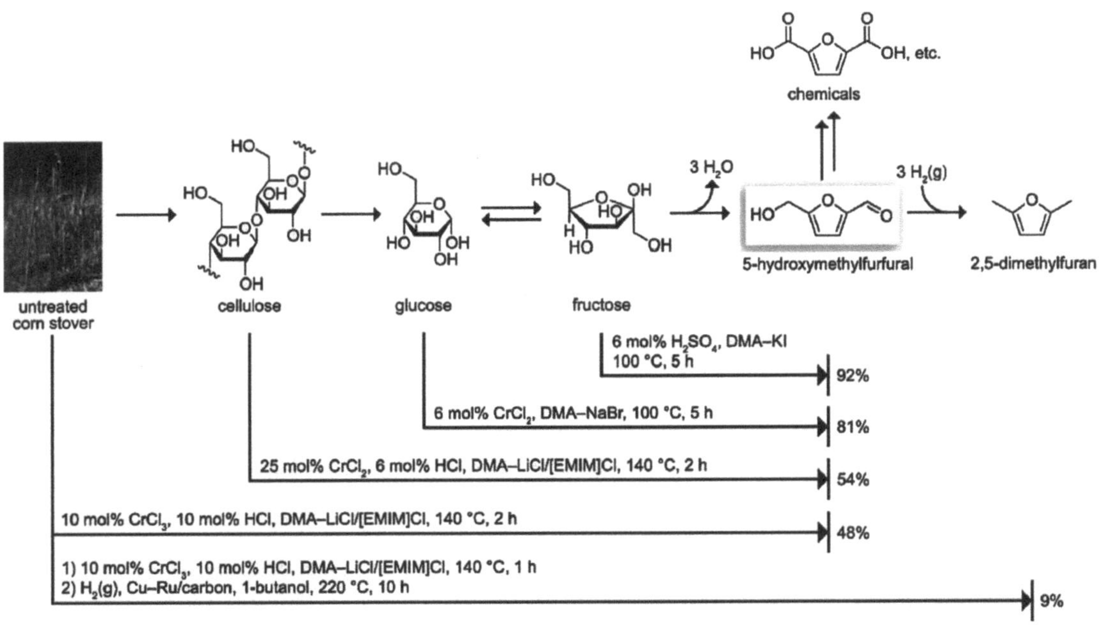

Figure 5
Direct transformation of untreated biomass into different chemicals [106].

3.4 The Transformation of CO_2 to Cyclocarbonates Using Ionic Liquids

CO_2 has been believed to be partly responsible for global warming and climate change as an important greenhouse gases. It is also a sustainable C_1 resource due to its nonflammable, nontoxic, economical, abundant and renewable characteristics. Therefore, the efficient transformation of CO_2 into useful chemicals is an important contribution from the viewpoint of environmental protection and resource utilization. In 1999, Brennecke et al. [107] found that when [C_6mim][PF_6] was exposed to the CO_2 atmosphere, its mass can increase 0.0881%. The Fourier transform infrared (FT-IR) measurements of CO_2 saturated ionic liquids showed the characteristic peaks of dissolved CO_2 at 2380 and 2400 cm^{-1}. Their subsequent researches [108−110] illustrated that although CO_2 can be dissolved in an ionic liquid phase with very low volume expansion, the solubilization of ionic liquids in CO_2 can be ignored. After that, numerous functional ionic liquids [111−115] were developed to increase the solubilization of CO_2, and the related results suggested that the increase in the basicity of both the cation and the anion benefits the dissolution of CO_2 in ionic liquids.

The high solubility of CO_2 in ionic liquids indicates that chemical transformation of CO_2 can be performed under ambient reaction conditions, which has attracted great attention in recent years. In 2001, Deng et al. [116] developed a procedure using [C_4mim][BF_4] as an efficient catalyst for the cycloaddition of CO_2 to propylene oxide with excellent selectivity. This method benefits from simple product separation by distillation and the ionic liquid catalyst can be recycled for another five cycles with a little activity decrease. Then, Kawanami et al. [117] studied the catalytic activity of different imidazolium-based ionic liquids for the cycloaddition of epoxide by CO_2 under supercritical conditions. It was found that the catalytic activity increased with the increase of the alkyl chain length on the imidazolium cations, and [C_8mim][BF_4] was found to have the highest activity. This catalytic system was also found to be suitable for the synthesis of various cyclocarbonates with satisfactory yields. Other Lewis acidic imidazolium ionic liquids were also found to be an efficient catalyst with an increased turnover number or turnover frequency [118−122] (Figure 5).

As the third generation of sustainable solvents, deep eutectic mixtures (DES), a special type of ionic liquids, have received great attention since Abbott et al. [123] found that when choline chloride was mixed with hydrogen bonding donors such as urea, the melting point of the mixture was much lower than that of the two components. They believed that the decrease of the melting point resulted from the interaction between the chloride and urea, and the existence of the associate compounds of chloride and urea was proven by electrospray ionization mass spectrometry. In 2007, Han et al. [124] found that the DES of choline chloride with urea has significant catalytic activity for the cycloaddition of CO_2 to epoxide, and this catalytic procedure benefits from biocompatibility and low cost of catalyst, no necessity of excessive CO_2, wide substrates tolerance and almost 100% conversion and selectivity. Importantly, the high polarity product, cyclocarbonate, cannot be dissolved in the DES, which makes it easy to separate the product from catalyst by decantation. Another important contribution of their work is the proposed hydrogen

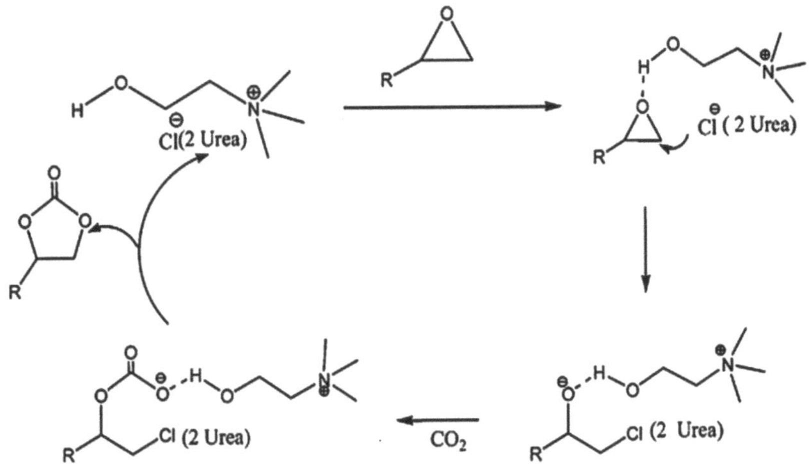

Figure 6
Activation effect of hydroxyl group on the cation of ionic liquids towards epoxides [124].

bond-assisted ring-opening mechanism. As illustrated in Figure 6, the authors suggested that the hydroxyl group on the choline cation could activate the epoxide as a hydrogen donor, and then the activated epoxide was attacked by the anion of ionic liquids to open the epoxy ring, react with CO_2 and form the target cyclocarbonates.

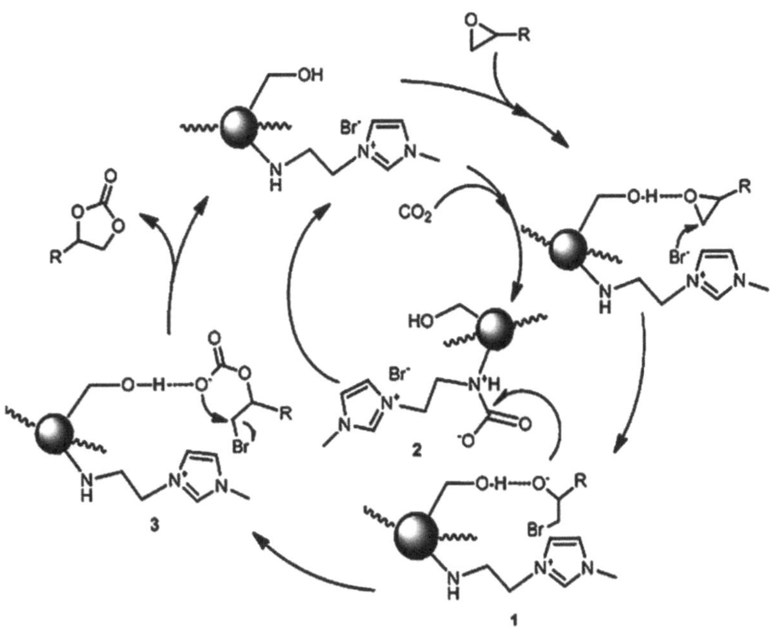

Figure 7
Chitosan-immobilized ionic liquids for the transformation of CO_2 [128].

Then the catalytic performance of functional ionic liquids for the chemical transformation of CO_2 received great attention. Zhang et al. [125–128] developed a series of hydroxyl functionalized imidazolium ionic liquids for this transformation, and they found that this type of ionic liquids could be easily bonded, by covalent bonds, to different immobilizers such as polystyrene resin and chitosan. The immobilized hydroxyl functionalized ionic liquids have high catalytic activity, and the product can be isolated easily. FT-IR measurements showed that there is no significant difference between the fresh and the reutilized ionic liquids. For the chitosan immobilization system, it was found that the functional groups like hydroxyl and amino group on chitosan have synergetic effect with ionic liquids on the catalytic process as illustrated in Figure 7. This transformation was accelerated by the dual activation from the hydrogen bond-assisted ring opening by the hydroxyl on chitosan and anion of ionic liquid, and the nucleophilic tertiary nitrogen-induced activation of CO_2 by the amino on the chitosan. DFT study also proved the existence of the hydrogen bonds between the hydroxyl groups on chitosan and the epoxide ring.

Park et al. [129,130] suggested that if the hydroxyl group on the cation of ionic liquids can activate epoxide as a hydrogen bond donor, then the carboxyl acid group which has higher hydrogen bond donor ability should have a higher activation effect on the epoxide ring. Thus, the incorporation of carboxylic acid moieties into ionic liquids could provide an effective method for improving the catalytic performance of ionic liquids. Based on this consideration, these authors synthesized a series of carboxylic acid group functionalized imidazolium ionic liquids and grafted them onto a silica gel surface, and then studied their catalytic performance for the cycloaddition reaction of CO_2 and epoxides. These ionic liquids showed enhanced catalytic activity compared with its hydroxyl group functionalized companions. The catalytic reaction mechanism is illustrated in Figure 8.

3.5 The Utilization of Functionalized Ionic Liquids in Sustainable Catalysis

Ionic liquids can be used as sustainable catalysts when their cations or anions have been functionalized with acidic or basic groups. The acidic or basic functional ionic liquids inherit the superiority of ionic liquid such as excellent and adjustable solubility, very low vapour pressure and high thermostability. Therefore, these ionic liquids can be easily recovered and recycled after the reaction, and the risk of exposure to the acidic or basic atmosphere can be efficiently avoided. In addition, the reactivity and selectivity may be remarkably increased through the synergetic effect of cation and anion of the ionic liquids [25b,124]. In addition, it is known that some acidic or basic functionalized ionic liquids are able to stabilize reaction intermediates; thus, such systems are useful to understand the reaction mechanism, and then to inspire the further design of catalyst [131].

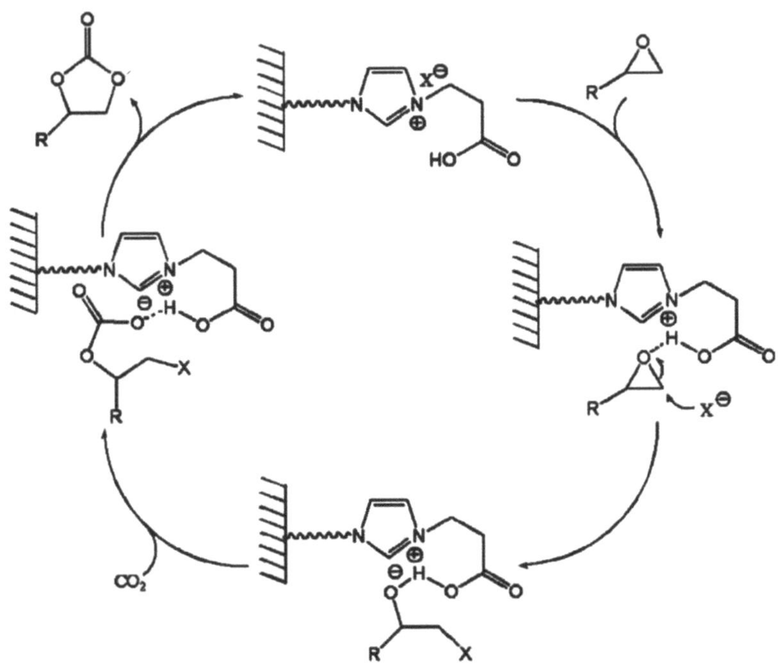

Figure 8
Silica immobilized carboxyl-functionalized ionic liquids for the transformation of CO_2 to cyclo-carbonate [129].

3.5.1 The Reactions Catalysed by Acidic Functionalized Ionic Liquids

Acid catalysis is an important technology and has been widely utilized in different types of chemical reactions. However, numerous environmental problems have arisen due to the use of strong acids and the generation of wastes during the related procedures. Therefore, with the development of the first generation of ionic liquids, the possibility of using ionic liquids as more environmentally sustainable alternatives for acid catalysis has attracted the attention of scientists. The strong acid catalysed electrophilic substitution reactions, traditionally using HF or concentrated sulphuric acid as the catalyst, are the earliest examples investigated in acidic ionic liquids. For example, Wilks et al. [132] developed a procedure for the Friedel–Crafts alkylations or acylation of benzene with halogenoalkanes or acetyl chloride in chloroaluminated ionic liquids. It was found that although $AlCl_3$ itself has low catalytic activity for the direct Friedel–Crafts reactions of benzene, the ionic liquid can promote this reaction remarkably, and the catalytic activity increases with the content of $AlCl_3$ when the mole fraction of $AlCl_3$ in chloroaluminate melts was >0.5. In this catalytic procedure, the ionic liquid can act as a catalyst to promote the generation of carbonium ions, and it can simultaneously act as the reaction medium to provide satisfactory stability for the carbonium ions. After that, some groups developed different catalytic procedures for the Lewis acid catalysed organic reactions and studied the catalytic mechanism. Typically, Salunkhe et al. [133] investigated the Friedel–Crafts sulphonylation reaction of benzene with 4-methyl benzesulphonyl chloride by using 1-butyl-3-methylimidazolium chloroaluminate ionic liquid as reaction media and Lewis acid as catalyst. At the same time [27], Al-NMR spectroscopy was used to examine the details of the reaction mechanism. It was shown that after 4-methyl benzesulphonyl chloride was reacted with aromatic hydrocarbon, the initially predominant $[Al_2Cl_7]^-$ turned to $[AlCl_4]^-$, due to the predominant interaction of the Lewis acidic species $[Al_2Cl_7]^-$ of the ionic liquid with the formed HCl during the sulphonylation reaction: $[Al_2Cl_7]^- + HCl \rightarrow 2[AlCl_4]^- + H^+$. They concluded that the high catalytic activity of chloroaluminate ionic liquids was resulted from the Lewis acid present in acidic ionic liquids, which can act as a chloride sink and then provide the driving force to enhance the acidity of HCl.

Kou et al. [134] developed a procedure using $[C_4mim]Cl/AlCl_3$ as the catalyst for the synthesis of monodedecyl diphenyl oxides. It was shown that the ionic liquid has a much higher catalytic activity than that of $AlCl_3$, and the ionic liquid catalytic system benefits from significant simplification in product isolation. They also established the relationship between Lewis acidity and the catalytic activity for the ionic liquid catalyst in the alkylation reaction using acetonitrile as an IR spectroscopic probe. Whatever, the chloroaluminate-based ionic liquids have been utilized in several key industrial processes including the hydrosilylation of carbon double bounds by Degussa, the alkylation of aromatics by BP, and difasol processes by Institut Francais du Pétrole (IFP) [135].

The increased sustainability of Lewis acid chloroaluminate ionic liquids in acidic catalytic procedures promotes the emergence of other Lewis acidic ionic liquids with the general structure of $[C_nmim]Cl/M_xCl_y$. For example, the ionic liquid $[C_4mim]Cl/FeCl_3$ was found to be an efficient catalyst and reaction media for a wide diversity of transformations such as Friedel–Crafts alkylation and acylation [136], hydroxyl methylation [137], cross-coupling reactions [138], C–C bond formation reactions [139] among others [140]. Interestingly, the magnetic characteristic of this ionic liquid makes the related separation procedures very intriguing [141,142]. Zinc chloride can also form ionic liquid with pyridium [143], imidazolium [144,145] or quaternary ammonium salts such as choline chloride [146]. The ionic liquid $[C_4mim]Cl/ZnCl_2$ has been found to be an efficient immobilizer for Ru complex to promote the hydrogenation reactions of hexane with high turnover frequencies and easy reutilization of the catalyst [144]. The DESs of choline chloride and zinc chloride ($[CHCl][ZnCl]_2$) have also been found to be an efficient catalyst and solvent for the Diels–Alder reactions [146], Fischer indole annulations reactions [147], and selective acylation of cellulose [148].

In our laboratory, $[CHCl][ZnCl]_2$ has been investigated as an efficient catalyst and reaction media for the direct nucleophilic substitution reactions of various alcohols with different nucleophiles [131b]. This reaction system benefits from high selectivity, good to excellent isolated yields, easy separation and work-up procedures, recoverability of ionic liquids, no use of volatile organic solvents, high tolerance of water, and no need for excessive nucleophiles. The control reactions using different zinc-based ionic liquids with different cations suggested that the hydroxyl group on the cation is essential for the high selectivity of the target compound. Combined with phase transformation during the reaction procedure, it is believed that the special microstructure of reaction mixture induced by the hydroxyl group on the cation has the ability to modulate the distribution of the reactants, carbocation and the products (Figure 9), and then to increase the catalytic activity and selectivity. Therefore, the excellent performance of this ionic liquid was believed to be a synergetic result of phase modulation effect of hydroxyl group on the choline cation and the Lewis acidity of zinc chloride.

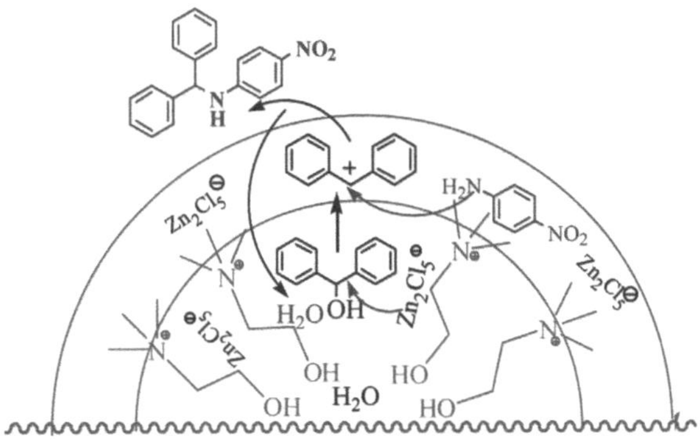

Figure 9
The special microstructure of the reaction mixture induced by the hydroxyl group on the cation of ionic liquid and their function on modulating the reactivity and selectivity [131b].

Besides the Lewis acid ionic liquids, many Bronsted acidic ionic liquids have been synthesized and found to be useful in sustainable catalysis procedures. The first strong Bronsted acidic ionic liquid developed by Davis et al. [149] was based on the alkane sulphonic acid group functionalized cation coupled with trifluoromethane sulphonic acid or *p*-toluenesulphonic acid anions. This type of ionic liquids was found to be a versatile solvent and catalyst for several classical acid-promoted organic reactions such as Fischer esterification, alcohol dehydrodimerization and the pinacol/benzopinacole rearrangement with a good product selectivity and a combination of the yields achievable using a homogeneous acid catalyst and the ease of catalyst/substrate separation using a heterogeneous catalyst. Deng et al. [150] synthesized four SO_3H-functionalized imidazolium ionic liquids with various anions such as trifluoromethane sulphonic acid, hydrosulphate, hydrophosphate and tetrafluoroborate, and determined their Hammet acidity functions using UV-vis spectroscopy. The correlation of their Hammet acidity with the catalytic activity for the Pechmann condensation suggested that the increased acidity favoured the catalytic activity, and this ionic liquid promoted system offers significant improvements for the synthesis of coumarins with high isolated yields of the products, simplicity in operation and no use of toxic catalysts and solvents. The same research group [151] also developed a series of protic pyridinium ionic liquids and studied the relationship between their catalytic activities for esterification of cyclic olefins with acetic acid and Hammet acidity function. Other Bronsted acid ionic liquids with carboxyl acid as the functional group were also synthesized and utilized in different acid-catalysed reactions. High catalytic activity, easy separation procedures and reusability of the ionic liquids [152–157] were observed, and the PKa values determined through titration against KOH have been used to evaluate the acidity of these ionic liquids [158].

3.5.2 The Reactions Catalyzed by Basic Functionalized Ionic Liquids

Compared with the widely used acidic ionic liquids in catalysis, the synthesis and application of basic ionic liquids are still in its infancy although exciting potential has been shown in basic catalysis with the advantages of easy separation, high catalytic activity and reusability. One of the basic ionic liquids, [C_4mim]OH, which has been utilized in different catalytic reactions including Michael addition [159], multicomponent reactions [160], Perkin reactions [161] and Mannich reactions [162], was not thought to be a really ionic liquid [163]. The reason is that [C_4mim]OH is prone to degradation in the absence of water, and can only be stable in water or polar organic solvents as a 1:1 mixture of N-heterocyclic carbine and solute.

Other kinds of basic ionic liquids mainly include a basic anion and a cation on which a basic site was incorporated. MacFarlane et al. [164] have given a specification of cations and anions involved in ionic liquids according to their basicity and acidity. MacNeil et al. [165] synthesized a series of basic ionic liquids with quaternized 1,4-diazabicycle[2,2,2]-octane as cation. The ionic liquids bearing short alkyl chains have been shown to have low toxicity to *Vibrio fischeri* [166] and used to catalyse Knoevenagel reactions [167] and facial synthesis of xanthenes derivatives [168] with high catalytic activity and reusability. Han et al. [169–171] developed a series of 1,1,3,3-tetramethylguanidium-based ionic liquids with different carboxyl acid anions, and these ionic liquids showed interesting catalytic activity for Henry reaction [169], Aldol reaction [170], Mannich reaction [171], Biginelli reaction [172] among others [173]. The catalytic activities of these ionic liquids showed a significant anion effect with the more basic anion and led to higher yields in a short time. The reactions between 2-butanone and aldehydes catalysed by this type of ionic liquids exhibited a different regioselectivity from traditional base-catalysed reactions. UV-vis spectroscopy study confirmed that the common enamine intermediate was formed and the anions affected generation of this intermediate. The observed regioselectivity was believed to result from the special structures of the intermediate involving the 1,1,3,3-tetramethylguanidium cation [170b].

Tertiary amino-functionalized imidazolium ionic liquid, 1-(N,N-dimethylaminoethyl)-2,3-dimethylimidazolium trifluoromethanesulphonate, has been synthesized and combined with the heterogeneous Ru catalyst in aqueous solution by Han and co-workers [174] to catalyse the hydrogenation of CO_2 for the production of formic acid with satisfactory activity and high selectivity. After the completion of the reaction, N_2 was allowed to pass through the reaction mixture at 130 °C and ambient pressure to remove the formic acid, and the catalytic system involving basic ionic liquid can be recycled and no significant turnover frequency decrease was observed after five runs. This suggests that both the catalyst and the ionic liquid can be reused after a simple separation process. Inductively coupled plasma atomic emission spectroscopy analysis indicated that there was no Ru loss after five runs, which indicates that ionic liquid is an ideal immobilizer for this complex. However, these authors [175] realized that only 1 mol of formic acid could be obtained per mole of the ionic liquid in a single reaction cycle. In order to increase the transformation efficiency, it is a wise choice to incorporate two or more basic groups in the ionic liquids. Therefore, in their subsequent work, 1,3-di(N,N-dimethylaminoethyl)-2-methylimidazolium trifluoromethanesulphonate was synthesized and used in the hydrogenation of CO_2 to prepare formic acid. The results demonstrated that 2 mol of formic acid could be obtained per mole of ionic liquid in a single reaction cycle, suggesting that the potential designability of ionic liquids can meet the special need of a specific process and increase the sustainability of the related reaction systems.

4. Outlook and Summary

As has been discussed above, ionic liquids have great promise in sustainable catalysis. Their utilization in catalysis can efficiently reduce the use of volatile organic solvents, simplify the operation and work-up procedures, increase the reactivity and selectivity and thus the atom efficiency, and lead to an increased sustainability of related processes. In addition, the convenient functionalization of ionic liquids makes it possible for us to prepare ionic liquids with specific functionality to meet the practical requirement of a particular process.

However, the fundamental physicochemical data such as melting point, glass transition temperature, polarity, solubility, density, viscosity and conductivity for ionic liquids and their mixtures with molecular solvent and electrolyte are still greatly limited and more data are required. The existence and complex interplay of Coulombic, hydrogen bond and van der Waals interactions among the ions of ionic liquids are far from being understood up to now, although they are highly important for the properties and functions of ionic liquids in catalysis. Also, the cost for the preparation of ionic liquids is often much higher than the traditional organic solvents. These factors weaken the utilization of ionic liquids in the industry and limit the rational design of ionic liquids for a particular catalytic system to a great extent. Therefore, it is still of great necessity to study the microstructures of ionic liquids in reaction systems and catalytic mechanisms of the related chemical reactions in ionic liquids, and then to establish the relationship between the structure of ionic liquids and their specific function in catalysis.

At this stage, it is wise to enrich the fundamental physicochemical data of ionic liquids and related multicomponent systems, to study the interactions between cations and anions of ionic liquids and those of the cations and anions with the other components in catalytic reaction systems, and to determine the catalytic mechanism of the ionic liquid-based catalytic systems. Based on these, an attempt is made to understand the relationship between the microstructure of ionic liquids and their catalytic function. Then, this relationship can be directly used to guide the rational design of ionic liquid-based catalytic systems. On the other hand, great attention should be paid to the development of biocompatible ionic liquids from renewable raw materials. These ionic liquids are particularly important for enzyme reactions and biological processes. We believed that with the rapid development of ionic liquid chemistry, ionic liquid-based catalysis will certainly make a great contribution to the research and application of sustainable catalysis.

References

[1] [a] Walden P. Molecular weights and electrical conductivity of several fused salts. Bull Acad Imp Sci (St Petersburg) 1914;8:405–22;
[b] Sugden S, Wilkins H. The parachor and chemical constitution. Part XII. Fused metals and salts. J Chem Soc 1929:1291–8.

[2] [a] Robinson J, Osteryoung RA. An electrochemical and spectroscopic study of some aromatic hydrocarbons in the room temperature molten salt system aluminum chloride-n-butylpyridinium chloride. J Am Chem Soc 1979;101:323–7;
[b] Appleby D, Hussey CL, Seddon KR, Turp JE. Room-temperature ionic liquids as solvents for electronic absorption spectroscopy of halide complexes. Nature 1986;323:614–6.

[3] Wilks JS, Zaworotko MJ. Air and water stable 1-ethyl-3-methylimidazolium based ionic liquids. J Chem Soc Chem Commun 1992:965–7.

[4] Fransisco M, van den Bruinhorst A, Kroon MC. Low-transition-temperature mixtures (LTTMs): a new generation of designer solvents. Angew Chem Int Ed 2013;52:3074–85.

[5] Bideau JL, Ducros JB, Soudan P, Guyomard D. Solid-state electrode materials with ionic-liquid properties for energy storage: the lithium solid-state ionic-liquid concept. Adv Funct Mater 2011;21:4073–8.

[6] de Maria PD. Ionic liquids in biotransformations and organocatalysis: solvents and beyond. John Wiley & Sons; 2012.

[7] Hallett JP, Welton T. Room-temperature ionic liquids: solvents for synthesis and catalysis. 2. Chem Rev 2011;111:3508–76.

[8] Han X, Armstrong W. Ionic liquids in separations. Acc Chem Res 2007;40:1079–86.

[9] Gordon JE, SubbaRao GN. Fused organic salts. 8. Properties of molten straight-chain isomers of tetra-n-pentylammonium salts. J Am Chem Soc 1978;100:7445–54.

[10] Holbery JD, Seddon KR. The phase behaviour of 1-alkyl-3-methylimidazolium tetrafluoroborates; ionic liquids and ionic liquid crystals. J Chem Soc Dalton Trans 1999;13:2133–40.

[11] Ngo HL, LeCompte K, Hargens L, McEwan AB. Thermal properties of imidazolium ionic liquids. Thermochim Acta 2000;357–358:97–102.

[12] Angell CA, Byrne N, Belieres J-P. Parallel developments in aprotic and protic ionic liquids: physical chemistry and applications. Acc Chem Res 2007;40:1228−36.

[13] Zhu AL, Li QQ, Li LJ, Wang JJ. One-pot synthesis of 3,4-dihydro-2(H)-pyrimidinones catalyzed by reusable acidic choline-based ionic liquids. Catal Lett 2013;143:463−8.

[14] Huddleston JG, Visser AE, Reichert WM, Willauer HD, Broker GA, Rogers RD. Characterization and comparison of hydrophilic and hydrophobic room temperature ionic liquids incorporating the imidazolium cation. Green Chem 2001;3:156−64.

[15] Chan BKM, Chang NH, Grimmett MR. The synthesis and thermolysis of imidazole quaternary salts. Aust J Chem 1977;30:2005−13.

[16] [a] Wasserscheid P, Keim W. Ionic liquids-new solutions for transition metal catalysis. Angew Chem Int Ed 2000;39:3772−89;
[b] Hyun B, Dzyuba SV, Bartsch RA, Quitevis EL. Intermolecular dynamics of room-temperature ionic liquids: Femtosecond optical Kerr effect measurements on 1-alkyl-3-methylimidazolium bis((trifluoromethyl)sulfonyl)imides. J Phys Chem A 2002;106:7579−85.

[17] Branco LC, Rosa JN, Moura RJJ, Afonso CAM. Preparation and characterization of new room temperature ionic liquids. Eur J Chem 2002;8:3671−7.

[18] Dzyuba SV, Bartsch RA. Influence of structural variations in 1-alkyl(aralkyl)-3-methylimidazolium hexafluorophosphates and bis(trifluoromethyl-sulfonyl)imides on physical properties of the ionic liquids. ChemPhysChem 2002;3:161−6.

[19] Backer SN, Baker GA, Kane MA, Bright FV. The cybotactic region surrounding fluorescent probes dissolved in 1-butyl-3-methylimidazolium hexafluorophosphate: effects of temperature and added carbon dioxide. J Phys Chem B 2001;105:9663−8.

[20] Shirota H, Castner EW. Why are viscosities lower for ionic liquids with $-CH_2Si(CH_3)_3$ vs $-CH_2C(CH_3)_3$ substitutions on the imidazolium cations? J Phys Chem B 2005;109:21576−85.

[21] [a] Siqueira LJA, Rebeiro MCC. Alkoxy chain effect on the viscosity of a quaternary ammonium ionic liquid: molecular dynamics simulations. J Phys Chem B 2009;113:1074−9;
[b] Chen ZJ, Xue T, Lee JM. What causes the low viscosity of ether-functionalized ionic liquids? Its dependence on the increase of free volume. RSC Adv 2012;2:10564−74.

[22] Reichardt C. Solvents and solvent effects in organic chemistry. 3rd ed. Weinheim, Germany: Wiley-VCH; 2003.

[23] [a] Wakai C, Oleinikova A, Ott M, Weingärtner H. How polar are ionic liquids? Determination of the static dielectric constant of an imidazolium-based ionic liquid by microwave dielectric spectroscopy. J Phys Chem B 2005;109:17028−30;
[b] Daguenet C, Dyson PJ, Krossling I, Oleinkova A, Slattery J, Wakai C, Weingärtner H. Dielectric response of imidazolium-based room-temperature ionic liquids. J Phys Chem B 2006;110:12682−8;
[c] Weingärtner H. Understanding ionic liquids at the molecular level: facts, problems, and controversies. Angew Chem Int Ed 2008;47:654−70.

[24] Huang M-M, Weingärtner H. Protic ionic liquids with unusually high dielectric permittivities. ChemPhysChem 2008;9:2172−3.

[25] [a] Wu Y, Sasaki T, Kazushi K, Seo T, Sakurai K. Interactions between spiropyrans and room-temperature ionic liquids: photochromism and solvatochromism. J Phys Chem B 2008;112:7530−6;
[b] Zhu AL, Liu RX, Li LJ, Li LY, Wang L, Wang JJ. Dual functions of N, N-dimethyl ethanolamnium-based ionic liquids for the Knoevenagel reactions at room temperature. Catal Today 2013;200:17−23.

[26] [a] Karmakar R, Samanta A. Steady-state and time-resolved fluorescence behavior of C153 and PRODAN in room-temperature ionic liquids. J Phys Chem A 2002;106:6670−5;
[b] Karmakar R, Samanta A. Dynamics of solvation of the fluorescent state of some electron donor−acceptor molecules in room temperature ionic liquids, [BMIM][$(CF_3SO_2)_2N$] and [EMIM][$(CF_3SO_2)_2N$]. J Phys Chem A 2003;107:7340−6;
[c] Mandal PK, Samanta A. Fluorescence studies in a pyrrolidinium ionic liquid: polarity of the medium and solvation dynamics. J Phys Chem B 2005;109:15172−717.

[27] Lungwitz R, Friederich M, Linert W, Spange S. New aspects on the hydrogen bond donor (HBD) strength of 1-butyl-3-methylimidazolium room temperature ionic liquids. New J Chem 2008;32:1493–9.
[28] Liu Y, Wang SS, Liu W, Wan QX, Wu HH, Gao GH. Transition-metal catalyzed carbon-carbon couplings mediated with functionalized ionic liquids, supported-ionic liquid phase, or ionic liquid media. Curr Org Chem 2009;13:1322–46.
[29] Lee JW, Shin JY, Chun YS, Bin Jang H, Song CE, Lee SG. Toward understanding the origin of positive effects of ionic liquids on catalysis: formation of more reactive catalysts and stabilization of reactive intermediates and transition states in ionic liquids. Acc Chem Res 2010;43:985–94.
[30] Zhao D, Dyson PJ, Laurenczy G, McIndoe JS. On the catalytic activity of cluster anions in styrene hydrogenation: considerable enhancements in ionic liquids compared to molecular solvents. J Mol Catal A 2004;214:19–25.
[31] Chauvin Y, Mussman L, Olivier H. A novel class of versatile solvents for two-phase catalysis: hydrogenation, isomerization, and hydroformylation of alkenes catalyzed by rhodium complexes in liquid 1,3-dialkylimidazolium salts. Angew Chem Int Ed 1995;34:2698–700.
[32] [a] Dupont J, Suarez PAZ, Umpierre AP, de Souza RF. Pd(II)-dissolved in ionic liquids: a recyclable catalytic system for the selective biphasic hydrogenation of dienes to monoenes. J Braz Chem Soc 2000;11:293–7;
[b] Consorti CS, Umpierre AP, de Souza RF, Dupont J, Suarez PAZ. Selective hydrogenation of 1,3-butadiene by transition metal compounds immobilized in 1-butyl-3-methyl imidazolium room temperature ionic liquids. J Braz Chem Soc 2003;14:401–5;
[c] Umpierre MG, Fecher GH, Morais J, Dupont J. Selective hydrogenation of 1,3-butadiene to 1-butene by Pd(0) nanoparticles embedded in imidazolium ionic liquids. Adv Synth Catal 2005;347:1404–12.
[33] Huang J, Jiang T, Gao HX, Han BX, Liu ZM, Wu WZ, et al. Pd nanoparticles immobilized on molecular sieves by ionic liquids: heterogeneous catalysts for solvent-free hydrogenation. Angew Chem Int Ed 2004;43:1397–9.
[34] Rossi LM, Machado G, Fichtner PFP, Teixeira SR, Dupont J. On the use of ruthenium dioxide in 1-n-butyl-3-methylimida-zolium ionic liquids as catalyst precursor for hydrogenation reactions. Catal Lett 2004;92:149–55.
[35] Yao Q. OsO_4 in ionic liquid [Bmim]PF_6: a recyclable and reusable catalyst system for olefin dihydroxylation. Remarkable effect of DMAP. Org Lett 2002;4:2197–9.
[36] Branco LC, Serbanovic A, Ponte MN, Afonso CAM. Clean osmium-catalyzed asymmetric dihydroxylation of in and supercritical CO_2 product recovery. Chem Commun 2005:107–9.
[37] Ritter S. New route to adipic acid avoids nitrous oxide production. Chem Eng News 2014;92:6.
[38] Sato K, Soki M, Noyori R. A "green" route to adipic acid: direct oxidation of cyclohexenes with 30 percent hydrogen peroxide. Science 1998;281:1646–7.
[39] Gui J, Liu D, Cong X, Zhang X, Jiang H, De Z, et al. Clean synthesis of adipic acid by direct oxidation of cyclohexene with H_2O_2 catalysed by $Na_2WO_4 \cdot 2H_2O$ and acidic ionic liquids. J Chem Res 2005:520–2.
[40] Vafaeezadeh M, Hashemi MM, Shakourian-Fard M. Catalytic roles of metals and supports on hydrodeoxygenation of lignin monomer guaiacol. Catal Commun 2012;26:54–8.
[41] Zhao GY, Jiang T, Wu WZ, Han BX, Liu ZM, Gao HX. Electro-oxidation of benzyl alcohol in biphasic system consisting of supercritical CO_2 and ionic liquids. J Phys Chem B 2004;108:13052–7.
[42] Sun N, Hou Y, Wu W, Niu M, Wang W. Electro-oxidation of benzyl alcohol to benzaldehyde in supercritical CO_2 with ionic liquid. Electrochem Commun 2013;28:34–6.
[43] [a] Moniruzzaman M, Nakashima K, Kamiya N. Recent advances of enzymatic reactions in ionic liquids. Biochem Eng J 2010;48:295–314;
[b] van Rantwijk F, Sheldon RA. Biocatalysis in ionic liquids. Chem Rev 2007;107:2757–85.

[44] [a] Constantinescu D, Herrmann C, Weingärtner H. Protein denaturation by ionic liquids and the Hofmeister series: a case study of aqueous solutions of ribonuclease A. Angew Chem Int Ed 2007;46:8887–9;
[b] Constatinescu D, Herrmann C, Weingartner H. Patterns of and aggregation in ionic liquids. Phys Chem Chem Phys 2010;12:1756–63.
[45] [a] Yang Z. Hofmeister effects: an explanation for the impact of ionic liquids on biocatalysis. J Biotechnol 2009;144:12–22;
[b] Fujita K, MacFarlane DR, Forsyth M, Yoshizawa - Fujita M, Murata K, Nakamura N, Ohno H. Solubility and stability of cytochrome c in hydrated ionic liquids: effect of oxo acid residues and kosmotropicity. Biomacromolecules 2007;8:2080–6;
[c] Lou WY, Zong MH, Liu YY, Wang JF. Efficient enantioselective hydrolysis of D,L-phenylglycine methyl ester catalyzed by immobilized *Candida antarctica* lipase B in ionic liquid containing systems. J Biotechnol 2006;125:64–74.
[46] [a] Lou WY, Zong MH, Smith TJ, Wu H, Wang JF. Impact of ionic liquids on papain: an investigation of structure–function relationships. Green Chem 2006;8:509–12;
[b] Lou WY, Zong MH. Efficient kinetic resolution of (R,S)-1-trimethylsilylethanol via lipase-mediated enantioselective acylation in ionic liquids. Chirality 2006;18:814–21.
[47] [a] Hernandez-Fernandez FJ, de los Rios AP, Rubio M, Gimez D, Villora G. Recent advances in supported ionic liquid membrane technology. J Chem Technol Biotechnol 2007;82:882–7;
[b] de los Rios AP, Fernandez FJH, Afonso FT, Gomez D, Villora G. Synthesis of esters in ionic liquids: the effect of vinyl esters and alcohols. Process Biochem 2008;43:892–5;
[c] Iwai N, Tanaka T, Kitazume T. Utility of ionic liquid for improvement of fluorination reaction with immobilized fluorinase. J Mol Catal B Enzym 2009;59:131–3.
[48] Kaar JL, Jesionowski AM, Berberich JA, Moulton R, Russell AJ. Impact of ionic liquid physical properties on lipase activity and stability. J Am Chem Soc 2003;125:4125–31.
[49] Ropel L, Belveze LS, Aki SNVK, Stadtherr MA, Brennecke JF. Octanol–water partition coefficients of imidazolium-based ionic liquids. Green Chem 2005;7:83–90.
[50] Lee SH, Doan TTN, Ha SH, Koo YM. Using ionic liquids to stabilize lipase within sol–gel derived silica. J Mol Catal B Enzym 2007;45:57–61.
[51] Lee SH, Koo YM, Ha SH. Influence of ionic liquids under controlled water activity and low halide content on lipase activity. Korean J Chem Eng 2008;25:1456–62.
[52] Lange C, Patil G, Rodolph R. Ionic liquids as refolding additives: N'-alkyl and N'-(ω-hydroxyalkyl) N-methylimidazolium chlorides. Protein Sci 2005;14:2693–701.
[53] Ou G, Yang J, He B, Yuan Y. Buffer-mediated activation of *Candida antarctica* lipase B dissolved in hydroxyl-functionalized ionic liquids. J Mol Catal B Enzy 2010;68:66–70.
[54] Tanaka H, Oka Y. Chaotropic ions and multivalent ions activate sperm in the viviparous fish guppy *Poecilia reticulate*. Biochim Biophys Acta, Gen Subj 2005;1724:173–80.
[55] Al-Zuhair S. Production of biodiesel by lipase-catalyzed transesterification of vegetable oils: a kinetics study. Biotechnol Prog 2005;21:1442–8.
[56] Gamba M, Lapis AAM, Dupont J. Supported ionic liquid enzymatic catalysis for the production of biodiesel. Adv Synth Catal 2008;350:160–4.
[57] Zhao H, Baker GA, Holmes S. New eutectic ionic liquids for lipase activation and enzymatic preparation of biodiesel. Org Biomol Chem 2011;9:1908–16.
[58] Zhao H, Zhang C, Crittle TD. DNA–based asymmetric catalysis: role of ionic solvents and glymes. J Mol Catal B: Enzym 2013;85–86:243–7.
[59] Tavares APM, Rodriguez O, Macedo EA. Ionic liquids as alternative co-solvents for laccase: study of enzyme activity and stability. Biotech Bioeng 2008;101:201–7.
[60] Moniruzzaman M, Kamiya N, Nakashima K, Goto M. Water-in-ionic liquid microemulsions as a new medium for enzymatic reactions. Green Chem 2008;10:497–500.

[61] Hernández-Fernández FJ, de los Ríos AP, Rubio M, Gómez D, Víllora G. Enhancement of activity and selectivity in lipase-catalyzed transesterification in ionic liquids by the use of additives. J Chem Tech Biotech 2007;82:882–7.
[62] Lee SH, Ha SH, Lee SB, Koo Y-M. Adverse effect of chloride impurities on lipase-catalyzed transesterifications in ionic liquids. Biotech Lett 2006;28:1335–9.
[63] Annual energy review 2009. Report No. DOE/EIS-0384 (2010). Washington, DC: U.S. Department of Energy, Available Electronically at: http://www.eia.gov.
[64] Dhepe PL, Fukuoka A. Cellulose conversion under heterogeneous catalysis. ChemSusChem 2008;1:969–75.
[65] Zhou C-H, Xia X, Lin C-X, Tong D-S, Beltramini J. Catalytic conversion of lignocellulosic biomass to fine chemicals and fuels. Chem Soc Rev 2011;40:5588–617.
[66] van de Vyver S, Geboers J, Jacobs PA, Sels BF. Recent advances in the catalytic conversion of cellulose. ChemCatChem 2011;3:82–94.
[67] Updegraff DM. Semimicro determination of cellulose in biological materials. Anal Biochem 1969;32:420–4.
[68] Wang H, Gurau G, Rogers RD. Ionic liquid processing of cellulose. Chem Soc Rev 2012;41:1519–37.
[69] Heinze T, Liebert T. Unconventional methods in cellulose functionalization. Prog Polym Sci 2001;26:1689–762.
[70] Xu AR, Wang JJ, Wang HY. Effects of anionic structure and lithium salts addition on the dissolution of cellulose in 1-butyl-3-methylimidazolium-based ionic liquid solvent systems. Green Chem 2010;12:268–75.
[71] Swatloski RP, Spear SK, Holbrey JD, Rogers RD. Dissolution of cellulose with ionic liquids. J Am Chem Soc 2002;124:4974–5.
[72] Zhang Q, Zhang S, Deng Y. Recent advances in ionic liquid catalysis. Green Chem 2011;13:2619–37.
[73] Remsing RC, Swatloski RP, Rogers RD, Moyna G. Mechanism of cellulose dissolution in the ionic liquid 1-n-butyl-3-methylimidazolium chloride: a ^{13}C and $^{35/37}Cl$ NMR relaxation study on model systems. Chem Commun 2006:1271–3.
[74] Zhang JM, Zhang H, Wu J, Zhang J, He JS, Xiang JF. NMR spectroscopic studies of cellobiose solvation in EmimAc aimed to understand the dissolution mechanism of cellulose in ionic liquids. Phys Chem Chem Phys 2010;12:1941–7.
[75] Liu HB, Sale KL, Holmes BM, Simmons BA, Singh S. Understanding the interactions of cellulose with ionic liquids: a molecular dynamics study. J Phys Chem B 2010;114:4293–301.
[76] Vitz J, Erdmenger T, Haensch C, Schubert US. Extended dissolution studies of cellulose in imidazolium based ionic liquids. Green Chem 2009;11:417–24.
[77] Erdmenger T, Haensch C, Hoogenboom R, Schubert US. Homogeneous tritylation of cellulose in 1-butyl-3-methylimidazolium chloride. Macromol Biosci 2007;7:440–5.
[78] Pinkert A, Marsh KN, Pang SS, Staiger MP. Ionic liquids and their interaction with cellulose. Chem Rev 2009;109:6712–28.
[79] Remsing RC, Hernandez G, Swatloski RP, Massefski WW, Rogers RD, Moyna G. Solvation of carbohydrates in n,n'-dialkylimidazolium ionic liquids: a multinuclear NMR spectroscopy study. J Phys Chem B 2008;112:11071–8.
[80] Novoselov NP, Sashina ES, Petrenko VE, Zaborsky M. Study of dissolution of cellulose in ionic liquids by computer modeling. Fibre Chem 2007;39:153–8.
[81] de Vyver SV, Geboers J, Jacobs PA, Sels BF. Recent advances in the catalytic conversion of cellulose. ChemCatChem 2011;3:82–94.
[82] Rinaldi R, Schüth F. Acid hydrolysis of cellulose as the entry point into biorefinery schemes. ChemSusChem 2009;2:1096–107.
[83] Xiang Q, Lee Y, Pettersson P, Torget R. Heterogeneous aspects of acid hydrolysis of alpha-cellulose. Appl Biochem Biotechnol 2003;107:505–14.

[84] Suganuma S, Nakajima K, Kitano M, Yamaguchi D, Kato H, Hayashi S, et al. Hydrolysis of cellulose by amorphous carbon bearing SO_3H, COOH, and OH Groups. J Am Chem Soc 2008;130:12787–93.
[85] Takagaki A, Tagusagawa C, Domen K. Glucose production from saccharides using layered transition metal oxide and exfoliated nanosheets as a water-tolerant solid acid catalyst. Chem Commun 2008:5363–5.
[86] Li C, Zhao ZK. Efficient acid-catalyzed hydrolysis of cellulose in ionic liquid. Adv Synth Catal 2007;349:1847–50.
[87] Li C, Wang Q, Zhao ZK. Acid in ionic liquid: an efficient system for hydrolysis of lignocelluloses. Green Chem 2008;10:177–82.
[88] Vanoye L, Fanselow M, Holbrey JD, Atkins MP, Seddon KR. Kinetic model for the hydrolysis of lignocellulosic biomass in the ionic liquid, 1-ethyl-3-methyl-imidazolium chloride. Green Chem 2009;11:390–6.
[89] Tao F, Song H, Chou L. Hydrolysis of cellulose by using catalytic amounts of $FeCl_2$ in ionic liquids. ChemSusChem 2010;3:1298–303.
[90] Rinaldi R, Palkovits R, Schuth F. Depolymerization of cellulose using solid catalysts in ionic liquids. Angew Chem Int Ed 2008;47:8047–50.
[91] Zhang ZH, Zhao ZBK. Solid acid and microwave-assisted hydrolysis of cellulose in ionic liquid. Carbohydr Res 2009;344:2069–72.
[92] Melero JA, Iglesias J, Morales G. Heterogeneous acid catalysts for biodiesel production: current status and future challenges. Green Chem 2009;11:1285–308.
[93] Villandier N, Corma A. One pot catalytic conversion of cellulose into biodegradable surfactants. Chem Commun 2010;46:4408–10.
[94] Ignatyev IA, Doorslaer CV, Mertens PGN, Binnemans K, De Vos DE. Reductive splitting of cellulose in the ionic liquid 1-butyl-3-methylimidazolium chloride. ChemSusChem 2010;3:91–6.
[95] Corma A, Iborra S, Velty A. Chemical routes for the transformation of biomass into chemicals. Chem Rev 2007;107:2411–502.
[96] Tong X, Ma Y, Li Y. Biomass into chemicals: conversion of sugars to furan derivatives by catalytic processes. Appl Catal A 2010;385:1–13.
[97] Horvath IT, Mehdi H, Fabos V, Boda L, Mika LT. γ-Valerolactone – a sustainable liquid for energy and carbon-based chemicals. Green Chem 2008;10:238–42.
[98] Rosatella AA, Simeonov SP, Frade RFM, Afonso CAM. 5-Hydroxymethylfurfural (HMF) as a building block platform: biological properties, synthesis and synthetic applications. Green Chem 2011;13:754–93.
[99] Zhao H, Holladay JE, Brown H, Zhang ZC. Metal chlorides in ionic liquid solvents convert sugars to 5-hydroxymethylfurfural. Science 2007;316:1597–600.
[100] Yong G, Zhang YG, Ying JY. Efficient catalytic system for the selective production of 5-hydroxymethylfurfural from glucose and fructose. Angew Chem Int Ed 2008;47:9345–8.
[101] Li CZ, Zhang ZH, Zhao ZBK. Direct conversion of glucose and cellulose to 5-hydroxymethylfurfural in ionic liquid under microwave irradiation. Tetrahedron Lett 2009;50:5403–4.
[102] Pidko EA, Degirmenci V, Santen RA, Hensen EJM. Glucose activation by transient Cr^{2+} dimers. Angew Chem Int Ed 2010;49:2530–4.
[103] Hu S, Zhang Z, Song J, Zhou Y, Han B. Efficient conversion of glucose into 5-hydroxymethylfurfural catalyzed by a common Lewis acid $SnCl_4$ in an ionic liquid. Green Chem 2009;11:1746–9.
[104] Chidambaramand M, Bell AT. A two-step approach for the catalytic conversion of glucose to 2,5-dimethylfuran in ionic liquids. Green Chem 2010;12:1253–62.
[105] Su Y, Brown HM, Huang X, Zhou X, Amonette JE, Zhang ZC. Single-step conversion of cellulose to 5-hydroxymethylfurfural (HMF), a versatile platform chemical. Appl Catal A 2009;361:117–22.
[106] Binder JB, Raines RT. Simple chemical transformation of lignocellulosic biomass into furans for fuels and chemicals. J Am Chem Soc 2009;131:1979–85.
[107] Blanchard LA, Hancu D, Beckman EJ, Brennecke JF. Green processing using ionic liquids and CO_2. Nature 1999;399:28–9.

[108] Scurto AM, Aki SNVK, Brennecke JF. CO_2 as a separation switch for ionic liquid/organic mixtures. J Am Chem Soc 2001;124:10276–7.

[109] Blanchard LA, Gu Z, Brennecke JF. High-pressure phase behavior of ionic liquid/CO_2 systems. J Phys Chem B 2001;105:2437–44.

[110] Anthony JL, Maginn EJ, Brenncke JF. Solubilities and thermodynamic properties of gases in the ionic liquid 1-n-butyl-3-methylimidazolium hexafluorophosphate. J Phys Chem B 2002;106:7315–20.

[111] Bates ED, Mayton RD, Ntai I, Davis Jr JH. CO_2 capture by a task-specific ionic liquid. J Am Chem Soc 2002;124:926–7.

[112] Fukumoto K, Yoshizawa M, Ohno H. Room temperature ionic liquids from 20 natural amino acids. J Am Chem Soc 2005;127:2398–9.

[113] Li XY, Hou MQ, Zhang ZF, Han BX, Yang GY, Wang XL, et al. Absorption of CO_2 by ionic liquid/polyethylene glycol mixture and the thermodynamic parameters. Green Chem 2008;10:879–84.

[114] Gurkan BE, De La Fuente JC, Mindrup EM, Ficke LE, Goodrich BF, Price EA, et al. Equimolar CO_2 absorption by anion-functionalized ionic liquids. J Am Chem Soc 2010;132:2116.

[115] Wang CM, Luo HM, Jiang DE, Li HR, Dai S. Carbon dioxide capture by superbase-derived protic ionic liquids. Angew Chem Int Ed 2010;49:5978–81.

[116] Peng JJ, Deng YQ. Cycloaddition of carbon dioxide to propylene oxide catalyzed by ionic liquids. New J Chem 2001;25:639–41.

[117] Kawanami H, Sasaki A, Matsui K, Ikushima Y. A rapid and effective synthesis of propylene carbonate using a supercritical CO_2–ionic liquid system. Chem Commun 2003:896–7.

[118] Kim HS, Kim JJ, Kim H, Jang HG. Imidazolium zinc tetrahalide-catalyzed coupling reaction of CO_2 and ethylene oxide or propylene oxide. J Catal 2003;220:44–6.

[119] Li FW, Xiao LF, Xia CG, Hu B. Chemical fixation of CO_2 with highly efficient $ZnCl_2$/[BMIm]Br catalyst system. Tetrahedron Lett 2004;45:8307–10.

[120] Sun JM, Fujita S, Zhao FY, Arai M. Synthesis of styrene carbonate from styrene oxide and carbon dioxide in the presence of zinc bromide and ionic liquid under mild conditions. Green Chem 2004;6:613–6.

[121] Palgunadi J, Kwon OS, Lee H, Bae JY, Ahn BS, Min NY, et al. Ionic liquid-derived zinc tetrahalide complexes: structure and application to the coupling reactions of alkylene oxides and CO_2. Catal Today 2004;98:511–4.

[122] Kim YJ, Varma RS. Tetrahaloindate (III)-based ionic liquids in the coupling reaction of carbon dioxide and epoxides to generate cyclic carbonates: H-bonding and mechanistic studies. J Org Chem 2005;70:7882–91.

[123] [a] Abbott AP, Capper G, Davies DL, Rasheed RK, Tambyrajah V. Novel solvent properties of choline chloride/urea mixtures. Chem Commun 2003:70–1;
[b] Abbott AP, Boothby D, Capper G, Davies DL, Rasheed RK. Deep eutectic solvents formed between choline chloride and carboxylic acids: versatile alternatives to ionic liquids. J Am Chem Soc 2004;126:9142–7.

[124] Zhu AL, Jiang T, Han BX, Zhang JC, Xie Y, Ma XM. Supported choline chloride/urea as a heterogeneous catalyst for chemical fixation of carbon dioxide to cyclic carbonates. Green Chem 2007;9:169–72.

[125] Sun J, Zhang SJ, Cheng WG, Ren JY. Hydroxyl-functionalized ionic liquid: a novel efficient catalyst for chemical fixation of CO_2 to cyclic carbonate. Tetrahedron Lett 2008;49:3588–91.

[126] Sun J, Cheng WG, Fan W, Wang YH, Meng ZY, Zhang SJ. Reusable and efficient polymer-supported task-specific ionic liquid catalyst for cycloaddition of epoxide with CO_2. Catal Today 2009;148:361–7.

[127] Sun J, Han LJ, Cheng WG, Wang JQ, Zhang XP, Zhang SJ. Efficient acid–base bifunctional catalysts for the fixation of CO_2 with epoxides under metal- and solvent-free conditions. ChemSusChem 2011;4:502–7.

[128] Sun J, Wang JQ, Cheng WG, Zhang JX, Li XH, Zhang SJ, et al. Chitosan functionalized ionic liquid as a recyclable biopolymer-supported catalyst for cycloaddition of CO_2. Green Chem 2012;14:654–60.

[129] Han LN, Choi HJ, Choi SJ, Liu BY, Park DW. Ionic liquids containing carboxyl acid moieties grafted onto silica: synthesis and application as heterogeneous catalysts for cycloaddition reactions of epoxide and carbon dioxide. Green Chem 2011;13:1023–8.
[130] Han LN, Choi SJ, Park MS, Lee SM, Kim YJ, Kim MI, et al. Carboxylic acid functionalized imidazolium-based ionic liquids: efficient catalysts for cycloaddition of CO_2 and epoxides. React Kinet Mech Catal 2012;106:25–35.
[131] [a] Chakraborti AK, Roy SR. On catalysis by ionic liquids. J Am Chem Soc 2009;131:6902–3; [b] Zhu AL, Li LJ, Wang JJ, Zhuo KL. Direct nucleophilic substitution reaction of alcohols mediated by a zinc-based ionic liquid. Green Chem 2011;13:1244–50.
[132] Boon JA, Levisky JA, Pflug JL, Wilkes JS. Friedel-Crafts reactions in ambient-temperature molten salts. J Org Chem 1986;51:480–3.
[133] Nara SJ, Harjani JR, Salunkhe MM. Friedel-Crafts sulfonylation in 1-butyl-3-methylimidazolium chloroaluminate ionic liquids. J Org Chem 2001;66:8616–20.
[134] Piao LY, Fu X, Yang YL, Tao GH, Kou Y. Alkylation of diphenyl oxide with dodecene catalyzed by the ionic liquids. Catal Today 2004;93–95:301–5.
[135] Plechkova NV, Seddon KR. Applications of ionic liquids in the chemical industry. Chem Soc Rev 2008;37:123–50.
[136] Valkenberg MH, deCastro C, Holderich WFJ. Friedel-Crafts acylation of aromatics catalysed by supported ionic liquids. Appl Catal A 2001;215:185–90.
[137] Bica K, Gaertner P. Metal-containing ionic liquids as efficient catalysts for hydroxymethylation in water. Eur J Org Chem 2008;20:3453–6.
[138] Bica K, Gaertner P. An iron-containing ionic liquid as recyclable catalyst for aryl Grignard cross-coupling of alkyl halides. Org Lett 2006;8:733–5.
[139] Maria V, Peter G, Katharina B. Iron catalyzed Michael addition: chloroferrate ionic liquids as efficient catalysts under microwave conditions. Sci China Chem 2012;55:1614–9.
[140] Bica K, Leder S, Gaertner P. From solvent to sustainable catalysis – chloroferrate ionic liquids in synthesis. Curr Org Synth 2011;8:824–39.
[141] Hayashi S, Hamaguchi H. Discovery of a magnetic ionic liquid [bmim]$FeCl_4$. Chem Lett 2004;33:1590–1.
[142] Zhang Q, Yang J, Lu X, Gui J, Huang M. Studies on an ionic liquid based on $FeCl_3$ and its properties. Fluid Phase Equilib 2004;226:207–11.
[143] Koura N, Endo T, Idemoto Y. The electrodeposition of amorphous Co–Zn alloy from ambient temperature molten salt electrolytes. J Non-Cryst Solids 1996;205:650–5.
[144] Lin Y, Sun I. Electrodeposition of zinc from a Lewis acidic zinc chloride-1-ethyl-3-methylimidazolium chloride molten salt. Electrochim Acta 1999;44:2771–7.
[145] Dupont J, Suarez PAZ, Umpierre AP, de Souza RF. Organo-zincate molten salts as immobilising agents for organometallic catalysis. Catal Lett 2001;73:211–3.
[146] Abbott AP, Capper G, Davis DL, Munro HL, Rasheed RK, Tambyrajah V. Preparation of novel, moisture-stable, Lewis-acidic ionic liquids containing quaternary ammonium salts with functional side chains. Chem Commun 2001:2010–1.
[147] Morales RC, Tambyrajah V, Jenkins PR, Davies DL, Abbott AP. The regiospecific Fischer indole reaction in choline chloride·$2ZnCl_2$ with product isolation by direct sublimation from the ionic liquid. Chem Commun 2004:158–9.
[148] Abbott AP, Bell TJ, Handa S, Stoddart B. O-acetylation of cellulose and monosaccharides using a zinc based ionic liquid. Green Chem 2005;7:705–7.
[149] Cole AC, Jensen JL, Ntai I, Tran KLT, Weaver KJ, Forbes DC, et al. Novel Brønsted acidic ionic liquids and their use as dual solvent–catalysts. J Am Chem Soc 2002;124:5962–3.
[150] Gu YL, Zhang J, Duan ZY, Deng YQ. Pechmann reaction in non-chloroaluminate acidic ionic liquids under solvent-free conditions. Adv Synth Catal 2005;347:512–6.

[151] Duan ZY, Gu YL, Zhang J, Zhu LY, Deng YQ. Protic pyridinium ionic liquids: synthesis, acidity determination and their performances for acid catalysis. J Mole Catal A Chem 2006;250:163–8.
[152] Thomazeau C, Olivier-Bourbigou H, Magna L, Luts S, Gilbert B. Determination of an acidic scale in room temperature ionic liquids. J Am Chem Soc 2003;125:5264–5.
[153] Greaves TL, Drummond CJ. Protic ionic liquids: properties and applications. Chem Rev 2008;108:206–37.
[154] Cui XJ, Zhang SG, Shi F, Zhang QH, Ma XY, Lu LJ, et al. The influence of the acidity of ionic liquids on catalysis. ChemSucChem 2010;3:1043–7.
[155] Srivastava R. Assessment of the catalytic activities of novel Brönsted acidic ionic liquid catalysts. Catal Lett 2010;139:17–25.
[156] Long JX, Li XH, Guo B, Wang LF, Zhang N. Catalytic delignification of sugarcane bagasse in the presence of acidic ionic liquids. Catal Today 2013;200:99–105.
[157] Rajkumar K, Rajendra S. Influence of $-SO_3H$ functionalization (N–SO3H or $N-R-SO_3H$, where R = alkyl/benzyl) on the activity of Brönsted acidic ionic liquids in the hydration reaction. Tetrahedron Lett 2012;53:3245–9.
[158] Fei ZF, Zhao DB, Geldbach TJ, Scopelliti R, Dyson PJ. Brønsted acidic ionic liquids and their zwitterions: synthesis, characterization and pKa determination. Chem Eur J 2004;10:4886–93.
[159] [a] Xu JM, Liu BK, Wu WB, Qian C, Wu Q, Lin XF. Basic ionic liquid as catalysis and reaction medium: a novel and green protocol for the Markovnikov addition of n-heterocycles to vinyl esters, using a task-specific ionic liquid, [bmIm]OH. J Org Chem 2006;71:3991–3;
[b] Ranu BC, Banerjee S, Jana R. Ionic liquid as catalyst and solvent: the remarkable effect of a basic ionic liquid, [bmIm]OH on Michael addition and alkylation of active methylene compounds. Tetrahedron 2007;63:776–82.
[160] [a] Ranu BC, Banerjee S, Roy S. A task-specific basic ionic liquid, [bmIm]OH-promoted efficient, green and one-pot synthesis of tetrahedrobenzo[b]pyran derivatives. Indian J Chem 2008;47B:1108–12;
[b] Gong K, Wang H-L, Fang D, Liu ZL. Basic ionic liquid as catalyst for the rapid and green synthesis of substituted 2-amino-2-chromenes in aqueous media. Catal Commun 2008;9:650–3.
[161] Li J, Sun H, Cai X-C, Dai LY. Application of basic ionic liquid [bmim]OH to Knoevenagel and Perkin reactions. Chin J Org Chem 2007;27:1296–9.
[162] Gong K, Fang D, Wang H-L, Liu ZL. Basic functionalized ionic liquid catalyzed one-pot Mannich-type reaction: three component synthesis of β-amino carbonyl compounds. Monatsh Chem 2007;138:1195–8.
[163] Yuen AKL, Masters AF, Maschmeyer T. 1,3-Disubstituted imidazolium hydroxides: dry salts or wet carbenes? Catal Today 2013;200:9–16.
[164] MacFarlane DR, Pringle JM, Johansson KM, Forsyth SA, Forsyth M. Lewis base ionic liquids. Chem Commun 2006:1905–17.
[165] Wykes A, MacNeil SL. Synthesis of new Lewis basic room-Temperature ionic liquids by monoquaternization of 1,4-diazabicyclo[2.2.2]octane (DABCO). Synlett 2007:107–10.
[166] Pretti C, Renzi M, Focardi SE, Giovani A, Monni G, Melai B, et al. Acute toxicity and biodegradability of N-alkyl-N-methylmorpholinium and N-alkyl-DABCO based ionic liquids. Ecotox Environ Safe 2011;74:748–53.
[167] Xu D-Z, Liu YJ, Shi S. A simple, efficient and green procedure for Knoevenagel condensation catalyzed by [C4dabco][BF_4] ionic liquid in water. Green Chem 2010;12:514–7.
[168] Zare-Bidaki A, Davoodnia A. A DABCO derived ionic liquid based on tetrafluoroborate anion: preparation, characterization and evaluation of its catalytic activity in the synthesis of 14-aryl-14H-dibenzo[a, j]xanthenes. Bull Kor Chem Soc 2012;33:1154–8.
[169] Jiang T, Gao HX, Han BX, Zhao GY, Chang YH, Wu WZ, et al. Ionic liquid catalyzed Henry reactions. Tetrahedron Lett 2004;45:2699–701.
[170] [a] Zhu AL, Jiang T, Wang D, Han BX, Liu L, Huang J, et al. Direct aldol reactions catalyzed by 1,1,3,3-tetramethylguanidine lactate without solvent. Green Chem 2005;7:514–7;

[b] Zhu AL, Jiang T, Han BX, Huang J, Zhang JC, Ma XM. Study on guanidine-based task-specific ionic liquids as catalysts for direct aldol reactions without solvent. New J Chem 2006;30:736–40.
[171] Gao HX, Han BX, Li JC, Jiang T, Liu ZM, Wu WZ, et al. Preparation of room-temperature ionic liquids by neutralization of 1,1,3,3-tetramethylguanidine with acids and their use as media for Mannich reaction. Synth Commun 2004;34:3083–9.
[172] Shaabani A, Rahmati A. Ionic liquid promoted efficient synthesis of 3,4-dihydropyrimidin-2-(1H)-ones. Catal Lett 2005;100:177–9.
[173] [a] Chavan SS, Degani MS. Ionic liquid mediated one-pot synthesis of 6-aminouracils. Green Chem 2012;14:296–9;
[b] Chavan SS, Degani MS. Ionic liquid catalyzed 4,6-disubstituted-3-cyano-2-pyridone synthesis under solvent-free conditions. Catal Lett 2011;141:1693–7;
[c] Gharnati L, Walter O, Arnold U, Doering M. Guanidinium-based phosphotungstates and ionic liquids as catalysts and solvents for the epoxidation of olefins with hydrogen peroxide. Eur J Inorg Chem 2011:2756–62.
[174] Zhang ZF, Xie Y, Li WJ, Hu SQ, Song JL, Jiang T, et al. Hydrogenation of carbon dioxide is promoted by a task-specific ionic liquid. Angew Chem Int Ed 2008;47:1127–9.
[175] Zhang ZF, Hu SQ, Song JL, Li WJ, Yang GY, Han BX. Hydrogenation of CO_2 to formic acid promoted by a diamine-functionalized ionic liquid. ChemSusChem 2009;2:234–8.

CHAPTER 4

Catalysis for the Production of Sustainable Chemicals and Fuels from Biomass

Md Imteyaz Alam[1], Basudeb Saha[2,3]

[1]*Renewable Energy and Chemicals Laboratory, Department of Chemical Engineering, Indian Institute of Technology, Delhi, India;* [2]*Department of Chemistry, Purdue University, West Lafayette, Indiana, USA;* [3]*Laboratory of Catalysis, Department of Chemistry, University of Delhi, Delhi, India*

1. Introduction

The global consumption of liquid fuel is projected to increase by 38% (119 million barrels per day) in 2040, which was about 87 million barrels per days in 2010 [1]. These liquid fuels are generally obtained from non-renewable petroleum sources that are the major source of global atmospheric pollution including rise in green house gas (GHG) emission. Depleting fossil fuel reserves and increasing demand of chemical and energy products have necessitated efforts to explore alternate renewable resources, such as biomass, because of its abundance (~498 million tons) as a carbon neutral source. Interestingly, carbohydrate ($C_nH_{2n}O_n$), a major component of biomass, has nearly similar C and H contents as those of hydrocarbons ($C_nH_{2n} \pm X$, where $X = 0$ or 2) but a high oxygen content. Effective catalytic pathways and mechanistic insights are desirable for removing oxygen from biomass to produce advanced fuel. Some processes like hydrolysis, dehydration, hydrogenolysis, hydrogenation, decarbonylation and decarboxylation have been explored to remove oxygen functionalities. C−C coupling reaction can also be performed to get jet range fuel directly from dehydrated products. Conversion of biomass resources into fuels and chemicals can be facilitated by effective catalytic routes.

2. Biomass and Biomass Compositions

The term *biomass* means organic materials such as wood, grass, algae, agricultural crops and their residues and wastes, including some animal waste [2]. These materials are abundant (~170 billion metric tons/per year) in nature. Cellulose, hemicelluloses and lignin (Figure 1) [3] are the major components of biomass that contain 35−50% of cellulose, 20−35% of hemicellulose and 10−25% of lignin [4]. Cellulose, a polymer of glucose units

Sustainable Catalytic Processes. http://dx.doi.org/10.1016/B978-0-444-59567-6.00004-2

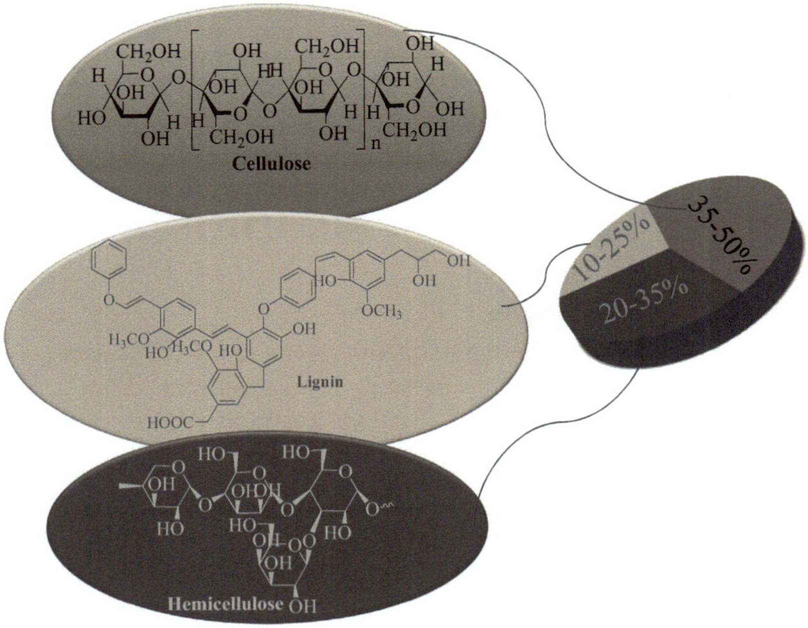

Figure 1
Lignocellulose composition: cellulose, hemicellulose and lignin.

linked together with β-glycosidic bonds, can be transformed into platform chemicals (e.g. 5-hydroxymethyl furfural (HMF) and levulinic acid (LA)) via hydrolysis followed by dehydration, and rehydration of HMF. Lignin is the most complex and rigid component of biomass with a three-dimensional aromatic network that provides structural rigidity to the plant. Out of large amount of available biomass, only 3–4% is used by humans for food and nonfood purposes. Utilizing the remaining component, especially lignocellulose, provides an effective and low cost replacement of fossil fuel to supply sustainable raw material for chemical and fuel production [5,6]. However, a major challenge in converting biomass into chemicals and hydrocarbon fuels is the cleavage of glycosidic linkage within the biomass feedstock to reduce both oxygen content and degree of polymerization.

3. *Strategy for Biomass Conversion*

The conversion of biomass resources into platform chemicals and fuels is strategically important to improve the sustainability in energy. In biorefinery processes, polysaccharides undergo hydrolysis to their constituent monomers, which yield furanic compounds like HMF or furfural upon dehydration over an acid catalyst. HMF and Furfural can be produced with high selectivity (e.g. 90%) from fructose and xylose using C_6 and C_5

sugars, respectively, as feedstock [7]. However, HMF yield is lower in the case of glucose, probably due to the requirement of an isomerization step. Extensive studies on various catalysts and their properties [8,9] suggest that the concentration and nature of the acid sites are important to achieve a high conversion of biomass and selectivity in products. These catalysts are categorized into five groups: organic acids, inorganic acids, salts, Lewis acids and others, which are discussed in Section 6.

Lignocellulosic biomass can be valorized into fuels and chemicals by three important pathways, namely, gasification, pyrolysis and pretreatment/hydrolysis. Gasification and pyrolysis are aimed to convert lignocellulose into syngas ($CO + H_2$) and liquid fractions (bio-oils) as important intermediates for fuels and chemical production at a high temperature (>300 °C). Such conditions can limit the process by making it difficult to control the reaction chemistries and problems in product purification due to high degrees of impurities. In case of hydrolysis, lignocellulose is converted to hexose and pentose sugars which undergo dehydration to produce HMF and furfural as platform chemicals. Disaccharides or polysaccharides, for example, cellobiose, inulin or cellulose, can be used as starting materials for HMF production, but these processes require a hydrolysis step for complete depolymerization to glucose, which is then converted to HMF in the presence of acid catalysts. Upon acid catalysis at high temperature and longer reaction time, monomer units decompose into side products [10]. Recently, Saha's group has developed various catalysts and solvents for the direct one-pot conversion of monosaccharides, disaccharides, polysaccharides and raw biomass (e.g. sugarcane bagasse and weeds) into furanic molecules. These reactions were carried out using single or combined metal chloride [11,12], ionic liquid (ILs) [13] and solid acid catalysts [14−17].

4. Biomass to Value-Added Chemicals

Two possible routes for the conversion of carbohydrates to chemicals are fermentation and catalysis (chemical transformation). In this chapter, we will discuss only chemical transformation of carbohydrates and plant biomass to produce value-added chemicals. The most convenient method for the preparation of HMF is the acid-catalysed dehydration of fructose, which is obtained by acid-catalysed hydrolysis of carbohydrates followed by selective isomerization of glucose. Many types of catalysts have been used for the dehydration of hexoses [8,9]. HMF can be transformed into other high value products such as 5-hydroxymethylfuranoic acid, 2,5-furandicarboxylic acid (FDCA), 2,5-bis(hydroxymethyl)furan and 2,5-furandicarboxaldehyde, LA and 2,5-bis(aminomethyl)furan (Scheme 1). Because of versatile applications, the US Department of Energy [18] has listed HMF as the top 10 biomass-derived platform chemical.

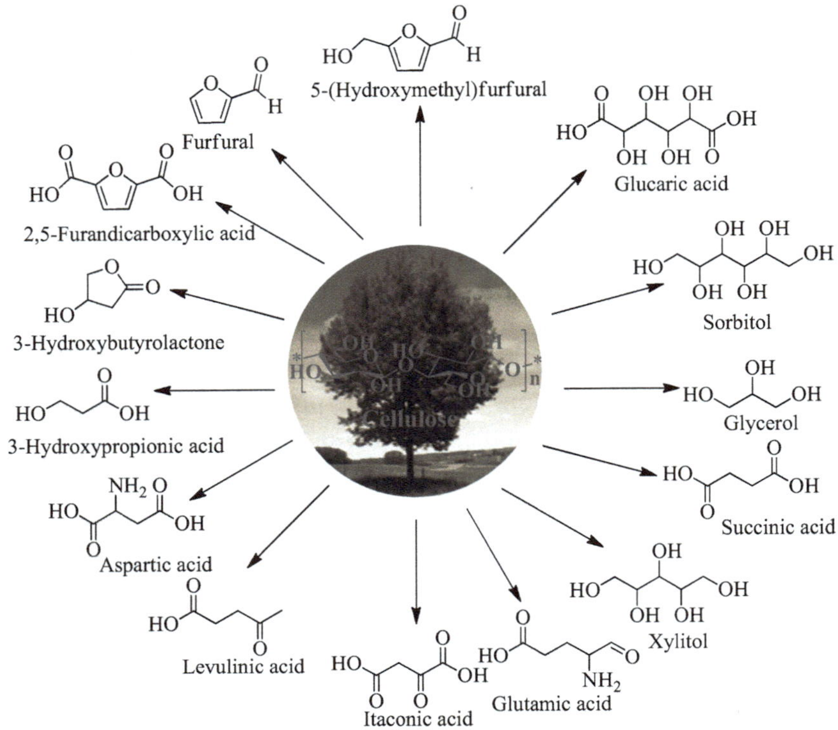

Scheme 1

Top platform chemicals from biomass.

5. 5-Hydroxymethyl Furfural

HMF is a dehydration product of hexose sugars; it is yellow in colour and highly soluble in water. The molecule contains a furan ring, having both aldehyde (http://en.wikipedia.org/wiki/Aldehyde) and alcohol functionalities. It can be found in natural products such as honey and a variety of heat-processed food products including milk, fruit juices, spirits and honey [19].

HMF is a valuable intermediate for fine chemicals, pharmaceuticals, fuels and furan-based polymers because it contains two different functional groups, hydroxyl and aldehyde, which allows its chemical transformation to a variety of products via oxidation, hydrogenolysis and condensation. It has been identified that HMF and its furan derivative compounds are the most suitable starting material for production of fuels, many thermo-resistant polymers and complex macrocycles (Scheme 2). HMF can be efficiently transformed into LA, fuels such as 2,5-dimethylfuran (DMF) and long chain hydrocarbons by hydrogenation, and 5-ethoxymethyl furfural (EMF) by esterification. Polyester building blocks, such as 2,5-furandicarbaldehyde, FDCA can be synthesized by selective oxidation

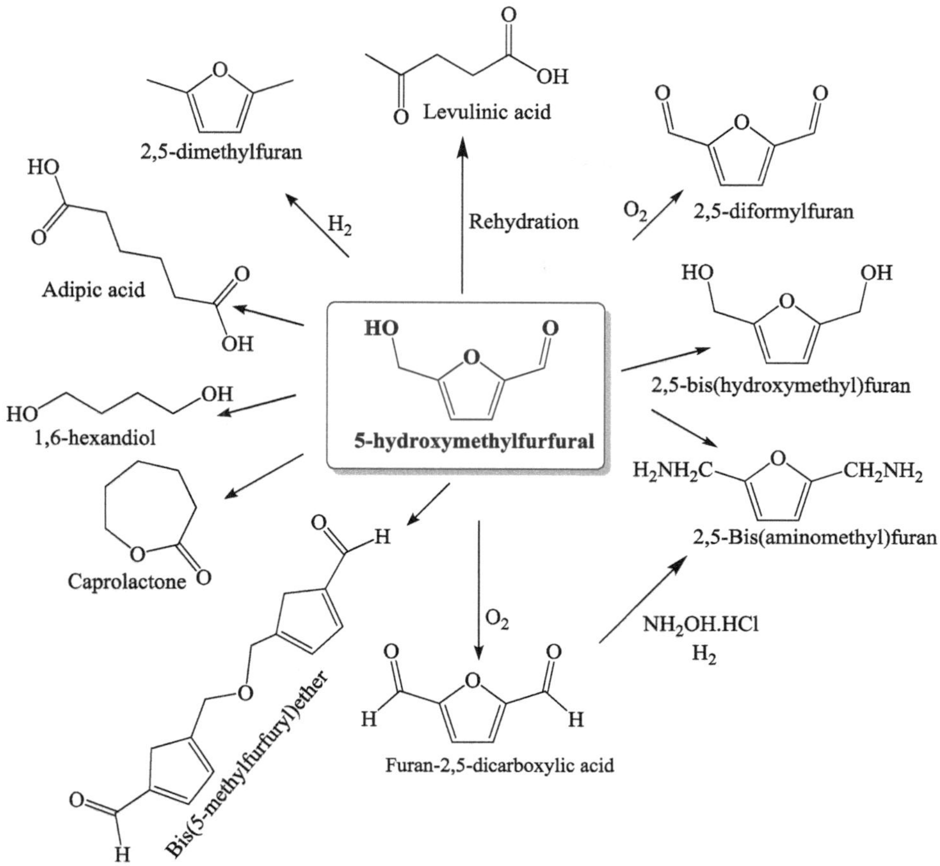

Scheme 2
Transformation of high potential value-added chemicals from HMF.

of HMF, which can further be used as a monomer for biodegradable polymers. Selective reduction of HMF can produce 2,5-dihydroxymethylfuran and 2,5-bis(hydroxymethyl) tetrahydrofuran (THF, tetrahydrafuran), which can substitute alcohol components in the production of polyester. A brief outline of the chemicals obtained from HMF is shown in Scheme 2 [18,20].

5.1 Aspects of HMF Synthesis

HMF production involves acid-catalysed triple dehydration of C_6 sugars, obtained from oligosaccharides and polysaccharides. The dehydration process produces high selective HMF when ketohexoses are used as starting substrates when compared to aldohexoses. The reason is aldohexose (glucose) enolyses to a very low degree and the enolization is a rate-determining step of the HMF formation from glucose. Glucose can also form cross-

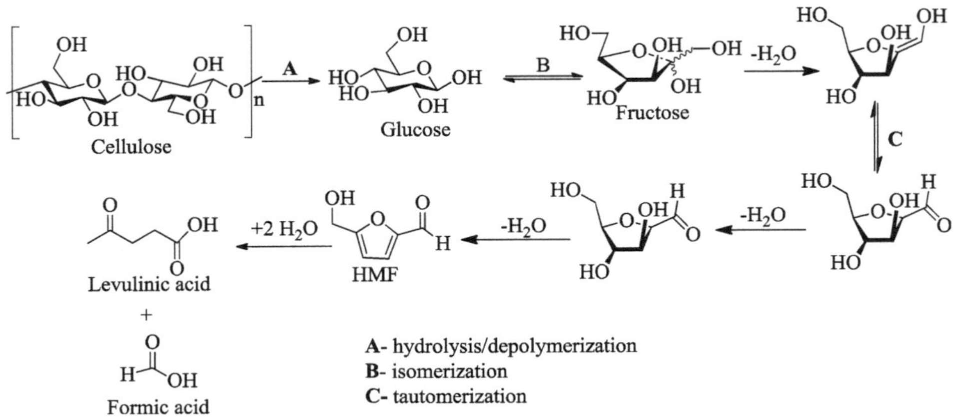

Scheme 3

Cyclic pathway showing cellulose conversion to HMF and its subsequent rehydration to LA.

polymerized materials with HMF. However, glucose is a low cost base feedstock in comparison to fructose. Conversion of abundant and low cost alternate substrates, such as bagasse [11], weeds [12], rice straw [21], pine wood [21] and corn stover [22], is desirable to make the process more cost effective. A general mechanism for acid-catalysed HMF production is shown in Scheme 3, which shows that cellulose in biomass initially hydrolyses/depolymerizes into glucose by an acidic catalyst and subsequently isomerizes to fructose. HMF is formed by the elimination of three molecules of water from fructose. Rehydration of HMF with water produces LA under acidic conditions.

The mechanism of HMF formation shows different reactivities for fructose and glucose conversions. Fructose is highly reactive and selective for HMF in comparison to glucose. Glucose has a stable ring structure, which hinders its ability to form the acyclic enediol intermediate. The cyclic mechanism as shown in Scheme 3, requires glucose to isomerize into fructose prior to dehydration. The dehydration reaction can occur in either aqueous or nonaqueous media, generally in the presence of acid catalysts. The dehydration of hexoses to HMF in terms of HMF yield is high when carried out in nonaqueous media. The higher HMF yields in nonaqueous solvents is due to the inhibition of the degradation of HMF to LA and cross-polymerized by-products. Oligosaccharides and polysaccharides including plant biomass are also effective substrates for HMF. Monophasic solvents such as dimethylsulphoxide (DMSO), *N,N*-dimethyl acetamide with lithium chloride (DMA−LiCl), isopropanol, alkylphenol, DMF, acetone, dioxane, polyglycol ether and n-butanol have been extensively used as the reaction media for the synthesis of HMF. The dehydration of fructose is more effective in DMSO [23] and DMA−LiCl [24], even in the absence of any catalyst. However, the use of a high boiling solvent (e.g. DMSO) poses a challenge in the separation and purification of HMF. Another disadvantage of DMSO is the possibility of formation of toxic S-containing by-products [11,25]. DMA−LiCl is a better solvent in terms of product yields and conversions without the formation of toxic by-products and is comparatively easier for product separation. ILs have also been used as solvents and catalysts with improved yield of HMF and furfural from C_6 and C_5 sugars, respectively, in a short reaction time. However, product separation from high boiling point and high cost of ILs remains a challenge for researchers. Recently, a biphasic reaction system using low boiling point organic solvents and water, including water–THF, water–butanol, water–methyl isobutyl ketone (MIBK) have been developed, which eliminates drawbacks associated with DMSO, IL and other high boiling solvents. In the latter process, organic solvents can be reused after product separation.

5.2 By-products in HMF Production

A major challenge in the HMF synthesis is the occurrence of side reactions [26–28]. Fructose dianhydrides, aromatics, organic acids and cross-polymeric materials, including humins (insoluble polymer) are the reported by-products. In addition to this, the formation of retroaldol products, for example, erythrose, pyruvaldehyde, glyceraldehyde, dihydroxyacetone, lactic acid and glycolaldehyde are also reported as possible by-products. The nature and quantity of the by-products in HMF synthesis depend on the nature of substrates, catalysts and process conditions.

It has been accepted that humins are formed from polymeric condensation of HMF with sugars. A similar effect has also been found in furfural production from pentoses where the addition of furfural to the pentose feed accelerated the formation of humin-like materials [29]. Akien and coauthors studied the formation of humins by using ^{13}C nuclear magnetic resonance and proposed that these by-products are formed through a 2,6-anhydro-β-D-fructofuranose intermediate [30]. Because of complex structure and composition, the nature of humins and their intermediates has not been well characterized. Recently, IR studies indicated the presence of furan, hydroxymethyl groups from HMF and the aromatic ring of benzaldehyde in humins [31]. Accordingly, a proposed pathway for the humin formation is depicted in Scheme 4.

6. Catalyst for Biomass Conversion

In the previous section, we have discussed that an acid catalyst is necessary for cellulose hydrolysis and dehydration to form HMF. Mineral acids, such as HCl, H_2SO_4 and H_3PO_4, have been effectively employed in the past to achieve carbohydrate conversion and HMF yield. Since mineral acids are corrosive and environmentally toxic, organic acids, such as oxalic acid, LA and maleic acid, were chosen by the researchers. However, these homogeneous catalysts also pose a challenge in the effective separation of the desired products and disposal of the spent catalysts. Consequently, solid acid catalysts (transition metal oxides, phosphates and other acidic materials) have been developed that can be

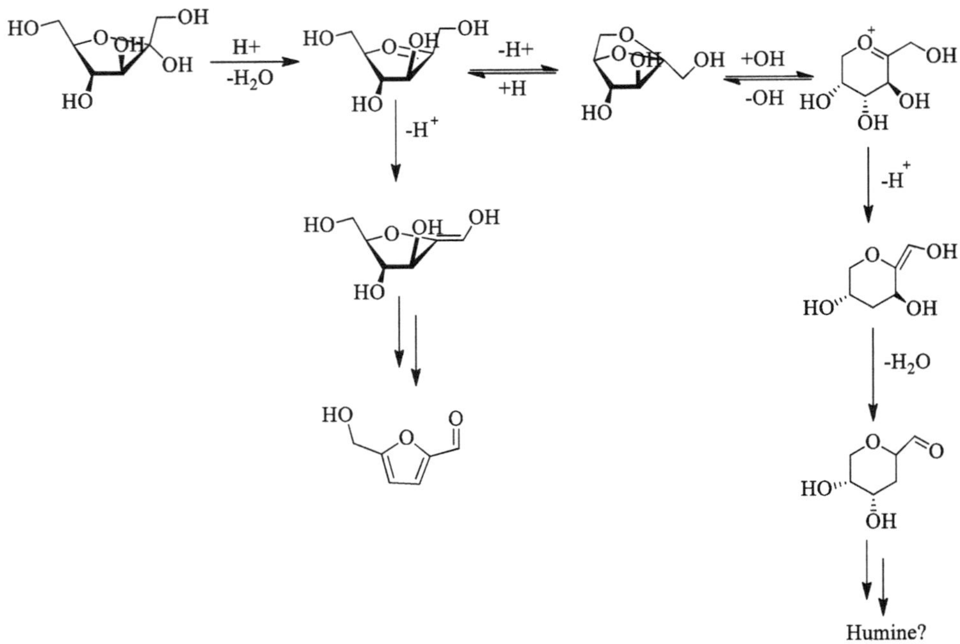

Scheme 4
Proposed pathway for humin formation [76].

easily separated and reused. Solid catalysts can also minimize corrosion of the reactor and environmental issues related to disposal of used catalysts [32]. The use of different solid acid catalysts for the conversion of lignocellulose to chemicals and fuels has been summarized in the recent literature [9,33,34]. Cottier and Descotes have classified acid catalysts into organic acids, mineral acids, salts, Lewis acids and solid catalysts [35] (Table 1).

Table 1: Classification of Acid Catalysts.

Organic acids	Oxalic acid, LA, maleic acid, *p*-toluenesulphonic acid, trifluoroacetic acid
Mineral acids	Phosphoric acid, Sulphuric acid, hydrochloric acid, iodine or hydroiodic acid, hydrobromic acid
Salts	$(NH_4)_2SO_4/SO_3$, pyridine/PO_4^{3-}, pyridine/HCl, aluminium salts, Th and Zr ions, zirconium phosphate, Cr, Al, Ti, Ca, in ions, $ZrOCl_2/CrCl_3$, $ZrOCl_2$, $VO(SO_4)TiO_2$, V, Zr, Cr, Ti, porphyrins
Lewis acids	$ZnCl_2$, $AlCl_3$, BF_3, $B(OH)_3$, $Sc(OTf)_3$, $Yb(OTf)_3$, $Ho(OTf)_3$, $Sm(OTf)_3$, $Nd(OTf)_3$ and $La(OTf)_3$
Solid catalyst	Ion exchange resins, zeolites, oxides of Ti, Zr, Nb, Ta, phosphates of Al, Ti, Zr, Nb, Ta, carbon-supported *p*-toluenesulphonic acid, $H_3PW_{12}O_{40}$, $Ag_3PW_{12}O_{40}$, HT-Amberlyst-15

7. Significance of 5-Hydroxymethyl Furfural as a Platform Chemical

7.1 Oxidation of HMF to FDCA

Oxidation of HMF produces a variety of chemicals such as 2,5-diformylfuran (DFF), hydroxymethyl furan carboxylic acid, formylfuran carboxylic acid and FDCA. Among these chemicals, oxidation of HMF to DFF and FDCA (Scheme 5) has great importance.

FDCA is a potential biorenewable substitute for petroleum-derived terephthalic acid used in the manufacturing of poly(ethylene terephthalate) (PET) [36]. Recently, FDCA has evolved as a suitable replacement for terephthalic acid in engineering plastics such as poly(ethylene furancarboxylate) (PEF), poly(butylene furancarboxylate) and mixed PEF.PET [37]. Different catalysts, processes and reaction chemistries for FDCA synthesis have been reported [9]. Generally, FDCA is synthesized via the oxidation of HMF using noble metals (e.g. Pt, Pd and Au)-based catalyst [38,39]. Because of the high cost of metals, these processes are yet to be commercialized. Recently, Saha and co-workers have developed an inexpensive metal-based catalyst for FDCA production. In their process, DFF was the only product in the absence of trifluoroacetic acid (HTFA) when $Co(OAC)_2$/$Mn(OAC)_2$/NaBr was used as a catalyst. In the presence of HTFA, HMF oxidation achieved a 60% FDCA yield at 90 °C for 3 h under 1 bar O_2 [40]. The same group has also developed an efficient Fe(III)-based catalyst for HMF oxidation to FDCA in water under mild conditions [41].

7.2 Oxidation of HMF to DFF

DFF is another versatile chemical obtained from HMF oxidation. DFF can be converted to a host of new chemicals for several applications including monomers, pharmaceutical intermediates and antifungal agents [42]. DFF may be further converted to furan, which hydrogenates to form tetrahydrofuran, a common laboratory solvent. Due to its high applicability, Saha and coauthors have recently developed a direct catalytic process for the production of DFF from glucose and fructose via an HMF intermediate [9,43].

7.3 HMF to LA

LA has been listed as a primary biorefinery building block and platform chemical because of its versatile applications (Scheme 6) [44]. It is formed by the rehydration of

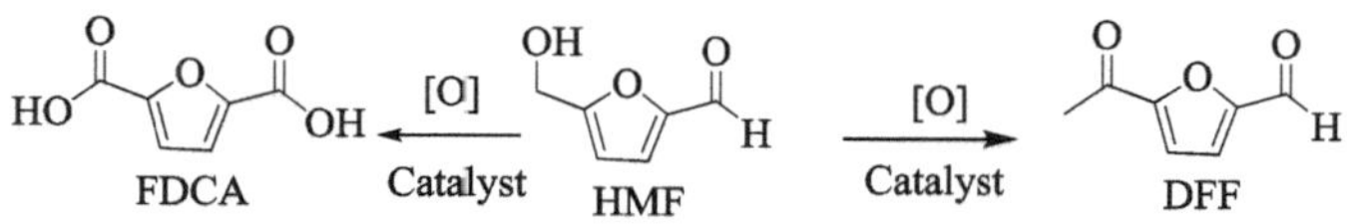

Scheme 5
Oxidation of HMF to DFF and FDCA.

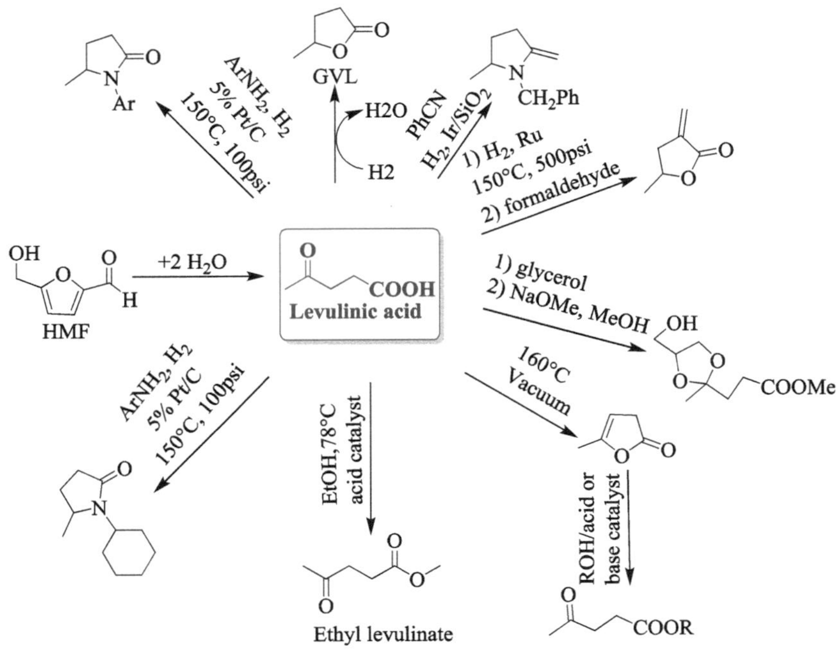

Scheme 6

Versatile applications of LA.

HMF with water in which the furan ring of HMF is cleaved in the presence of an acid catalyst and an equivalent amount of formic acid is also formed. The cyclic route for the transformation of glucose to LA via HMF is represented in Scheme 3. Mineral acids, solid acids, such as Amberlite IR-120, LZY-zeolite, $S_2O_8^{2-}/ZrO_2-SiO_2-Sm_2O_3$, $CrCl_3$/HY zeolite, MFI-type zeolite and Nafion, and Lewis acidic metal salts, such as $CrCl_3$ and $AlCl_3$ have been used as catalysts for the conversion of carbohydrates and HMF to LA. The reaction chemistries and process conditions for these methods are extensively described in the literature [9].

8. Second-Generation Biofuels from Biomass

Biofuels in many forms such as solid (fuelwood, charcoal and wood pellets), liquid (ethanol, biodiesel and pyrolysis oils) and gas (biogas like methane) are obtained from biomass. These are broadly divided into primary and secondary biofuels. Primary biofuels are used in an unprocessed form (mainly for heating, cooking or electricity production such as fuel wood, wood chips and pellets) while secondary biofuels are produced by catalytic and non-catalytic processing of biomass. Secondary biofuels are further classified as first, second and third generations on the basis of raw materials and technologies used for their production (Figure 2).

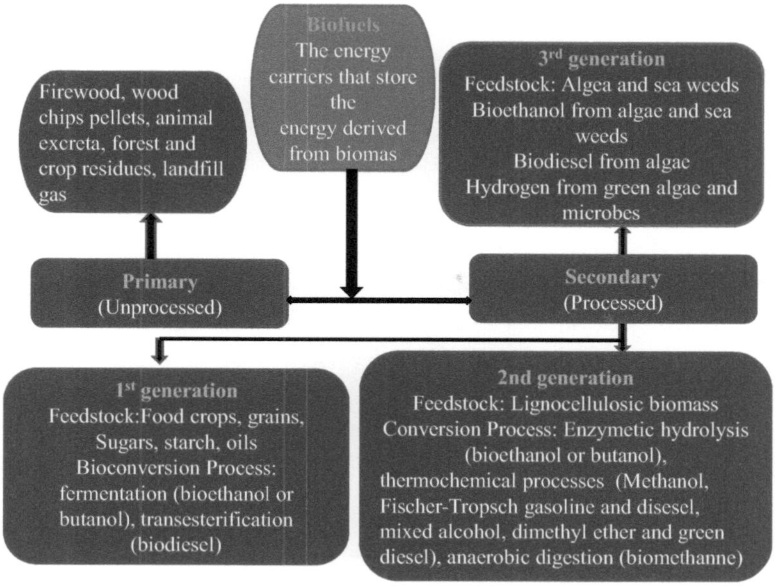

Figure 2
A short introduction of biofuel.

The second-generation biofuels are fuels produced from sustainable feedstock. Sustainability of a feedstock depends upon its availability, impact on GHG emissions and impact on biodiversity and land use. The second-generation biofuel is better than the first-generation biofuel, because of the abundance of nonfood lignocellulosic biomass. Scheme 7 shows a range of fuels and fuel additives that are produced from lignocellulose via catalytic routes.

8.1 Cellulosic Ethanol

The use of ethanol as a motor fuel has been steadily increasing across the globe. The reasons for the increasing demand of ethanol as a fuel are that it (1) decreases the dependence on foreign oil, (2) reduces trade deficits, (3) creates jobs in rural areas, (4) reduces air pollution and (5) carbon emissions. Ethanol, unlike gasoline, is an oxygenated fuel that contains 35% oxygen, which reduces particulate and NO_x emissions from combustion.

Ethanol can be made synthetically from petroleum or by microbial conversion of biomass materials through fermentation. In 1995, about 93% of the ethanol in the world was produced by the fermentation method and about 7% by the synthetic method. The fermentation method generally uses three steps: (1) production of fermentable sugars, (2) fermentation of sugars to ethanol and (3) separation and purification of the ethanol, usually by distillation.

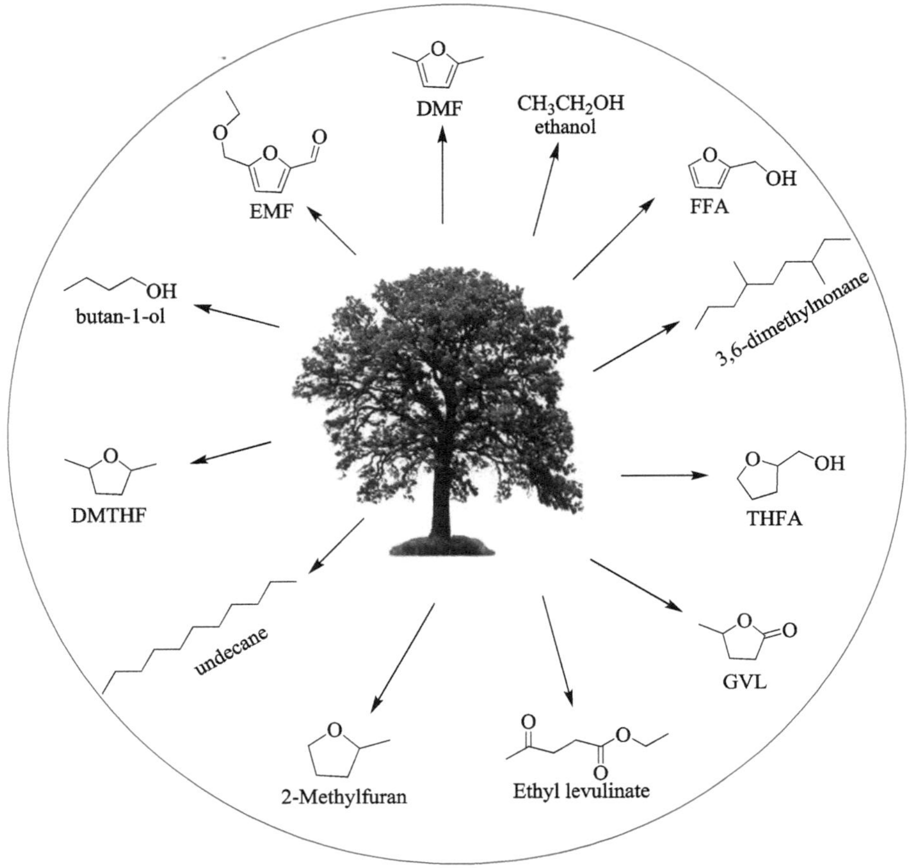

Scheme 7
Liquid biofuel components obtained from lignocellulosic biomass.

8.2 5-Ethoxymethyl Furfural

EMF is a heterocyclic compound containing a furan ring with an aldehyde and ester group. It is a colourless to pale yellow liquid that is considered to be a potential biofuel. It is obtained either by ethanolysis of chloromethylfurfural/bromomethylfurfural or etherification of HMF with ethanol (Scheme 8) [43]. The physical and chemical properties of this ether (liquid form at room temperature, high cetane number, good oxidation stability) compare well with those of standard gasoline. The energy density of EMF (8.7 kWh/L) is similar to that of standard gasoline (8.8 kWh/L), and diesel (9.7 kWh/L), and is significantly higher than that of ethanol (6.1 kWh/L) [45]. Recently, Gruter et al. used EMF-blended commercial diesel in engine tests, which significantly impacted the reduction of soot (fine particulates) and SO_x emission [46]. EMF can also be converted to 5-(ethoxymethyl)furfural alcohol via hydrogenation, which is miscible in diesel and has a similar combustion profile to that of ethanol. Because of the above reasons, EMF is an accepted fuel or fuel additive.

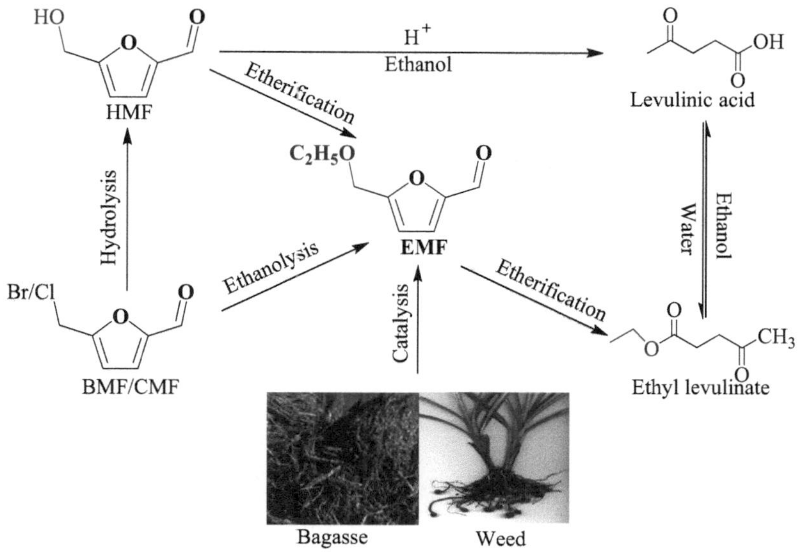

Scheme 8
Reaction route for the transformation of lignocelluloses, HMF, BMF and CMF into EMF and ethyl levunate.

In addition to being a fuel, EMF has been extensively recognized as a flavour and aroma component additive in wines and beers, due to its low toxicity [47].

A number of catalytic routes have been developed for the synthesis of EMF. It was previously synthesized from sugars via 5-chloromethylfurfural (CMF), in which CMF was produced by using concentrated hydrochloric acid. In that process, nucleophilic substitution of chlorine in CMF with ethanol resulted in EMF. Using H_2SO_4 as a catalyst, the EMF yield from fructose was approximately 60%. However, mineral acids (HCl and H_2SO_4) [48] are corrosive and pose serious safety concerns. To address these challenges, $AlCl_3$ [49], Imidazolium propanesulphonic acids [50], $Fe_3O_4@SiO_2^-H_3PW_{12}O_{40}$ [51], SnBEA and Amberlyst-131 [52], $H_3PW_{12}O_{40}$ [53] and graphene oxide (GO) [54] catalysts have been developed for EMF synthesis. Recently, Abu-Omar and coauthors have synthesized EMF from glucose, inulin and sucrose in one pot via etherification [53]. Lew and co-workers have used a combined SnBEA and Amberlyst-131 acid catalyst for the one-pot synthesis of EMF from glucose with a 31% yield.

8.3 2,5-Dimethylfuran

DMF is a heterocyclic compound of furan with a molecular formula of C_6H_8O. It has a boiling point of 92–94 °C and energy density of 33.7 MJ/kg, which is 40% higher than that of ethanol. DMF is almost immiscible (1.47 g/L at 25 °C) in water [55]. The research octane number (RON) of DMF is 119 and latent heat of vaporization at 20 °C is

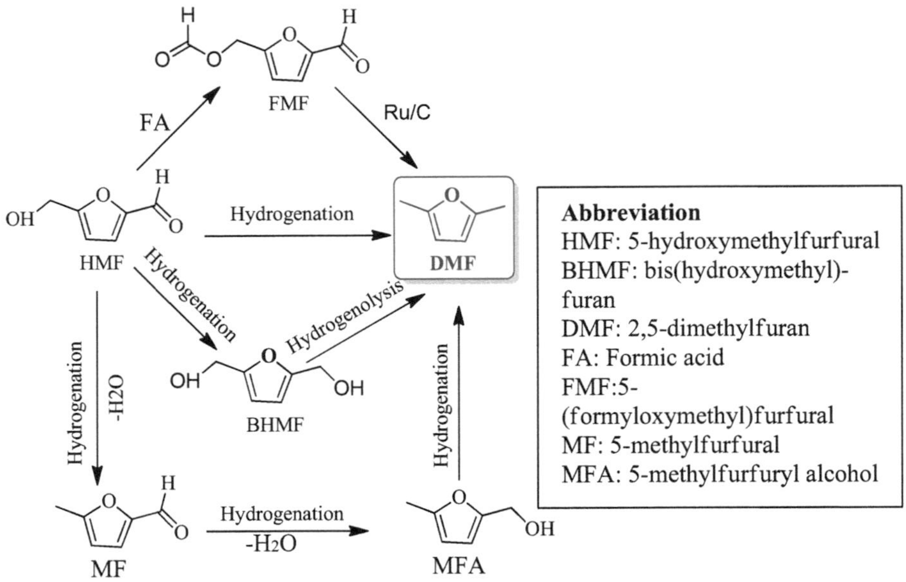

Scheme 9
Catalytic routes for DMF production from HMF.

31.91 kJ/mol [56]. Hu et al. have compared the antiwear and antifriction properties of DMF with petrol (gasoline) and confirmed that it can be used as an additive in petrol [57] Rothamer and Jennings also tested DMF-blended petrol for engine knocking and confirmed positive results of DMF-blended petrol over the unblended analogue [58]. DMF can be obtained from HMF via different hydrogenation routes in which BHMF, FMF, MF and MFA are proposed intermediates (Scheme 9). The fuel properties [58] of DMF against ethanol and standard gasoline are compared in Table 2.

Table 2: Comparison of Fuel Properties of DMF with Those of Ethanol and Gasoline.

Properties	Unit	DMF	Ethanol	Gasoline
Molecular formula	–	C_6H_8O	C_2H_6O	C_4–C_9
Molecular mass	kg/mol	96.13	46.07	100–105
Density at 20 °C	kg/m^3	889.7	790.7	744.6
Water solubility at 25 °C	g/L	1.47	~100	Insoluble
Gravimetric oxygen content	%	16.67	34.78	0
H/C ratio		1.33	3.00	1.865
O/C ratio		0.17	0.5	0
Stoichiometric air/fuel ratio		10.72	8.95	14.56
Gravimetric calorific value (LCV, liquid fuel)	MJ/kg	33.7	26.9	43.2
Volumetric calorific value (LCV, liquid fuel)	MJ/L	30	21.3	32.2
RON		119	110	95.8
Auto-ignition temperature	°C	285.85	423	257
Latent heat of vaporization at 20 °C	kJ/mol	31.91	43.25	38.51

LCV = Lower calorific value.

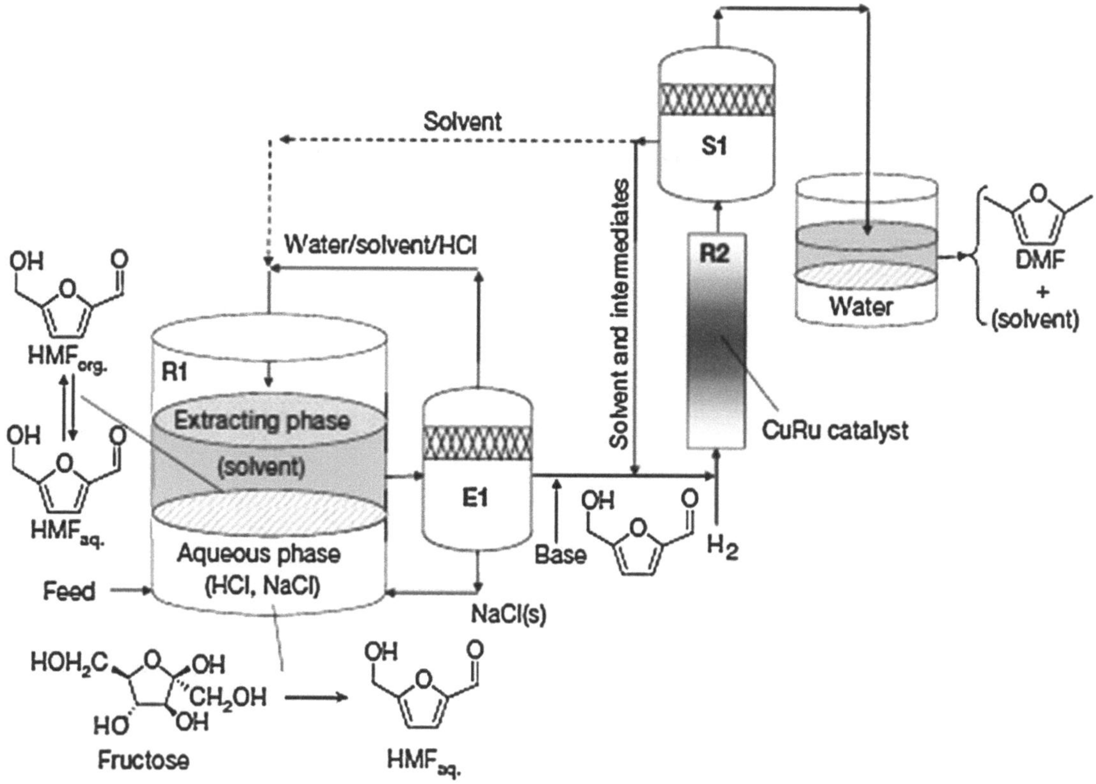

Figure 3
Schematic diagram of the process for conversion of fructose to DMF. *Adapted from Ref. [59].*

Dumesic group have developed a catalytic process for the production of DMF from fructose in biphasic medium (Figure 3) [59]. The aqueous layer in the biphasic reactor contained an acid catalyst and a sugar, while the extracting phase contained butanol for extraction of HMF as it is formed in the aqueous phase. The partitioning of HMF into the extracting phase improved in the presence of an inorganic salt. HMF was subsequently hydrogenated to DMF (71% yield) over a CuRu catalyst under H_2. Raines et al. developed a method for DMF from untreated corn stover [24]. This two-step method involved the $CrCl_3$–HCl-catalysed transformation of corn stover into HMF, followed by the hydrogenation–hydrogenolysis of HMF to DMF by a Cu-Ru/C catalyst in the presence of H_2. In this process, a toxic chromium salt along with a mineral acid was used as the catalyst for the degradation of corn stover into HMF. Chidambaram and Bell reported a two-step approach for the catalytic conversion of glucose to DMF in IL ([EMIM]Cl) Cl and acetonitrile at 120 °C under 62 bar H_2 pressure over a Pd/C, which gave a maximum 47% conversion of glucose with 32% DMF selectivity [60]. However, a potential drawback of this method was that IL decreased the solubility of H_2. Hence, a high

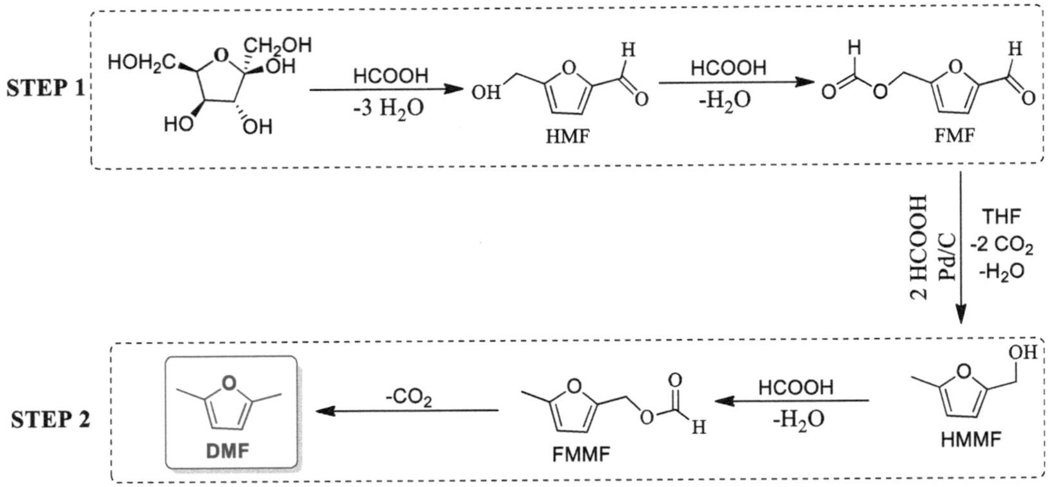

Scheme 10
One-pot DMF production from fructose using formic acid as a hydrogen source.

pressure of H_2 (62 bar) was required, which made the process energy intensive. Under similar reaction conditions, the Ru/C catalyst was ineffective for HMF hydrodeoxygenation. In the same year, Thananatthanachon and Rauchfuss found Pd/C as an efficient catalyst for the one-pot conversion of fructose to DMF with a maximum 51% overall DMF yield by heating a solution of HMF in refluxing tetrahydrofuran in the presence of formic acid, H_2SO_4, and Pd/C catalyst. Another method reported a maximum 81% yield of 2,5-dimethyltetrahydrofuran with a small amount of DMF from fructose by using $RhCl_3$ catalyst in H_2O/toluene/HI mixture [61].

Recently, De et al. produced DMF and EMF in one pot from a range of readily available biomass substrates such as carbohydrate, algal biomass and sugarcane bagasse (Scheme 10). The first step of this reaction involved the in situ HMF synthesis using formic acid as a catalyst for fructose conversion and $[DMA]^+[CH_3SO_3]^-$ (DMA=*N*,*N*-dimethyl acetamide) as a catalyst for cellulose and untreated biomass conversions. In the subsequent step, HMF was transformed into DMF by hydrogenation and hydrogenolysis reactions using FA and Ru/C, which produced maximum 32% DMF from fructose and 27% from agar [62]. Recently, Zu et al. have developed an Ru/Co_3O_4 catalyst to achieve maximum 93% DMF yield at 130 °C and 7 bar H_2 [63]. Vlachos et al. used Ru/C catalyst and 2-propanol as a hydrogen source to achieve maximum 81% DMF from HMF at 190 °C for 6 h [64].

8.4 Long Chain Hydrocarbons

8.4.1 Sylvan (2-Methylfuran) Diesel Process

Sylvan is a heterocyclic compound containing a furan ring and a methyl group. It is a key intermediate in the Sylvan diesel process. It has the unique property to react with diverse

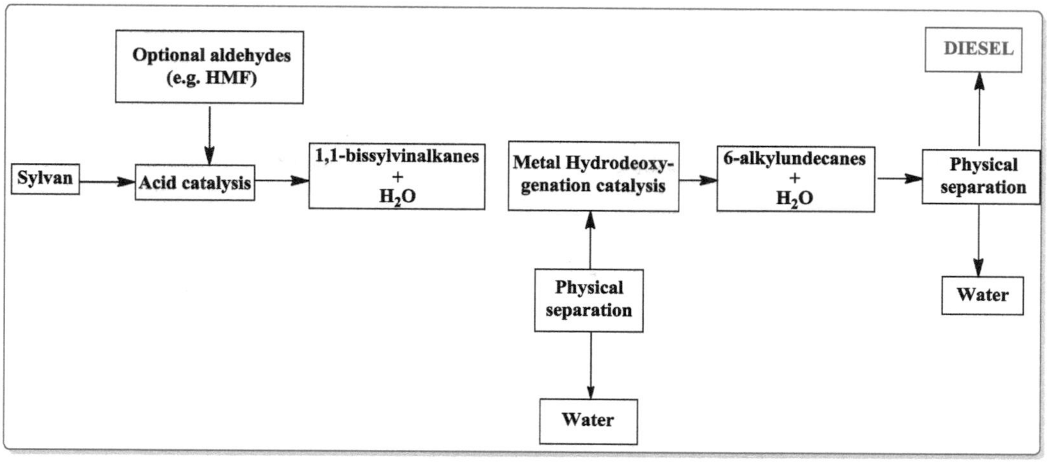

Figure 4
Flow diagram of the Sylvan process for diesel synthesis from lignocellulosic biomass.

molecules, mostly aldehydes and ketones, to give a precursor with an adequate carbon atom number that can be hydrogenated. This process is very flexible and can use molecules derived from different lignocelluloses, which undergo catalytic hydroxyalkylation/alkylation and hydrodeoxygenation. In the hydroxyalkylation/alkylation step, two Sylvan molecules were coupled with an aldehyde or a ketone to produce a C_{12}-oxygenated precursor [65,66]. Subsequent catalytic hydrodeoxygenation of the precursor led to the formation of premium diesel [67]. Water is the only by-product in this process, which can be easily separated without requiring energy-intensive distillation [65,66]. Figure 4 shows the schematic diagram of the process. The literature surveys on the hydroxyalkylation/alkylation and hydrodeoxygenation steps for the synthesis of long chain hydrocarbons are summarized below.

Catalytic hydroxyalkylation/alkylation involves coupling of two Sylvan molecules with an aldehyde or a ketone to yield oxygenated intermediates. This step can be carried out using different homogeneous and solid catalysts. Under solvent-free conditions, either catalyst can perform hydroalkylation/alkylation of 2-MF with butanal. The main reason for choosing butanal as a potential linker of two Sylvan molecules is that (1) it is a biomass-derived molecule that can be obtained by the selective oxidation of 1-butanol produced from biomass fermentation with engineered enzymes [68] and (2) the final hydrogenated product contains 14 carbon atoms and fits perfectly within the boiling point range of diesel fuel. Formaldehyde and acetaldehyde can also be used as linkers that are obtained from biomass pyrolysis and fast pyrolysis [69] or from Fischer–Tropsch synthesis as by-products [70]. Using 2.5 wt% of *p*-toluenesulphonic acid (*p*-TosOH) as a catalyst and a 2:1 M ratio (stoichiometric ratio) of 2-MF and butanal, 85% conversion of 2-MF with 91% selectivity in an oxygenated C_{14} precursor has been achieved. Among solid acids,

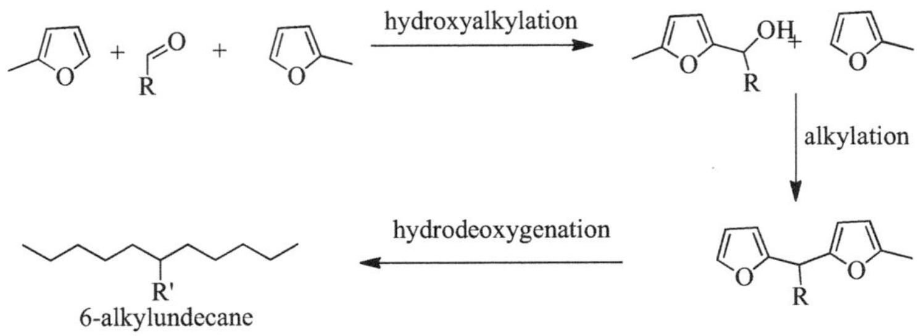

aldehyde (RCHO)	R	R' (after hydrodeoxygenation)
butanal	$-CH_2-CH_2-CH_3$	$-CH_2-CH_2-CH_3$
5-hydroxymethylfurfural	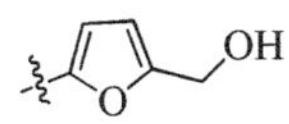	$-CH_2CH_2CH_2CH_2CH_2$
5-methylfurfural	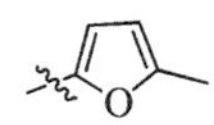	$-CH_2-CH_2-CH_3$

Scheme 11

Formation of 1,1-bisylvylalkanes by hydroxyalkylation/alkylation from Sylvan and an aldehyde with subsequent hydrodeoxygenation to 6-alkyl undecane. *Adapted from Ref. [65].*

delaminated zeolites have been identified as promising catalysts. Trimerization of Sylvan takes place in the presence of a sulphuric acid catalyst, in which 4-oxopentanal was formed as an intermediate via the furan ring opening of 2-MF. The hydroxyalkylation/alkylation of 2-MF with 4-oxopentanal is fast; thus, coupling products with aldehyde were not detected.

Subsequent hydrodeoxygenation of the oxygenated precursors for removing oxygen takes place at moderate temperatures and high hydrogen pressures and produce the corresponding mono-branched paraffinic diesel product with an 87% yield. The final alkane products are suitable for high-quality diesel fuel application. The progression of reaction sequence is shown in Scheme 11.

8.4.2 Cascade Process for Jet Fuels

HMF and furfural are used as building blocks for the production of liquid fuels by means of a cascade process involving dehydration, hydrogenation and aldol–condensation reactions as shown in Scheme 12. The process begins with acid-catalysed depolymerization of polysaccharides (e.g. starch, cellulose or hemicellulose) or raw biomass such as rice husk, weeds and bagasse to yield pentose and hexose sugar monomer units, which then undergo dehydration to form furan compounds (furfural and HMF). In the next step, the CO group in the furan compounds acts as a reactive centre for C–C coupling through aldol-condensation reactions with carbonyl molecules such

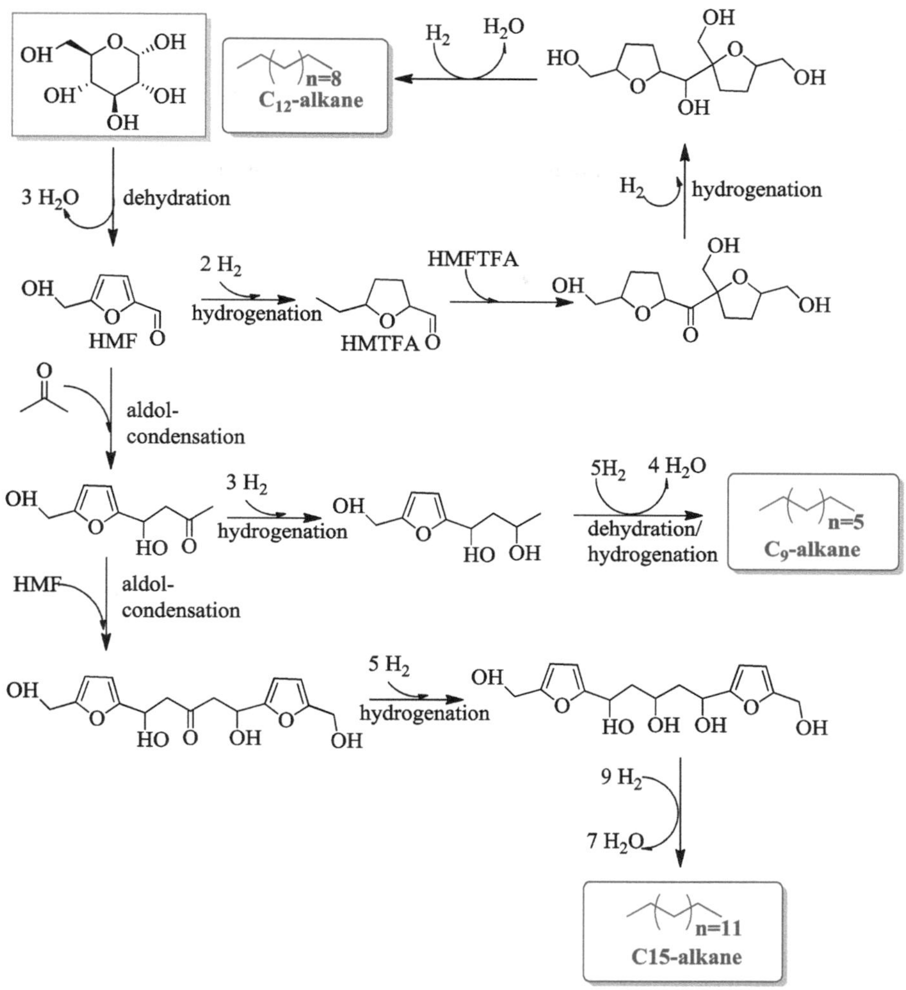

Scheme 12

Reaction routes for the transformation of biomass-derived glucose into liquid alkanes via HMF. *Adapted from Ref. [71].*

as acetone. These condensations are base catalysed (e.g. NaOH, Mg–Al oxides) and are typically carried out in polar solvents like water or biphasic solvent like water THF. As a result of the aldol condensation, long chain C–C compounds containing unsaturated C=C and C=O bonds (i.e. aldol adduct) are formed. The carbon chain of the adduct can be further increased by condensation with another molecule of furan substrate. The unsaturated C=C and C=O bonds in aldol adducts is subsequently hydrogenated over metal catalysts such as Pt, Pd or Ru to form alkanes. Both aldol condensation and hydrogenation steps can be coupled in one step by using appropriate bifunctional catalyst, for example, Pd/$MgO-ZrO_2$. Finally, the complete removal of

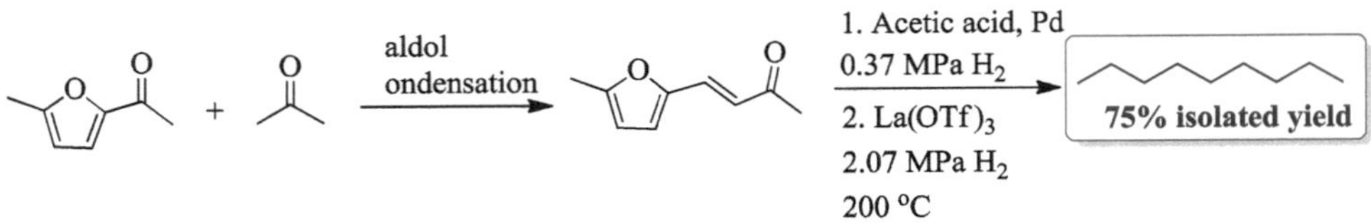

Scheme 13
Synthesis of higher alkane jet fuel by organocatalytic condensation and hydrodeoxygenation. *Adapted from Ref. [72].*

oxygen is performed via aqueous-phase dehydration/hydrogenation (APD/H) reactions to form liquid alkanes. Oxygen is progressively removed from the water-soluble adducts over a bifunctional metal–acid catalyst by cycles of dehydration and hydrogenation reactions. Huber et al. improved APD/H reactions using $Pt-SiO_2-Al_2O_3$ in a four-phase reactor involving aqueous solution of adducts, a hydrogen gas inlet stream, a hexadecane sweep stream and a solid catalyst [71]. Bifunctional $Pt/NbPO_4$ catalyst produced a pure organic stream of liquid hydrocarbon fuels with high molecular weights (C_9-C_{15}), without the use of a hexadecane sweep stream. However, this approach increased the carbon chain length in a nonselective manner by cross-condensation reactions. The resulting deoxygenation of these molecules provided a multifaceted mixture of molecules ranging from C_1 to C_{15} under high H_2 pressures and temperatures.

More recently, Sutton and co-workers have developed a new strategy (Scheme 13) for making a long chain alkane under relatively mild conditions using organocatalysts. These catalysts can eliminate the need for large quantities of caustic base or flammable organic solvents [72]. By this new and efficient one-pot process, a large number of linear alkanes can be produced selectively in good to excellent yields from biomass-derived furanic substrates via organoaldol chemistry. This method provides a great synthetic approach for the production of higher linear alkanes C_nH_{2n+2} with $n \geq 9$ for fuel applications (Scheme 13).

8.4.3 LA to Transportation Fuels

As we have already discussed in the previous section, LA can be transformed into a variety of chemicals and fuels (Figure 5). In this section, we discuss catalytic routes to produce long chain hydrocarbon fuels. A catalytic process has been reported for the conversion of LA to liquid hydrocarbon fuels of different classes (Figure 5). This pathway involves oxygen removal from LA-derived oxygenated intermediates and decarboxylation reactions, combined with C–C coupling processes such as ketonization, isomerization and oligomerization that are required to increase the molecular weight and to adjust the structure of the final hydrocarbon product. As a

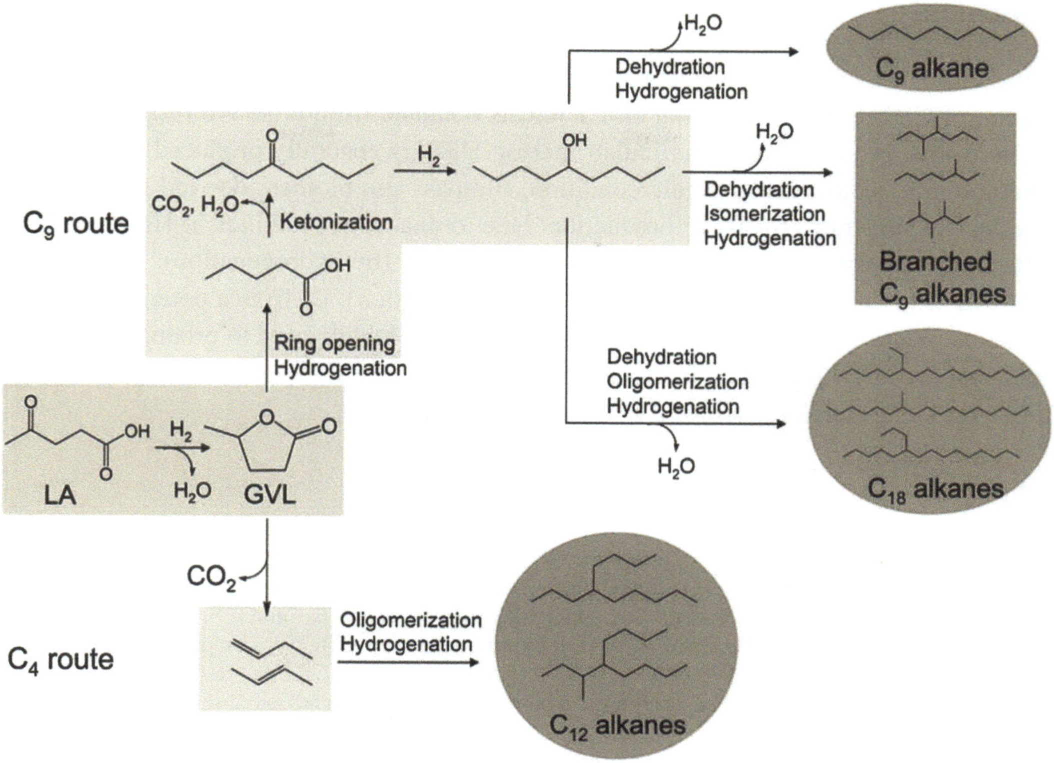

Figure 5
Catalytic pathway for the transformation of LA and γ-valerolactone (GVL) into liquid hydrocarbon transportation fuels. *Adapted from Ref. [75].*

first step, aqueous LA is hydrogenated to water-soluble GVL using a low acidic catalyst like Ru/C at low temperatures. This water-soluble GVL can then be upgraded to liquid hydrocarbon fuels by following two major pathways: the C_9 route and the C_4 routes (Figure 5). In the first route, GVL is converted to 5-nonanone via pentanoic acid over a water-tolerant multifunctional Pd/Nb_2O_5 catalyst. The GVL is first transformed into hydrophobic pentanoic acid by means of ring-opening (on acid sites) and hydrogenation reactions (on metal sites) at reasonable temperatures and pressures. Subsequently, Pentanoic acid ketonized to 5-nonanone, and reaction conditions can be monitored to permit this conversion reaction to take place on the same Pd/Nb_2O_5 reactor with a maximum carbon yield of 70% [73]. A multifunctional catalyst (Pd/Nb_2O_5 + $Ce_{0.5}Zr_{0.5}O_2$) has been developed to improve yields of the alkenes up to 90% [74] by progressive deoxygenation of biomass. Recently, Bond and co-workers have developed a promising route to upgrade aqueous solutions of GVL into jet fuels through the formation of C_4 alkenes [75].

9. Summary

Biomass is a natural sustainable source for chemicals and fuels with minimal or no emission of GHG emission. Among all chemicals obtained from biomass, furfurals and their derivative products are of particular interest. HMF is generally produced by the dehydration of fructose under acidic condition. Biomass and biomass-derived glucose are also effective substrates for HMF production. Due to dual functionalities in HMF structure, it is considered as a precursor to produce DMF (by hydrogenation), EMF (by esterification), FDCA (by oxidation) and LA (by rehydration). DMF is a potential biogasoline while FDCA can substitute fossil-based terephthalic acid to produce plastic bottles and textiles. Long chain hydrocarbon (C_9−C_{18}) can also be produced from furfural and LA via C−C coupling followed by hydrodeoxygenation.

References

[1] U.S. Energy Information Administration. International energy outlook. 2014.
[2] Definition of biomass. In: Basu P, editor. Biomass gasification and pyrolysis. Elsevier; 2011. p. 325−6.
[3] Carrier M, Loppinet-Serani A, Denux D, Lasnier J-M, Ham-Pichavant F, Cansell F, et al. Thermogravimetric analysis as a new method to determine the lignocellulosic composition of biomass. Biomass Bioenergy 2011;35:298−307.
[4] Sun Y, Cheng J. Hydrolysis of lignocellulosic materials for ethanol production: a review. Bioresour Technol 2002;83:1−11.
[5] Gallezot P. Conversion of biomass to selected chemical products. Chem Soc Rev 2012;41:1538−58.
[6] Lange JP, van der Heide E, van Buijtenen J, Price R. Furfural—a promising platform for lignocellulosic biofuels. ChemSusChem 2012;5:150−66.
[7] Chheda JN, Roman-Leshkov Y, Dumesic JA. Production of 5-hydroxymethylfurfural and furfural by dehydration of biomass-derived mono- and poly-saccharides. Green Chem 2007;9:342−50.
[8] Hu L, Zhao G, Hao W, Tang X, Sun Y, Lin L, et al. Catalytic conversion of biomass-derived carbohydrates into fuels and chemicals via furanic aldehydes. RSC Adv 2012;2:1184−206.
[9] van Putten RJ, van der Waal JC, Jong Ed de, Rasrendra CB, Heeres HJ, de Vries JG. 5-Hydroxymethyl furfural: a versatile platform chemical made from renewable resources. Chem Rev 2013;113:1499−597.
[10] Antal MJ, Mok WSL, Richards GN. Mechanism of formation of 5-(hydroxymethyl)-2-furaldehyde from d-fructose and sucrose. Carbohydr Res 1990;199:91−109.
[11] Dutta S, De S, Alam Md I, Abu−Omar MM, Saha B. Direct conversion of cellulose and lignocellulosic biomass into chemicals and biofuel with metal chloride catalysts. J Catal 2012;288:8−15.
[12] De S, Dutta S, Saha B. Microwave assisted conversion of carbohydrates and biopolymers to 5-hydroxymethylfurfural with aluminium chloride catalyst in water. Green Chem 2011;13:2859−68.
[13] Alam Md I, De S, Dutta S, Saha B. Solid-acid and ionic-liquid catalyzed one-pot transformation of biorenewable substrates into a platform chemical and a promising biofuel. RSC Adv 2012;2:6890−6.
[14] Dutta S, De S, Patra AK, Sasidharan M, Bhaumik A, Saha B. Microwave assisted rapid conversion of carbohydrates into 5-hydroxymethylfurfural catalyzed by mesoporous TiO_2 nanoparticles. Appl Catal A Gen 2011;409−410:133−9.
[15] Dutta S, Patra AK, De S, Bhaumik A, Saha B. Self-assembled TiO_2 nanospheres by using a biopolymer as a template and its optoelectronic application. ACS Appl Mater Interfaces 2012;4:1560−4.
[16] Mazzotta MG, Gupta D, Saha B, Patra AK, Bhaumik A, Abu-Omar MM. Efficient solid acid catalyst containing Lewis and Brønsted acid sites for the production of furfurals. ChemSusChem 2014;7:2342−50.

[17] Alam Md I, De S, Singh B, Saha B. Titanium hydrogen phosphate: an efficient dual acidic catalyst for 5-hydroxymethylfurfural (HMF) production. Appl Catal A Gen 2014;486:42–8.

[18] Werpy TA, Petersen G. Top value added chemicals from biomass. Golden, CO: US Department of Energy (DOE); 2004. DOE/GO-102004–1992.

[19] Boisen A, Christensen TB, Fub W, Gorbanev YY, Hansen TS, Jensen JS, et al. Process integration for the conversion of glucose to 2,5–furandicarboxylic acid. Chem Eng Res Des 2009;87:1318–27.

[20] Bozell JJ, Petersen GR. Technology development for the production of biobased products from biorefinery carbohydrates: the US Department of Energy's "Top 10" revisited. Green Chem 2010;12:539–54.

[21] Zhang Z, Zhao ZK. Microwave-assisted conversion of lignocellulosic biomass into furans in ionic liquid. Bioresour Technol 2010;101:1111–4.

[22] Mascal M, Nikitin EB. High-yield conversion of plant biomass into the key value-added feedstocks 5-(hydroxymethyl)furfural, levulinic acid, and levulinic esters via 5-(chloromethyl)furfural. Green Chem 2010;12:370–3.

[23] Musau RM, Munavu RM. The preparation of 5-hydroxymethyl-2-furaldehyde (HMF) from d-Fructose in the presence of DMSO. Biomass 1987;13:67.

[24] Binder JB, Raines RT. Simple chemical transformation of lignocellulosic biomass into furans for fuels and chemicals. J Am Chem Soc 2009;131:1979–85.

[25] Bicker M, Hirth J, Vogel H. Dehydration of fructose to 5-hydroxymethylfurfural in sub- and supercritical acetone. Green Chem 2003;5:280–4.

[26] Kabyemela BM, Adschiri T, Malaluan RM, Arai K. Glucose and fructose decomposition in subcritical and supercritical water: detailed reaction pathway, mechanisms, and kinetics. Ind Eng Chem Res 1999;38:2888–95.

[27] Aida TM, Sato Y, Watanabe M, Tajima K, Nonaka T, Hattori H, et al. Dehydration of d-glucose in high temperature water at pressures up to 80 MPa. Supercrit Fluids 2007;40:381–8.

[28] Kabyemela BM, Adschiri T, Malaluan RM, Arai K. Degradation kinetics of dihydroxyacetone and glyceraldehyde in subcritical and supercritical water. Ind Eng Chem Res 1997;36:2025–30.

[29] Zeitsch KJ. The chemistry and technology of furfural and its many by-products. The Netherlands: Elsevier; 2000.

[30] Akien GR, Qi L, Horvath IT. Molecular mapping of the acid catalysed dehydration of fructose. Chem Commun 2012;48:5850–2.

[31] Patil SKR, Lund CRF. Formation and growth of humins via aldol addition and condensation during acid-catalyzed conversion of 5-hydroxymethylfurfural. Energy Fuels 2011;25:4745–55.

[32] Arata K. Organic syntheses catalyzed by superacidic metal oxides: sulfated zirconia and related compounds. Green Chem 2009;11:1719–28.

[33] De S, Dutta S, Patra AK, Bhaumik A, Saha B. Self-assembly of mesoporous TiO_2 nanospheres via aspartic acid templating pathway and its catalytic application for 5-hydroxymethyl-furfural synthesis. J Mater Chem 2011;21:17505–10.

[34] Dutta S, De S, Saha B, Alam MI. Advances in conversion of hemicellulosic biomass to furfural and upgrade to biofuels. Catal Sci Tecnol 2012;2:2025–36.

[35] Cottier L, Descotes G. 5-(hydroxymethyl)furfural syntheses and chemical transformation. Trends Heterocycl Chem 1991;2:233–48.

[36] Gandini, Belgacem MN. Furans in polymer chemistry. Prog Polym Sci 1997;22:1203–379.

[37] Dutta S, De S, Saha B. A brief summary of the synthesis of polyester building block chemicals and biofuels from 5-Hydroxymethylfurfural. ChemPlusChem 2012;77:259–72.

[38] Davis SE, Houk LR, Tamargo EC, Datye AK, Davis RJ. Oxidation of 5-hydroxymethylfurfural over supported Pt, Pd and Au catalysts. Catal Today 2011;160:55–60.

[39] Davis SE, Zope BN, Davis RJ. On the mechanism of selective oxidation of 5-hydroxymethylfurfural to 2,5-furandicarboxylic acid over supported Pt and Au catalysts. Green Chem 2012;14:143–7.

[40] Saha B, Dutta S, Abu-Omar MM. Aerobic oxidation of 5-hydroxylmethylfurfural with homogeneous and nanoparticulate catalysts. Catal Sci Techn 2012;2:79–81.

[41] Saha B, Gupta D, Abu–Omar MM, Modak A, Bhaumik A. Porphyrin-based porous organic polymer-supported iron (III) catalyst for efficient aerobic oxidation of 5-hydroxymethyl-furfural into 2, 5-furandicarboxylic acid. J Catal 2013;299:316–20.
[42] Sharma RV, Yadav GD. Preparation of 2, 5-diformylfuran from 5-hmf with a heterogeneous mesoporous silver containing manganese catalyst. 2012. WO2012073251 A1.
[43] Lew CM, Rajabbeigi, Tsapatsis NM. One-pot synthesis of 5-(Ethoxymethyl)furfural from glucose using Sn-BEA and Amberlyst catalysts. Ind Eng Chem Res 2012;51:364.
[44] Dunlop AP, Madden JW. Process of preparing γ-valerolactone. 1957. US Patent 2786852.
[45] Mascal M, Nikitin EB. Direct, high-yield conversion of cellulose into biofuel. Angew Chem Int 2008;47:7924.
[46] Gruter GJM, Dautzenberg F. Method for the synthesis of 5-hydroxymethylfurfural ethers and their use. 2011. U.S. Patent Appl. 2011/0082304 A1.
[47] Silva HO, de Pinho PG, Machado BP, Hogg T, Marques JC, Camara JS, et al. Impact of forced-aging process on madeira wine flavor. J Agric Food Chem 2008;56:11989.
[48] Balakrishnan M, Sacia ER, Bell AT. Etherification and reductive etherification of 5-(hydroxymethyl) furfural: 5-(alkoxymethyl)furfurals and 2,5-bis(alkoxymethyl)furans as potential bio–diesel candidates. Green Chem 2012;14:1626–34.
[49] Liu B, Zhang Z, Huang K, Fang Z. Efficient conversion of carbohydrates into 5-ethoxymethylfurfural in ethanol catalyzed by $AlCl_3$. Fuel 2013;113:625–31.
[50] Kraus GA, Guney T. A direct synthesis of 5-alkoxymethylfurfural ethers from fructose via sulfonic acid-functionalized ionic liquids. Green Chem 2012;14:1593–6.
[51] Wang S, Zhang Z, Liua B, Li J. Silica coated magnetic Fe_3O_4 nanoparticles supported phosphotungstic acid: a novel environmentally friendly catalyst for the synthesis of 5-ethoxymethylfurfural from 5-hydroxymethylfurfural and fructose. Catal Sci Technol 2013;3:2104–12.
[52] Lew CM, Rajabbeigi N, Tsapatsis M. One-pot synthesis of 5-(ethoxymethyl)furfural from glucose using Sn-BEA and Amberlyst catalysts. Ind Eng Chem Res 2012;51:5364–6.
[53] Yang Y, Hu C, Abu-Omar MM. Conversion of glucose into furans in the presence of $AlCl_3$ in an ethanol–water solvent system. Bioresour Technol 2012;116:190–4.
[54] Wang H, Deng T, Wang Y, Cui X, Qi Y, Mu X, et al. Graphene oxide as a facile acid catalyst for the one-pot conversion of carbohydrates into 5-ethoxymethylfurfural. Green Chem 2013;15:2379–83.
[55] http://www.ecofriendlymag.com/sustainable-transporation-and-alternative-fuel/direct-injection-engine-study-finds-that-dmf-is-a-promising-biofuel-with-combustion-performance-and-regulated-emissions-comparable-to-gasoline/.
[56] Luque R, Herrero-Davila L, Campelo JM, Clark JH, Hidalgo JM, Luna D, et al. Biofuels: a technological perspective. Energ Environ Sci 2008;1:542–64.
[57] Hu E, Hu X, Wang X, Xu Y, Dearn KD, Xu H. On the fundamental lubricity of 2,5-dimethylfuran as a synthetic engine fuel. Tribol Int 2012;55:119–25.
[58] Rothamer DA, Jennings JH. Study of the knocking propensity of 2,5-dimethylfuran–gasoline and ethanol–gasoline blends. Fuel 2012;98:203–12.
[59] Roman–Leshkov Y, Christopher J, Barrett Z, Liu Y, Dumesic JA. Production of dimethylfuran for liquid fuels from biomass-derived carbohydrates. Nature 2007;447:982–6.
[60] Chidambaram M, Bell AT. A two-step approach for the catalytic conversion of glucose to 2,5-dimethylfuran in ionic liquids. Green Chem 2010;12(12):1253–62.
[61] Yang W, Sen A. One-step catalytic transformation of carbohydrates and cellulosic biomass to 2,5-dimethyltetrahydrofuran for liquid fuels. ChemSusChem 2010;3:597–603.
[62] De S, Dutta S, Saha B. One-pot conversions of lignocellulosic and algal biomass into liquid fuels. ChemSusChem 2012;5:1826–33.
[63] Zu Y, Yang P, Wang J, Liu X, Ren J, Lu1 G, et al. Efficient production of the liquid fuel 2,5-dimethylfuran from 5-hydroxymethylfurfural over Ru/Co_3O_4 catalyst. Appl Catal B Env 2014;146:244–8.

[64] Jae J, Zheng W, Lobo RF, Vlachos DG. Production of dimethylfuran from hydroxymethylfurfural through catalytic transfer hydrogenation with ruthenium supported on carbon. ChemSusChem 2013;6:1158–62.
[65] Corma A, de la Torre O, Renz M, Villandier N. Production of high–quality diesel from biomass waste products. Angew Chem Int 2011;50:2375–8.
[66] Corma A, de la Torre O, Renz M. Production of high quality diesel from cellulose and hemicellulose by the Sylvan process: catalysts and process variables. Energy Environ Sci 2012;5:6328–44.
[67] Gandini A, Belgacem MN. Prog Polym Sci 1997;22:1203–379.
[68] Kumar M, Gayen K. Developments in biobutanol production: new insights. Appl Energy 2011;88:1999–2012.
[69] Oasmaa A, Solantausta Y, Arpiainen V, Kuoppala E, Sipila K. Energy Fuels 2010;24:1380–8.
[70] Nel RJJ, de Klerk A. Ind Eng Chem Res 2009;48:5230–8.
[71] Huber GW, Chheda J, Barret C, Dumesic JA. Production of liquid alkanes by aqueous-phase processing of biomass derived carbohydrates. Science 2005;308:1446–50.
[72] Sutton AD, Waldie FD, Schlaf RW, 'Pete' Silks III MLA, Gordon JC. The hydrodeoxygenation of bioderived furans into alkanes. Nat Chem 2013;5:428–32.
[73] Serrano-Ruiz JC, Wang D, Dumesic JA. Catalytic upgrading of levulinic acid to 5-nonanone. Green Chem 2010;12:574–7.
[74] Serrano-Ruiz JC, Braden DJ, West RM, Dumesic JA. Conversion of cellulose to hydrocarbon fuels by progressive removal of oxygen. Appl Catal B 2010;100:184–9.
[75] Serrano-Ruiz JC, Dumesic JA. Catalytic routes for the conversion of biomass into liquid hydrocarbon transportation fuels. Energy Environ Sci 2011;4:83–99.
[76] Assary RS, Redfern PC, Hammond JR, Greeley J, Curtiss LA. Computational studies of the thermochemistry for conversion of glucose to levulinic acid. J Phys Chem B 2010;114:9002–9.

CHAPTER 5

Lignin Deconstruction: Chemical and Biological Approaches

Saikat Dutta
Department of Chemical Engineering, National Taiwan University, Taipei, Taiwan

1. Introduction

Human beings are desperate to replace sources of energy and materials, that is, petroleum oil with biomass, which has been solely used so far by other living organisms for the production of high-energy compounds, such as adenosine triphosphate (ATP), and a wide range of polymers, such as DNA, proteins and polysaccharides, for millions of years. In the context of being a key component in plant cell walls and most abundant renewable aromatic polymer, lignin has yet to play a significant role as a feedstock. Lignin is composed of approximately 25% of lignocellulosic biomass and it is an amorphous natural polymer consisting of a three-dimensional (3D) arrangement of methoxylated phenyl propane [1]. The complex structure of lignin is amorphous, polyaromatic and incorporated with numerous ether linkages, −OH and methoxy groups. Lignin is regarded as a cross-linked macromolecule composed of three types of monolignols, including *p*-coumaryl alcohol, coniferyl alcohol and sinapyl alcohol with the proportions depending on the source [2]. For example, lignin from softwood consists largely of coniferyl units (4-(3-hydroxy-1-propenyl)-2-methoxyphenol), whereas hardwoods consist largely of syringol (2,6-dimethoxyphenol) units excluding exceptions. The *p*-coumaryl units without methoxy groups on the aromatic ring are also found in lignin. Moreover, coniferyl, syringol and *p*-coumaryl units are interconnected through various cross-linkages (C−O−C = β−O−4, α−O−4, 4−O−5) and C−C inter-unit linkages (β−1, β−5, β−β, 5−5) as formed during the biosynthesis of macromolecular lignin (Figure 1) of which the most abundant is the β−O−4 ether linkage [3]. It is essential to understand the details of linkages present in the macromolecular structure in order to design strategies for selective breaking of these linkages. The essential difference between lignin in hardwoods and softwoods is the number of methoxy groups on the aryl rings. Hardwood contains two or three methoxy groups per aromatic rings that originate from coniferyl and syringol alcohol units; however, softwoods contain only one methoxy group that originates from the

Sustainable Catalytic Processes. http://dx.doi.org/10.1016/B978-0-444-59567-6.00005-4

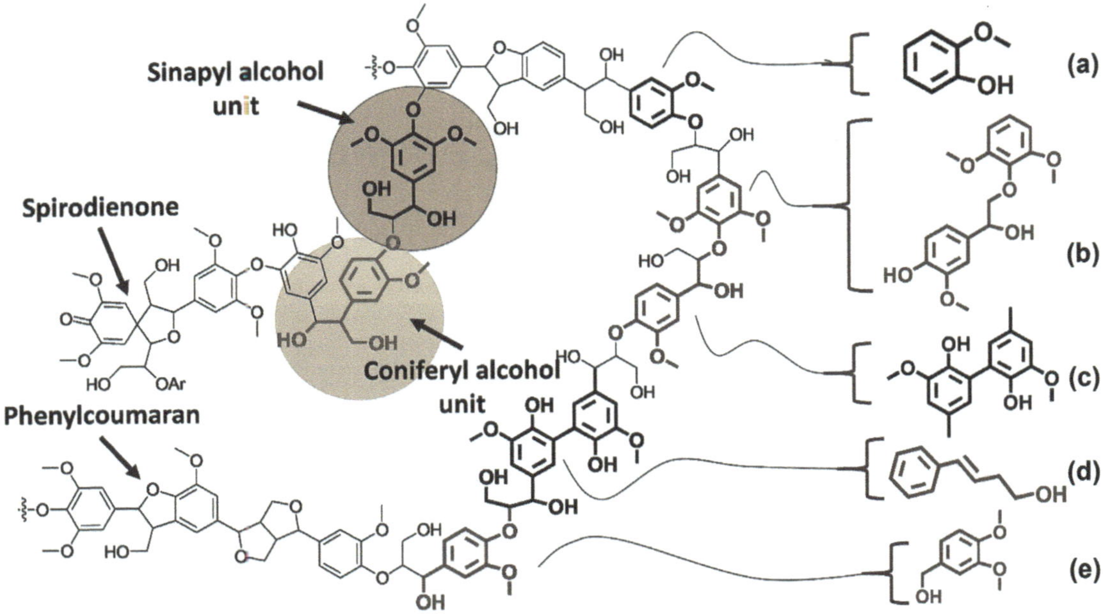

Figure 1

Schematic depiction of lignin, showing various linkages and lignin model compounds to model (a) phenol and methoxy functionality, (b) β–O–4 linkages, (c) 5–5′ linkages, (d) propyl side chain and (e) benzylic groups.

polymerization of coniferyl alcohol [4]. The common linkage between the monomers is the β–O–4, which represents approximately 50–65% of all inter-subunit bonds (Table 1).

The most abundant linkages are β–O–4 and 5–5′ constituting approximately 50–65% and 20–25% of structure, respectively [5]. Chemical conversion of cellulose and

Table 1: Types and Frequencies (%) of Inter-subunit Linkages in Softwood and Hardwood Lignin [5].

	Linkage	Softwood Lignin	Hardwood Lignin
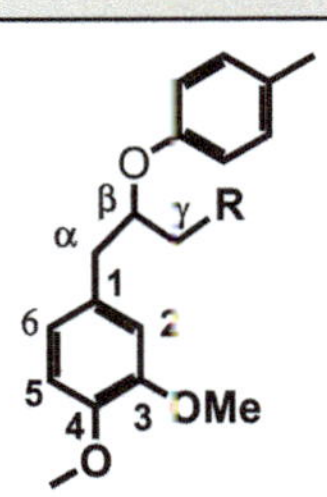	β–O–4	49–51	65
	α–O–4	6–8	–
	β–5	9–15	6
	β–1	2	15
	5–5	9.5	2.3
	4–O–5	3.5	1.5
	β–β	2	5.5

hemicellulose has been extensively studied, while that of lignin remains scarce albeit lignin being the most abundant renewable aromatic polymer, constituting up to 15–30% of the weight and 40% of the energy content of lignocellulosic biomass [6,7]. As we know, its rigid cross-linked structure renders lignin resistant to chemical degradation.

The catalytic valorization of lignin represents a potentially useful method to access bulk and fine chemicals and there are different approaches via chemical routes for developing such processes [6]. Significantly, disruption of linkages of lignin represents a potential route for the production of a wide range of aromatic compounds, which are otherwise accessible only from petroleum feedstocks. The utilization of lignin as feedstock for conversions to hydrocarbons also offers a significant opportunity for enhancing the overall operational efficiency, carbon conversion rate, economic viability and sustainability of biofuel production [8]. The challenge, however, is the propensity of the aromatic lignin macromolecular assembly to condense and degrade, thereby generating high amounts of relatively intractable solid residues in biorefinery, and paper and pulp industries. Therefore, depolymerization of lignin and its subsequent conversion to value-added products are vital to enhance the profitability of biorefineries [9]. Among three main biopolymers that make up the cell wall, lignin is the most recalcitrant. It comprises up to 3% of the plant's mass and is needed for support and water transport [10]. Furthermore, lignin's high aromatic content is responsible for its high energy content. A number of different linkages occur naturally in lignin's structure with the most abundant being the β–O–4 ether linkage (Figure 1) [2]. Therefore, methods to break apart the lignin 3D network while preserving its aromatic nature promises to provide a valuable stream of chemicals. Previous methods to depolymerize lignin include hydrolysis [11–13], pyrolysis [14,15], catalytic hydrogenation [16,17], oxidation [5] and hydrocracking [18]. Schinski et al. disclosed a process for the hydrotreatment of lignin to yield aromatic products, requiring the use of a hydrogen feedstock [8,19]. Lignin treatment and gasification using supercritical water ($T_c = 647.3$ K, $P_c = 221$ bar) have also been reported to primarily form light alkanes and hydrogen [20–22], although the disadvantages of these processes include the high reaction temperatures and pressures (often T 673 K and P 250 bar).

Lignin is ever underutilized and the bulk of produced lignin, representing about 70 million metric tons per year, is only employed as a combustible material for its high heat value [8]. Precise control over the techniques to selectively depolymerize lignin is expected to propel the growth of this field further into new realms of synthetic chemistry focused on utilizing lignin as feedstocks. Despite templating biomaterial being an active area for harnessing the structural complexity to mimic natural functions [23], disruption of lignin polymeric structure has been of remarkable interest due to following reasons: (1) large reservoir of aromatic hydrocarbons, (2) deoxygenation of ethereal oxygen to obtain hydrocarbons, (3) nanostructured carbon materials or hybrid materials by strategic carbonization of lignin in absence of structure/pore directing template. Considering the above facts, this book chapter focuses on describing structural features of recalcitrant lignin in woody biomass. Further, we picked the essence of emerging strategies of lignin valorization, mainly its chemical aspects, for example, hydrodeoxygenation (HDO), synthetic biological methods and carbonization for novel porous carbon materials. Major emphasis is placed on the chemical and material aspects of these techniques, which promise to create a significant impact on lignin valorization. Lignin model reaction to understand the underlying features of depolymerization and selective cleavage of a bond present is described with important recently reported cases in all sections of emerging lignin valorization. Future prospects of lignin valorization and possible new techniques are also proposed.

2. Lignin Valorization Techniques

2.1 Chemical Techniques

Considering lignin as a renewable reservoir of aromatic building blocks, breaking of complex cross-linked polymeric units via selective inter-unit C—O—C bond cleavage and subsequent HDO of the low-molecular-weight moieties to aromatic and aliphatic hydrocarbons is an emerging strategy. This is to create avenues through the chemical and biological conversion of lignin to building block chemicals and liquid fuels. Despite immense potentials, the conversion of lignin to liquid fuels particularly to aliphatic and aromatic hydrocarbons has always been a challenging task. Generally, propensity of the aromatic lignin macromolecular assembly condense and degrade via generating high amounts of relatively intractable solid residues in biorefinery processes. Among the recently developed processes, hydrogenolysis of C—O bonds results in the breakdown of high molecular weight components, and this process offers phenols as end products. However, in order to access liquid hydrocarbon, recently an HDO technique under hydrogenolytic conditions has gained tremendous importance. This technique required noble metal nanoparticle catalysts and, in most of the cases, biofunctional or bimetallic metal nanoparticles exhibited a remarkable performance as compared to their mononuclear counterparts. Liquid phase oxidation of lignin, which is mainly dependent on benzylic C—H or C—OH bond transformation to carbonyl groups, depends on metal catalysts [6,24]; however, metal-free catalysts are also known now [25]. Enzymatic hydrolysis is a promising strategy for cellulose degradation; similarly, enzymatic oxidation can be considered as a major future prospect for lignin valorization and for the production of aromatics. It would be essential to discover lignin sources that have the potential to provide ordered porous carbons directly, among which accessing 3D interconnected porous carbon is one of the major targets. This developing area of lignin as a major source of materials applied for advanced applications which offers ample scope to develop non-templating strategy of deriving materials from a renewable source.

2.1.1 Lignin Depolymerization into Monomers

Lignin depolymerization (LDP) into aromatic products under mild conditions is a desired approach for its valorization. So far, hydrolysis, oxidation and reduction are well-established methods, in which aqueous phase alkali hydroxide or carbonate catalysed hydrolysis of C—O—C linkages results in phenol derivatives [26—28]. Oxidative cleavage of C—H bonds and/or C—C bonds adjacent to C—O—C linkages produces vanillin and its analogues [29,30]. This reaction causes oxidative damage to the aromatic fragments of the lignin, leading to extreme oxidation to CO_x and H_2O. Besides, vanillin would repolymerize into oligomers leading to poor recovery of depolymerization products. Reductive depolymerization has been considered as a promising method of LDP to phenols. With hydrogenation or hydrogenolysis methods, C—O—C linkages are selectively cleaved into phenols [31]. The reduction method may partially avoid the condensation of phenol intermediates to oligomers, a key issue associated with the LDP reaction [32]. Kou et al. have reported that lignin is hydrogenated to monomeric phenols over noble metal catalysts mainly consisting of Pt, Ru, Pd and Rh supported on activated carbon under 4 MPa H_2 [33]. Zhang et al. have shown that woody lignin is catalytically hydrogenated to phenolic compounds like guaiacol and syringols [34]. Alternatives such as CuCr oxide [35], Co—Mo—S/Al_2O_3 [36], activated carbon-, alumina- or silica-supported Ru [37] or Pt [38] have also been examined to obtain monomeric phenols via the hydrogenation of many lignin or model compounds. A strategy for conversion of lignosulphonate into 4-ethylguaiacol and 4-propylguaiacol (PGol) over heterogeneous nickel catalysts revealed that aryl—O—alkyl bonds (C—O—C) and hydroxyl groups of lignin are hydrogenated to phenols and alkanes, respectively, while preserving the aromatic structure [39]. When insoluble solid lignin comes in contact with heterogeneous catalysts in common solvents, mass transfer becomes limited and may retard the process. Generally, heterogeneously catalysed conversion of native lignin failed to address the key chemical aspects of the process and thus the problem is far less elucidated. Fragmentation and hydrogenolysis are the key steps occurring in depolymerizing native lignin into monomeric phenols in alcoholic solvents over nickel-based catalysts (Figure 2) [40]. The best selectivity towards monomeric phenol products is >90% from about 50% conversion of birch wood lignin.

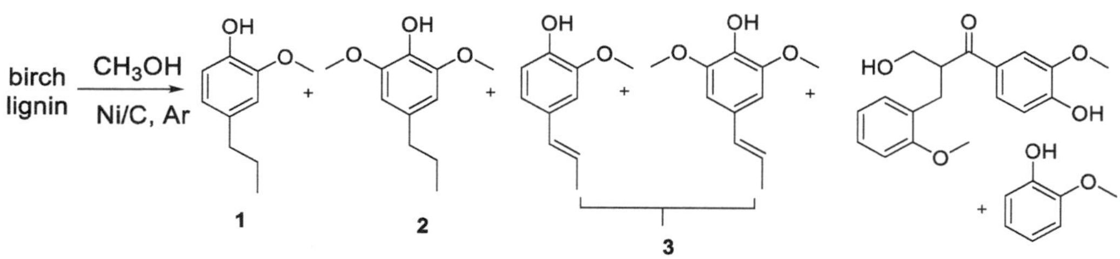

Figure 2
β—O—4 Major products in the lignin depolymerization reaction over Ni/C catalyst.

It is demonstrated that nickel-based catalysts are highly active and selective in native lignin conversion into monomeric phenols; these include propylguaiacol and propylsyringol. It is shown that alcohols act as nucleophilic reagents for C–O–C cleavage in alcoholysis and function as a source of active hydrogen in the medium. The results of this study confirmed that the conversion of smaller lignin fragments into monomeric phenols undergoes β-elimination dehydration coupled with hydrogenation and subsequent hydrogenolysis.

In the above-discussed depolymerization process, efficiency depends on the CH_3OH as solvent when it comes in contact with an active site of the catalyst to generate active hydrogen species, which is energetically favourable than the activation of molecular H_2. A maximum 54% lignin conversion in methanol can be achieved (Table 2, entry 3) with a total selectivity of 89% for the components containing PGol and 4-propylsyringol as major products. The Ni/C system for LDP also exhibits reusability of at least four times with about 50% lignin conversion in each cycle. Since Ni forms MeOMe from MeOH and reforms it to H_2, CO and CO_2 must be generated in this process as by-products. Ni catalyst can also be deactivated in the presence of water. LDP from intact biomass currently produces a largely heterogeneous slate of molecules, which makes lignin valorization an enormously complicated process. Even with an extensive research effort on lignin conversion in last one decade, a selective catalyst that narrows the product slate of LDP products down to something manageable is yet to be found. The exception is Song's Ni/C-based work, but deactivation of the Ni/C in the presence of water and limitation of

Table 2: Catalytic Results of the Lignin Depolymerization Reaction.

Entry	Catalyst	Solvent	Conversion (%)	Selectivity (%)			
				1	2	3	Others
1	No catalyst	CH_3OH	n.d.	–	–	–	–
2	AC	CH_3OH	n.d.	–	–	–	–
3	Ni/C	CH_3OH	54	22	67	1	10
4	Ni/C	EtOH	48	25	70	3	2
5	Ni/C	EG	50	25	72	1	2
6	Ni/C	i-PrOH	27	12	37	26	25
7	Ni/C	25% Glycerol + H_2O	16	16	67	–	17
8	Ni/C	1,4-Dioxane	15	–	6	93	<1
9	Ni/C	1% MOH + H_2O	9	31	63	–	6
10	Ni/C	25% MeOH + H_2O	22	16	74	–	11
11	Ni/C	Cyclohexane	<1	–	–	–	
12	Ni/C	MeOH	42	26	71	–	3
13	Ni/SBA-15	MeOH	27	9	18	72	<1
14	Ni/Al_2O_3	MeOH	19	–	18	81	1
15	Cu/C	MeOH	9	–	–	99	1
16	Cu-Cr oxide	MeOH	2	–	–	99	1

Reaction conditions: birch sawdust (2.0 g), catalyst (0.1 g), solvent (40 mL), 200 °C, 6 h, 1 atm Ar. n.d. = not detected. 1, Propylguaiacol; 2, Propylsyringol; 3, Propenylguaiacol.

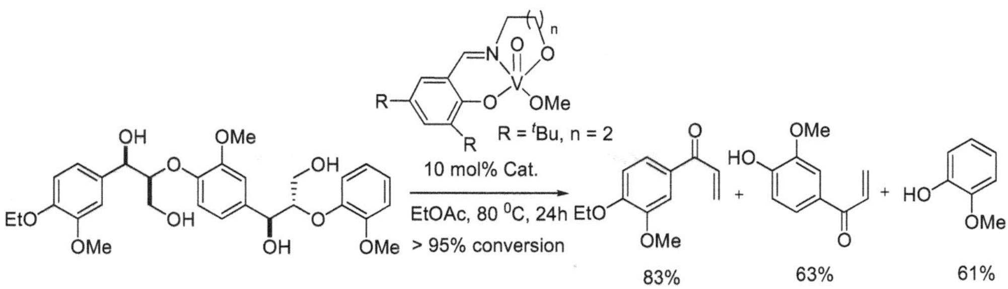

Figure 3
β—O—4 Vanadium-catalysed degradation of a trimeric lignin model.

strategy with a specific variety of sawdust birch wood are certain downsides of this process. Non-oxidative degradation involving cleavage of β—O—aryloxy bonds in lignin models has been a standard system to simplify the development for depolymerization of lignin. Recently, vanadium-catalysed β—O—aryloxy bond cleavage in a lignin model has been reported; this involves the formation of enones resulting from the C—O bond cleavage (Figure 3) [41]. This also reveals that pretreatment and isolation can alter the structure of lignin.

Production of phenols from lignin via depolymerization techniques clubbed with further conversion (hydrogenolysis or cracking) is a challenging task that has been solved via catalytic cracking and hydrogenolysis involving the breakdown of molecular components into small fragments via breaking of C—O bonds in water—alcohol medium. This has been suggested as the major reaction route. For these processes of phenol production, tungsten phosphide [42], ZrO_2—Al_2O_3—FeO_x [43] and iron oxide [44] play a major role, respectively. Depolymerization by catalytic alcoholysis of lignin can offer high-value small molecules of carbon C_6—C_{10} liquid products by using a nanostructured molybdenum carbide catalyst in supercritical ethanol; in this case, the advantage is that lignin fragments can be rapidly stabilized on the surface of the catalyst under supercritical conditions of ethanol [45]. This recent finding suggests that catalytic alcoholysis is an alternative route to produce value-added chemicals from lignin due to the fact that release of an active hydrogen atom from an alcohol, for example, methanol is more facile than that from a hydrogen molecule. This approach can be regarded as one of the key strategies to derive organic molecules from lignin degradation.

2.1.2 Upgradation of Lignin Monomers

Currently, HDO is receiving more attention, which involves the complete removal of oxygen atoms from the substrate under hydrogenolytic conditions [46]. Deoxygenation of polyols (e.g. glycerol) to useful organics such as allyl alcohol, propanal and acrolein is a well-established method [47,48]. Several catalytic upgrading routes that involve HDO have been widely studied using monomeric lignin model compounds to produce hydrocarbons [35].

HDO has been considered as the major tool to upgrade lignin-derived bio-oils and the HDO process involves reaction with hydrogen to produce hydrocarbons and water [49]. In order to highlight the importance of HDO reactions in bio-oil upgrading, HDO chemistry of lignin-derived compounds or raw lignin with comparison of activity of employed catalysts is essential. Solid surface that stabilizes water—oil emulsions and catalyses reactions at the liquid/liquid interface can lead to biphasic HDO using a deposited Pd onto carbon nanotube-inorganic oxide hybrids. This is useful for biomass refining as revealed when employed for lignin conversion [50]. The hybrid solid nanoparticle is capable of simultaneously stabilizing an emulsion and catalysing reactions in both aqueous and organic phases. In order to remove oxygenate functional groups in phenolic compounds derived from lignin, such as vanillin, the HDO of vanillin by nanohybrid emulsions on Pd was studied as a model reaction performed on the aqueous side of the interface where the Pd is deposited on the hydrophilic side. As illustrated schematically in Figure 4, a range of different products and phase migration processes occur due to the varying extents of hydrogenation, hydrogenolysis and decarbonylation of vanillin catalysed by Pd as the reaction conditions were modified.

The major advantage of this method is the solid-stabilized emulsions; a continuous process could be designed in which two homogeneous phases coexist with the emulsion in a layered configuration of oil/emulsion/water. Full conversion can be expected on both sides

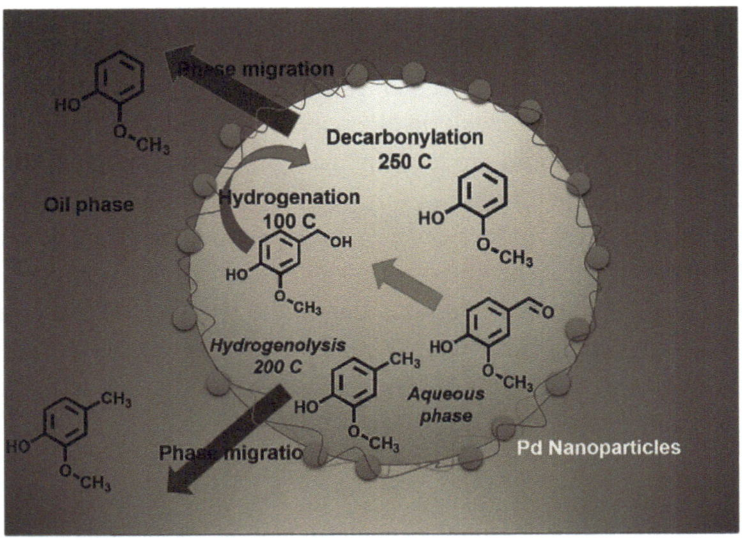

Figure 4

Schematic illustration of the reactions taking place at the water/oil interface in the solid-stabilized emulsions using vanillin as a lignin model. Depending on the reaction temperature, the prevailing reactions are hydrogenation, hydrogenolysis or decarbonylation, and depending on the relative solubility, the products remain in the aqueous phase or migrate to the oil phase.

of the emulsion with constant removal of oil soluble products from the top layer and water-soluble products from the bottom layer while the reaction is continuous in the emulsion.

Recently, various supported noble metals, such as Ru, Pd, Pt, Rh and their alloys, have been widely used in aqueous phase non-sulphide-based HDO process for conversion of lignin-based monomeric model phenolic substrates to hydrocarbons [51–53]. The process of hydrotreatment of pyrolysis oil fractions has been considered as a promising energy career; however, chemical aspects of the HDO reactions in fast pyrolysis oil were not investigated in detail even though it was found that the carbohydrate fraction has been very reactive in the HDO of pyrolysis oil. It is important to disclose the chemical aspects of the disassembly/'unzipping' process of lignin involving HDO, which offers production of value-added products. Recent investigations on model lignin compounds have demonstrated the utility of ruthenium hydrogenation catalysts and nickel carbene hydrogenation catalysts for the hydrogenolysis of the aromatic CO bonds in aryl ethers important for the generation of fuels and chemical feedstocks from lignin. Study on the lignin model compounds revealed the relative reactivity of ether substrates scale as Ar–O–Ar > Ar–OMe > $ArCH_2$–OMe (Ar, Aryl; Me, Methyl) [54]. Refractory aryl ether biopolymer to hydrocarbon conversion is a formidable challenge to synthetic chemists. Results from this work have demonstrated that the selective cleavage of aromatic C–O bonds in the presence of other C–O bonds can be conducted without reduction of the arene units, although the mechanistic issues are not yet resolved for this nickel system.

Performing HDO to remove covalently bonded oxygen from lignin offers opportunities to not only break apart lignin's complex structure but also to increase the overall energy density and value of the resulting products. The major complexity is that C–O and C–C bond strengths are comparable, making selective HDO versus aromatic hydrogenation challenging. There are efforts recently made for overcoming such critical issues. For example, a bifunctional catalyst constructed with Pd metal nanoparticle on carbon and $ZnCl_2$ that selectively cleaves C–O bonds in a variety of lignin models and lignin under relatively mild conditions (150 °C and 30–300 psi of H_2) [55]. When both Pd/C and Zn^{+2} are present, the benzyl alcohol and aldehyde groups can be selectively deoxygenated in good yields without hydrogenation of the phenyl ring. Most importantly, monomeric lignin surrogates can be deoxygenated depending on the synergy between the Pd/C and Zn^{+2}. The β–O–4 linkage of the lignin macromolecule is the most abundant repeating subunit and selective cleaving of such an ether linkage undergoing HDO by using the Pd/C–Zn^{+2} system provided a means of unfolding the complex polymeric structure of lignin into small molecules of fuel value (Figure 5, Eqn (1)). In this model compound, guaiacyglycerol–β–guaiacyl ether was hydrogenated and deoxygenated at 150 °C under 20 bars of hydrogen using Zn/Pd/C in MeOH to yield primarily two products, guaiacol and 2-methoxy-4-propylphenol (Figure 5, Eqn (1)). Catalyst performance was tested for the

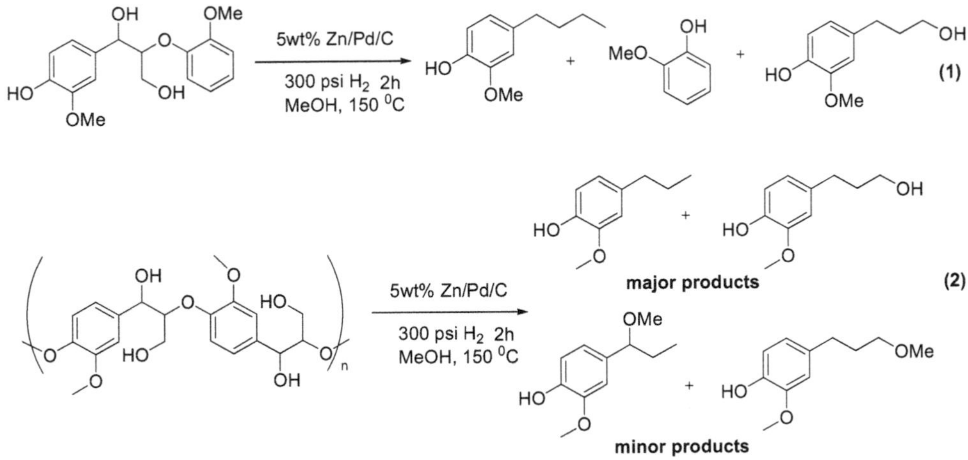

Figure 5
β–O–4 Dimeric and polymeric lignin cleavage and HDO to remove ether oxygen using the Zn/Pd/C system in methanol.

β–O4 synthetic lignin polymer (M_n 3390, DP_n 12.1) which undergoes cleavage and deoxygenation (Figure 5, Eqn (2)). Extended X-ray absorption fine structure (EXAFS) and X-ray absorption near edge structure (XANES) studies confirmed that the 30% of the metallic nanoparticles are oxidized, and EXAFS studies also confirmed that there is no existence of Pd–Zn bimetallic alloy. However, at higher temperatures (250 °C), XANES and Pd K-edge suggest a small amount of Pd–Zn alloy formation. The investigation of the chemical aspects of deoxygenation revealed that Zn adsorption on carbon, most likely onto –OH sites, and not forming a Pd–Zn alloy or direct Pd–Zn interaction was very different as revealed from XAS studies. The hypothesis given by the authors based on the evidence suggests that Zn^{+2} participates in one of two ways, it either activates the substrate via binding to OH groups inducing reactivity with Pd–H on the surface via hydrogen spillover or alternatively Zn^{+2} ions are desorbed into solution at the reaction temperature (150 °C). Zinc binds the substrate and activates its cleavage upon encountering Pd-H sites on the catalyst surface. Once the reaction mixture is cooled, the Zn^{+2} ions are readsorbed onto the C surface. However, Zn/Pd/C was found to be more effective for the cleavage of β–O–4 linkage of lignin molecules and subsequent HDO of aromatic fragments without any chemical change in aromatic functional groups. The system is capable of removing alcohol oxygen atoms on the alkyl chains and maintaining valuable aromaticity and is thus useful for cleaving recalcitrant ether bonds of model lignin dimer and polymers.

With the initial development of LDP chemistry, deoxygenation via the selective hydrogenolysis of C–O bonds has been realized as a key strategy. Dispersed and immobilized nanoscale metal catalysts with dual functionalities play a significant role in the hydrogenolysis of C–O bonds [56]. In this regard, understanding the nature of active sites in such a bifunctional catalyst and reaction pathways of C—O bond scission are essential to be addressed. How superior a bimetallic catalyst can act on hydrogenolysis of lignin C—O bond was witnessed in a β—O—4 type C—O bond hydrogenolysis of a lignin model compound 2-phenoxy-1-phenylethanol with an optimized 85% Ni and 15% Ru composition [57]. This bimetallic system contains near-surface Ru in the $Ni_{85}Ru_{15}$; however, X-ray absorption and photoelectron spectroscopic results suggest that electron-enriched Ni atoms preferentially occupies at the surface studies. Hydrogenolysis results at various temperatures and H_2 pressures led the authors to conclude that enhanced activation of H_2 and substrate due to electron-rich Ni at the surface was the major responsible factor for higher activity than for single component metal catalysts. Using the $Ni_{85}Ru_{15}$ system, organosolv lignin from *Betula platyphylla suk* shows that monomeric products can be achieved under the typical reaction conditions as noted in Table 3 in which case the entire process can be investigated by measuring the intensity of C—O bond vibration in the guaiacyl unit and nonconjugated C=O vibration using Fourier transform infrared spectroscopy (FTIR). This method of the NiRu bimetallic system at an optimized ratio of two metals suggests the importance of such species in other processes, which is required to prevent hydrogenation of the aromatic ring, in which one of the metal component is more active when used as a single metal component.

A further interesting feature was reported in the case of the NiAu catalyst for the hydrogenolysis of lignin into phenolic chemicals with a high selectivity for the β—O—4 type C—O bond hydrogenolysis, where an optimum Ni:Au 7:3 offers 99% conversion of the lignin model 2-phenoxy-1-phenylethanol with a maximum 87% monomer yield when

Table 3: Organosolv Lignin Depolymerization over Ni, Ru, $Ni_{85}Ru_{15}$ Catalyst in Aqueous Medium (Reaction Conditions: 50 mg of Organosolv Lignin, 3 mL Aqueous Solution Containing 0.22 mmol Metal and 0.44 mmol Polyvinylpyrrolidone (PVP), 10 bar H_2) [41].

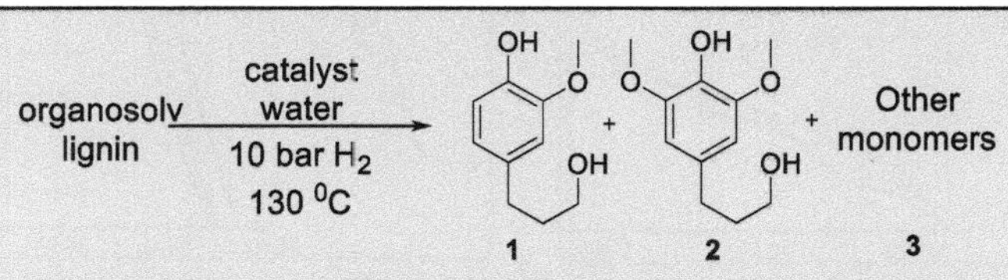

Catalyst	Time (h)	Yield (wt%) 1	2	3	Residual Lignin (wt%)
Ni	1	0	0	0	46
Ru	1	0.1	0.03	0.03	50
$Ni_{85}Ru_{15}$	1	0.1	0.6	0.1	58
$Ni_{85}Ru_{15}$	12	1.4	5.0	0.4	56
Ru	12	0.2	0.06	0.6	38

dimers were reaction intermediates [58]. The comparative Turn over frequency (TOF) of initial hydrogenolysis results confirm the more active Ni_7Au_3 catalysts with more active surface sites and increased dispersion. From the EXAFS, X-ray photoelectron spectroscopy (XPS), inductively coupled plasma mass spectrometry (ICP-MS), ultraviolet visible (UV-Vis) experiments, the authors proposed that the structure of the NiAu catalyst would be an Ni enriched NiAu shell encapsulates Au core (Figure 6(a)). The exceptional performance of the NiAu catalyst at an Ni:Au = 7:3 may be related to the unique electronic state of Ni where an electronic modification of bimetallic species Ni_7Au_3 can be revealed from XANES spectra (Figure 6(b)). Here, the white light intensity of Au increased with increasing Ni loadings, pointing to a decreased electron density of Au atoms while interacting with Ni which results in a concurrent shift of the Ni absorption edge towards a lower energy as Au content increases. In this case, Au acts as an electron donor in the NiAu catalyst, enabling neighbouring Ni atoms to be more electron enriched, which is useful for lignin hydrogenolysis. Upon applying for hydrogenolysis of organosolv lignin, the average molecular weight dramatically decreases due to the hydrogenolysis, exhibiting complete depolymerization. This case of manipulation of electronic properties

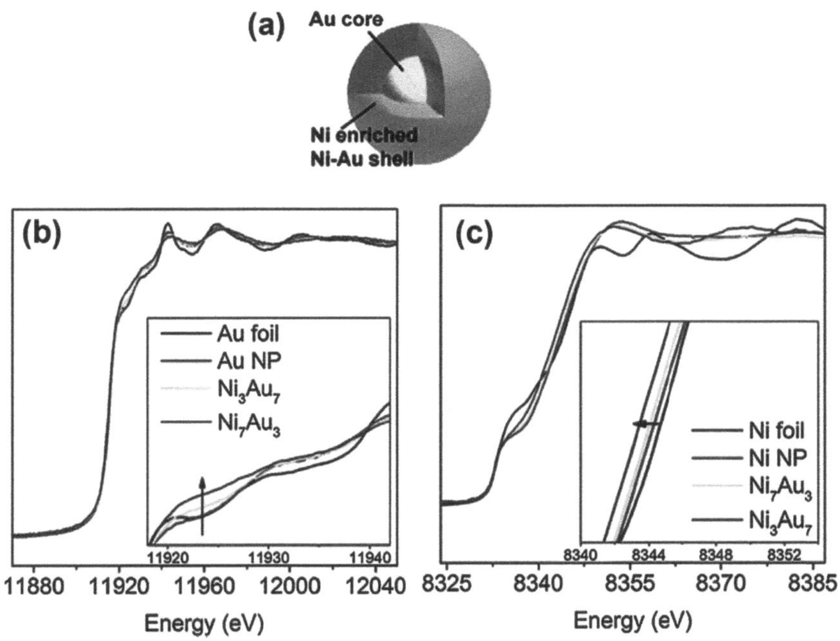

Figure 6

(a) Proposed structural representation of NiAu catalyst. (b) Normalized XAS spectra at the AuLIII edge of Au, Ni_3Au_7 and Ni_7Au_3 catalysts. A spectrum of Au foil was included as a reference; (c) normalized XAS spectra at the Ni K edge of Ni, Ni_3Au_7 and Ni_7Au_3 catalysts. A spectrum of Ni foil was included as a reference. *Reproduced with permission, Copyright Royal Society of Chemistry, 2014.*

of one component of a bimetallic system (Ni) in enhancing the overall activity revealed by spectroscopic studies has a larger implication on future system design for lignin valorization.

In addition to developing and exploring new combinations of catalysts capable of controlled HDO, studies on oligomeric lignin in order to identify new combinations of catalyst matrices with multi-catalytic functionalities suitable for efficient depolymerization of the lignin polymeric framework into a monomeric intermediate and subsequent removal of oxygen via the HDO process require analysis of inter-unit linkages. For example, a recent study revealed for technical grade lignin to produce C_7 to C_9 monomeric lignin intermediates under HDO conditions with 5% noble metal (Ru, Rh, Pt)/Al_2O_3 (or C)-zeolyst (e.g. NH_4^+ Z-Y 57277-14-1) catalyst matrix [59]. These multi-functional catalyst matrices showed promising selectivity in the conversion of lignin to aromatic hydrocarbons (alkyl benzene derivatives) from a wide variety of lignins under HDO conditions in aqueous media. Results show that Al_2O_3-supported noble metal catalysts are effective in reducing functional groups (e.g. methoxyl groups) in lignin aromatic molecules, but are rather inactive for the hydrogenation of the aromatic ring. Thus, under HDO conditions, Al_2O_3 supports are preferable to carbon supports under different HDO conditions with noble metals (Ru, Pt and Rh) and acidic zeolites (CBV 28014 CY 1.6 and NH^{4+} Z-Y 57277-14-1) in the presence of hydrogen to further improve the yields and selectivity for the production of aromatic hydrocarbons (alkyl benzene derivatives) for incorporation into commercial jet fuels to increase bulk density and lower the cloud point of the fuels. Replacing the acidic zeolite CBV 8014 CY 1.6 with NH^{4+} Z-Y 57277-14-1 was shown to enhance aromatic hydrocarbons and product selectivities with the noble metals Rh and Pt on Al_2O_3 supports under HDO conditions. The high reactivity of technical grade lignin based on their substructures (Figure 7) accounted for their high reactivity as substrates for HDO reactions. Lignin structure predominantly comprising of 8–O–4′ inter-unit linkages (C–O–C bonds) would undergo both hydrogenolysis and

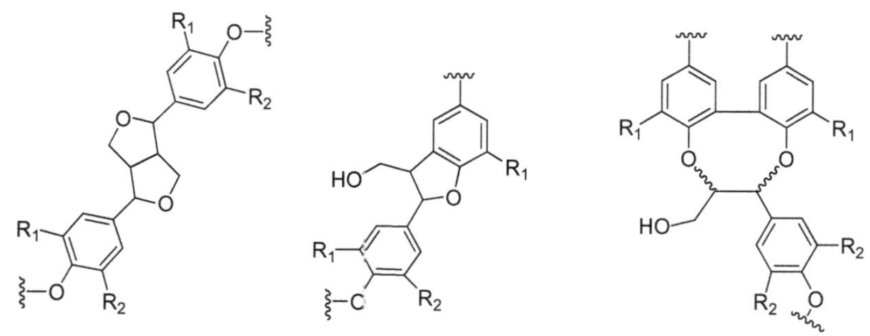

Resinol (8-8')   Phenylcoumaran (8-5')   Dibenzodioxacin (5-5'/b-O-4'/7-O-4')

Figure 7
Inter-unit linkage-based substructures of technical grade lignin.

deoxygenation via cleavage and simultaneous removal of oxygen in the presence of a noble metal-integrated zeolite catalyst matrix.

Various analyses of lignin model compounds and distribution of products obtained during lignin aqueous phase reforming revealed that LDP through the disruption of the abundant β–O–4 linkages and to a lesser extent the 5–5′ carbon–carbon linkages to form monomeric aromatic compounds is well known [60]. However, less efforts have been invested for developing selective HDO techniques to access hydrocarbons from lignin or lignin model compounds. In this context, a combined depolymerization and (hydro)-deoxygenation of lignin in a single step to access the phenolic product over a Pt catalyst with formic acid (H_2 source) can significantly reduce molecular weight and oxygen content [61]. However, this single-step disassembly of lignin offers monomeric cyclohexyl derivatives and aromatic products [62,63].

A combined depolymerization and HDO of lignin in two steps using Pt/Al_2O_3 (ethanol/water) and $CoMo/Al_2O_3$ under high pressure of H_2 was reported for the organosolv and kraft lignin. This involves liquid phase reforming, extraction of lignin oil and is further subject to an HDO step to access deoxygenated products, that is, monooxygenated phenolics (Figure 8) [64]. It happens that, without a pretreatment step, the HDO process with Mo_2C/CNF (carbon nanofibre) on organosolv lignin offers only a small percentage of monomeric phenolics (Figure 9). In spite of a positive temperature effect on reducing the oxygen content of HDO-derived products, the required level of success for hardwood lignin is yet to be achieved. This must include a convenient pretreatment step.

3. Enzymatic Techniques

The majority of the carbon in plant material remains trapped in structural biopolymers comprised of cellulose, hemicellulose and lignin rather than easily accessible sugars and starches. Lignin is the most chemically difficult of these feedstocks to deconstruct because of its aromatic monomer units and diverse set of C–C and C–O cross-linked units. Its eminent structural complexity consequently requires multiple families of enzymes to achieve its breakdown. The current model for lignin breakdown is derived from studies of

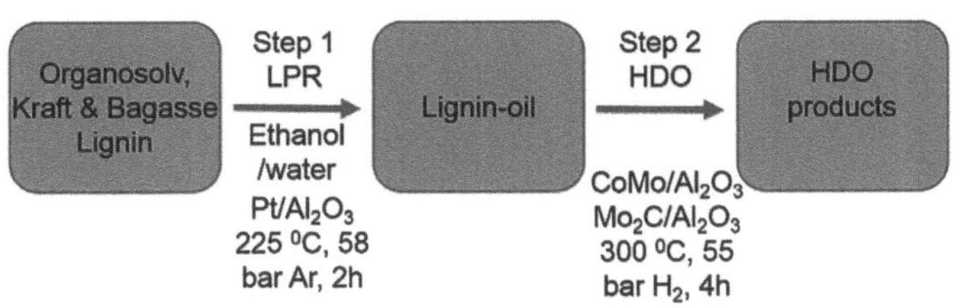

Figure 8
Two-step approach to valorization of lignin involving HDO.

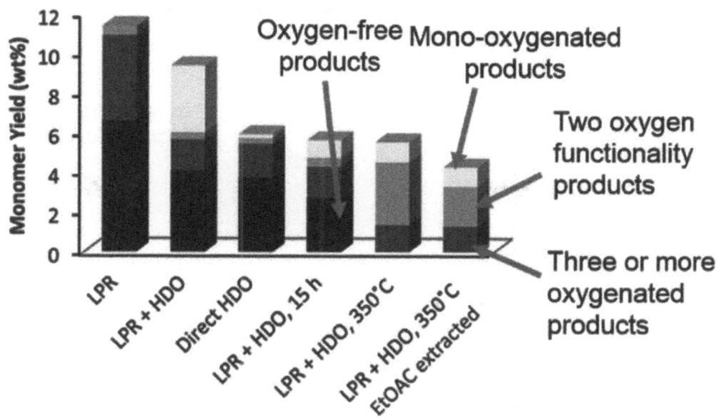

Figure 9
Yield of monomeric aromatic products after HDO of organosolv lignin oil over Mo_2C/CNF catalyst under different conditions.

fungal metalloenzymes and is based on an oxidative combustion of lignin through various radical-mediated paths, which essentially generate phenoxyl and phenyl radicals on the substrate followed by depolymerization (Figure 10) [65].

In order to explore new chemical strategies for lignin degradation, discovery and characterization of oxidative and accessory enzymes in lignin-reactive soil bacteria that exhibit rich activity towards depolymerization and utilization of biomass-derived carbon sources has been essential. In this regard, it is desired to devise a strategy for the rapid discovery and functional identification of new enzymes from unsequenced bacteria under lignin-reactive growth conditions, which is a general approach to any cultural microbes of interest. The enzyme is involved in a multistep transformation of complex substrates in whole organisms and this strategy is applied to study lignin reactivity. Amycol is an oxidative enzyme for phenol, rather for aromatic ring oxidation. However, the genome of

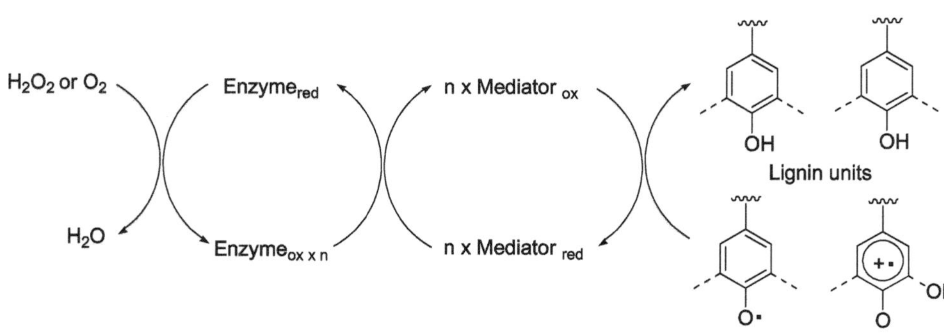

Figure 10
Oxidative reactions initiating the radical-dependent breakdown of lignin. Small-molecule mediators, including veratryl alcohol, Mn(II) and oxalate, are oxidized directly by fungal peroxidases and laccases, which utilize H_2O_2 or O_2, respectively.

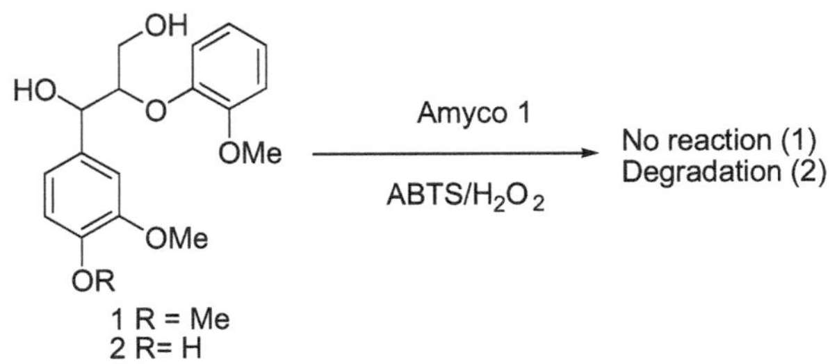

Figure 11
Reactivity of Amyco1 with lignin model dimers 1 and 2.

Z. sp. 75iv2 shows that it potentially contains demethylases for structurally similar lignin-related compounds (Figure 11), which can synergistically degrade biomass by uncapping phenolic sites. This discovery could pave a way towards exploring the function and reactivity of the full oxidative system of Amycol, including laccases and DyPs and their synergy with other enzyme families [65].

Generally, chemical methods to depolymerize lignin generate low yields of complex mixtures of products and also results in the formation of insoluble high molecular weight products. Initially, microbial degradation of lignin has been investigated primarily in white-rot and brown-rot fungi depending on the ability of these fungi to produce extracellular lignin peroxidase and Mn peroxidase enzymes for oxidation of lignin. It was revealed that fungi use iron and redox mediators to carry out Fenton oxidation to depolymerize lignin [66]. Although fungal lignin deconstruction has been studied since long, there is no commercial biocatalytic process involving this technique to date [67]. Several bacterial strains are also able to break down lignin, including actinobacteria such as *Streptomyces viridosporus*, which produces an extracellular peroxidase, certain pseudomonads, and the anaerobic bacterium *Enterobacter lignolyticus* SCFI. In *Rhodococcus jostii* RHA1, whose genome sequence has been determined, identification and characterization of peroxidase DypB drive its application towards oxidizing lignin models, Kraft lignin, lignocellulose and Mn^{2+} [68]. Two further bacterial peroxidase enzymes have been reported from *Amycolatopsis* sp. 75iv2: a haeme-containing enzyme capable of oxidizing lignin model compounds [65] and a Dyp2 peroxidase enzyme with a higher catalytic activity for Mn^{2+} oxidation [69]. Synthetic biology is currently finding a prominent status in designing new biosynthetic routes to alkanol, biodiesel and biofuels as a non-fermentative pathway [70]. It was previously observed that *R. Jostii* RHAI DypB (lignin oxidizing enzymes) is able to cleave a β−aryl ether lignin model compound to produce vanillin and a two-carbon fragment glycoaldehyde. The other aromatic metabolite, 5-carboxyvanillic acid (2), is a potential intermediate in the biphenyl catabolic pathway and can be decarboxylated to form vanillic acid for growth [71]. Thus, the vanillic acid catabolic pathway might be important for lignin breakdown.

For the vanillic catabolic pathway, one would desire the production of vanillin from the lignin component of lignocellulose using gene deletion. This process can be engineered for the production of aromatic compounds which offers an advantage of generating a small number of metabolites in a predictive way. This requires a robust organism for which a mutant strain growing to a high cell density also prevents the formation of toxic aldehyde metabolites. A gene deletion strain of *R. Jostii* RHAI in which vanillin dehydrogenase gene had been deleted, when grown on minimal medium containing 2.5% wheat straw lignocellulose and 0.05% glucose, offers vanillin as the product [72]. A hypothetical catabolic pathway in *R. jostii* RHA1 for breakdown of β−aryl ether and biphenyl components of lignin, based upon the structure of observed metabolites, is depicted in Figure 12.

A synthetic biological technique has been used in *Escherichia coli* to engineer new biosynthetic routes to alkanol and biodiesel, but thus far, there are limited applications towards biomass deconstruction and little use of other bacterial hosts for synthetic biology. Gene deletion mutant-based biocatalytic route to access aromatic chemicals like vanillin from wheat straw lignocellulose is a step significantly further towards a metabolic engineering-based approach [73]. Lignin biodegradation has been studied using an extracellular enzyme from the fungus *Trametes versicolor* [74]. Enzyme multicopper-containing laccase (E.C > 1.10.3.1) is mild and catalyses the oxidation of the substrate by

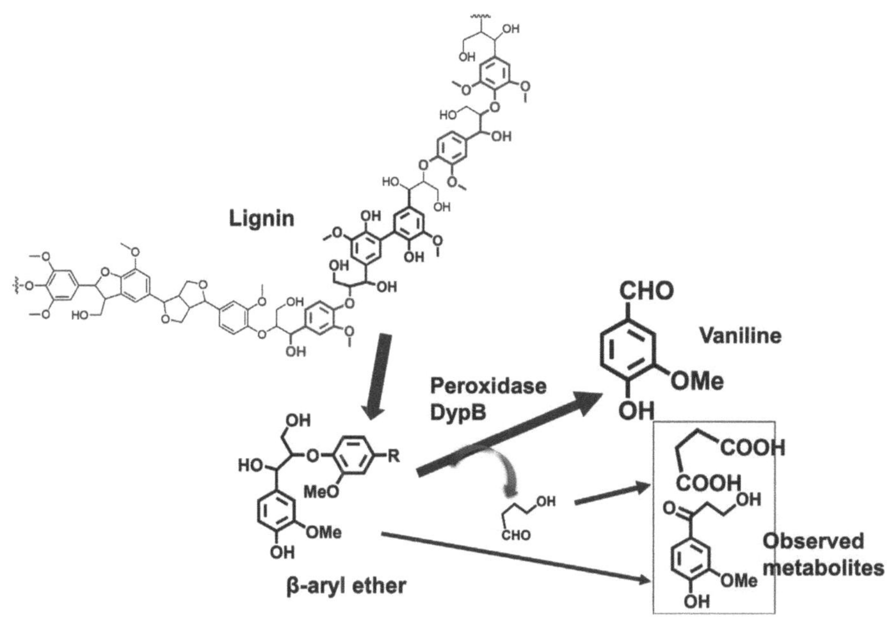

Figure 12

Hypothetical catabolic pathways in *R. jostii* RHA1 for the breakdown of β-aryl ether and biphenyl components of lignin, based upon the structures of observed metabolites.

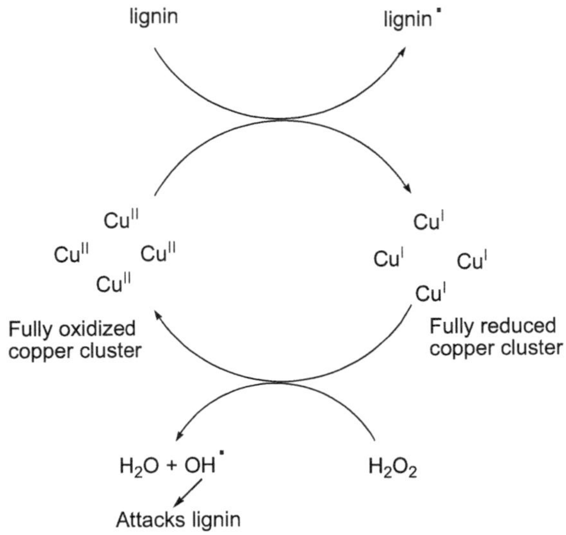

Figure 13
Possible catalytic cycles of laccase in the presence of H_2O_2.

accepting electrons from copper ions and then reducing molecular oxygen to water (Figure 13) [75]. However, laccase has limited bioremediation potential because of its specificity for the phenolic compounds in lignin, to the exclusion of other compounds. However, this can be improved by inclusion of a mediator such as 1-hydroxybenzotriazole, 2,2-azonobis(3-ethylbenzthiazoline-6-sulphonate), violuric acid, and 2,2′,6,6′-tetramethyl-piperidine-N-oxyl and extend the substrate range of laccase to non-phenolic subunits of lignin. However, these synthetic mediators are expensive and their subunits even can produce toxic residual compounds. It also interesting that a significant enhancement of degradation efficiency of trinitrotoluene by laccase from *Phanerochaete chrysosporium* was observed by the application of TiO_2-assisted photocatalytic pretreatment [76].

Laccases (EC 1.10.3.2) are a class of multicopper oxidases that oxidizes a range of substrates, including phenols, aromatic amines and non-phenolic substrates including aromatic and nonaromatic compounds containing hydroxyl and amine groups in the presence of atmospheric oxygen. Although laccases are believed to be involved in lignin synthesis in plants and in cell pigment formation, as well as metal oxidation in fungi and bacteria, their physiological roles are unclear. There is involvement of the small *Streptomyces* laccases in lignin degradation, acid precipitable polymeric lignin obtained in the presence of wild-type *Streptomyces coelicolor* A3(2) (SCWT) and its laccase-less mutant (SCΔLAC) in the presence of *Miscanthus x giganteus* lignocellulose [77]. When using a series of laccases (*S. coelicolor* A3(2), *Streptomyces lividans* TK24, *Streptomyces viridosporus* T7A and *Amycolatopsis* sp. 75iv2) using both lignin model compounds and ethanolsolv lignin where degradation of phenolic model compounds and non-phenolic compound in the presence of redox mediators. Their activities are the highest at pH 8.0 with a low K_{rel}/K_{app} for LM–OH, suggesting that the enzymes' natural substrates must be different in shape or chemical nature. This was more revealed from the crystal structure analysis of laccases from *S. viridosporus* T7A (SVLAC) and *Amycolatopsis* sp. 75iv2 both with and without the bound lignin substrate.

Another type of lignin-degrading enzyme comprises Amidohydrolase enzymes, termed Ligl, obtained from *Sphingomonas paucimobilis* and they catalyse the reversible hydrolysis of 2-pyrone-4,6-dicarboxylate to 4-oxalomesaconate and 4-carboxy-2-hydromuconate in lignin degradation. Ligl was never considered as a metalloenzyme in spite of containing residues that could bridge a binuclear metal centre at the C-terminus of β-strand 3 or 4. It is very important to clearly understand the crystal structure of Ligl in the presence and absence of the product bound on the active site.

Microbial systems provide molecular information on LDP as they have evolved to selectively break down lignin using metalloenzyme-dependent radical pathways. Both fungi and bacteria have been observed to metabolize lignin; however, their differential reactivity with this substrate suggests that they may utilize different chemical strategies for its breakdown (Figure 14) [78]. Recent advances in studying bacterial lignin degradation as an approach to exploring greater diversity in the environment have been an emerging area, which involves diffusible mediators that directly react with lignin to generate radical sites within the substrate and initiate a cascade of bond scissions that ultimately leads to decomposition to smaller aromatic compounds, CO_2 and water [79].

Recent advances in studying lignin degradation in bacteria and efforts to expand the scope of our understanding of environmental microbial processes have been made. This involves bacterial lignin metabolism and the action of bacterial haeme peroxidases. Bacterial systems have been found to be less oxidatively powerful as compared to lignolytic fungal

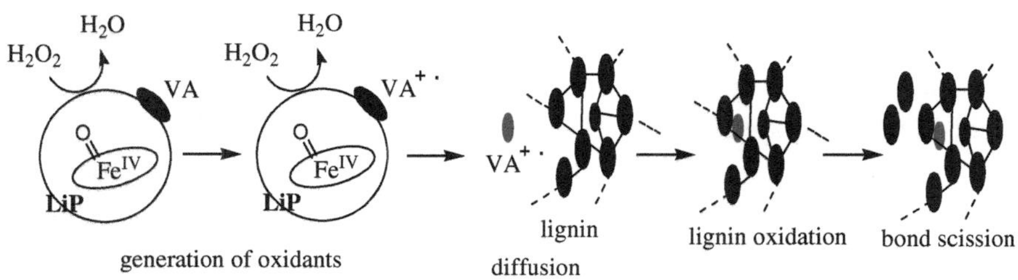

Figure 14
The current model for lignin degradation involves enzymatic generation of the radical mediator, which can then diffuse to the lignin substrate and transfer the oxidizing equivalent to the polymer. Upon formation of a lignin-based radical, bond scission reactions will ensue that lead to depolymerization.

systems to date but may provide a rich source for elucidating new accessory enzymes that act synergistically with the major oxidative enzymes to activate and uncap various sites, similar to that involved in cellulose degradation. Recent studies also demonstrated the existence of oxidative enzymes that are equipped for associating with lignin, which could indicate that this process is not completely mediated through secondary small molecules mediators. To this end, molecular detail for understanding how lignin degradation occurs in the environment would be essential to explain the differential reactivity profiles of lignin-processing enzymes.

Remarkably, non-radical ligninolytic enzymes, such as β-etherases, should deliver a more specific and effective alternative for lignin cleavage and valorization. Because the major intermolecular lignin linkage is the arylglycerol—β—aryl ether bond, one pathway proposed for biochemical lignin catabolism entails the enzymatic cleavage of such linkages. By using β-etherases, which cleave β-aryl ether bonds [80], valuable industrially useful low-molecular-mass lignins that retain aromatic rings would be selectively achieved. Surprisingly, until now, few studies have focused on the biocatalytic characterization and application of β-etherases in lignin valorization. Their natural occurrence has scarcely been reported, for example, for an ascomycete believed to be a member of the genus *Chaetomium* [81] and for the soil a-proteobacterium *Sphingobium paucimobilis* SYK-6. The extracellular fungal β-etherase was shown to cleave the β-aryl ether bonds in two lignin model dimers, guaiacylglycerol β-guaiacyl ether and guaiacylglycerol—β—O—4—methylumbelliferone, as well as in a polymeric synthetic lignin. However, the gene coding for this β-etherase has not been identified, so far. A key aspect for the successful implementation of β-etherases (and other enzymes) in lignin valorization is the complete dissolution of lignin polymers in the reaction media.

4. Carbonization

Currently, bulk synthesis of materials for energy storage from renewable sources using scalable technologies has been a major target. Lignin is a major by-product of chemical pulping and increasingly available for biofuel production. In spite of chemical heterogeneity and structural variations, lignin's high carbon content and phenolic structure make it an excellent alternative source of superior carbon materials (Figure 15). Additionally, the high mass retention characteristic of lignin in thermal processing has been witnessed in its conversion to carbon fibres (CFs) [82,83], porous activated carbon particles [84—86] and graphitic form of CFs and more. In addition, lignin-based CFs have comparable mechanical properties to those of commercial polyacrylonitrile (PAN)-based CFs for reinforcing composites. Novel microstructured carbon materials find a wide range of applications including adsorbents and anode materials for batteries or supercapacitors [87]. Thus, it is highly interesting to design a synthesis technique to access microstructured carbon materials directly from lignin involving fibre-processing and chemical modifications.

4.1 Carbon Fibres

The first report on the use of lignin as a precursor for CFs deals with the transformation of lignin to functional polymers to access spinnable materials and fibres after thermal stabilization followed by carbonization (Figure 15) [86]. In contrast to the approach based on the use of a binder polymer, a straightforward method of obtaining carbon microfibres and nanofibres by the co-electrospinning Alcell lignin solutions at room temperature followed by thermal stabilization and carbonization to produce CFs has been established [80]. Alcell lignin electrospun nanofibres were stabilized by thermal treatment in air at room temperature to 200 °C for 24 h. This produces fibres of a similar size and this was then carbonized at 900 °C to obtain carbon nanofibres. The carbon crystallites with a preferred orientation along the fibre axis offer a smooth fibre surface. The fibres remain in the glassy state ($T_g > T$) and the structural organization in the carbon nanofibre was revealed with Raman Spectroscopic study which confirms that a lower contribution of structural disorder in the higher-temperature and graphitic carbon formation (E_{2g}, 1575 cm^{-1}) confirmed the onset of structural organization. The fibre is oxidation resistant and highly microporous and a large micropore volume makes these nanofibre suitable for the CO_2 adsorption–desorption process. A rapid freezing process of aqueous lignin solution followed by sublimation of the resultant ice forms a uniform network that comprises individual interconnected lignin nanofibres and carbonization that offers a similarly organized network of carbon nanofibres [81]. The rapid freezing process resulted in the elimination of larger spaces and macropores due to the formation of smaller ice

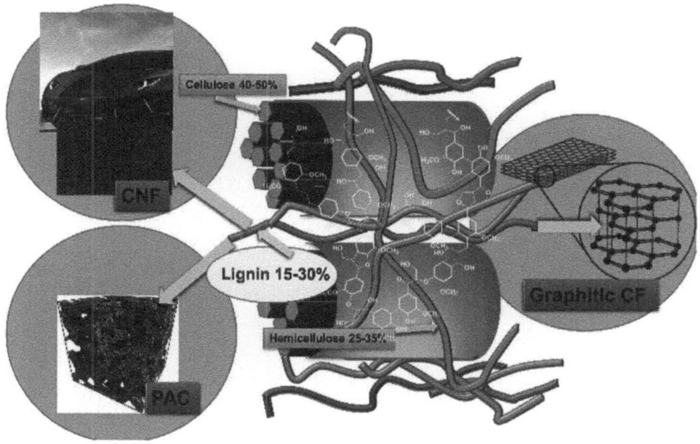

Figure 15
Pictorial depiction of lignin as a potential source for the various nanostructured carbon materials suitable for advanced technological applications.

crystals. Smaller and uniform fibre diameters were generated due to the more rapid heat removal and uniform cooling rates, which resulted in lignin and subsequently in individual CFs with a diameter <100 nm (Figure 16). Fibres tend to align preferentially along the direction of the freezing front, which depends on the concentration.

The conventional graphitic anode of lithium ion batteries can be replaced with lignin CFs (LCFs) when synthesized and applied as monolithic fused mats containing specific charge capacities [89]. LCFs carbonized at 1000 °C have specific charge capacities. Irreversible lithium insertion and extraction were witnessed in the more disordered LCFs carbonized at 1000 °C with the smallest nanoscale graphitic domains. The question is 'Why do we need a graphitic domain in a CF that can enhance the performance of the battery'? This happens due to the reason that carbonized structure is modulated to yield an active material with an internal mesostructure consisting of a high-density, nanoscale crystalline (graphitic) domain surrounded by a continuous, low-density, highly disordered carbon matrix. This amorphous carbon enables facile, isotropic lithium transport throughout the fibre structures, providing access to charge storage sites within the nanoscale graphitic domains [90]. It was found that matrix graphitization increased with an increase in carbonization temperature, that is, long-range graphite order with no preferential orientation along the fibre axis as shown in the SEM image of typical LCF-fused mats after carbonization and varying levels of fibre–fibre fusion (Figure 17) in which the mat densities can be controlled. The mats are excellent anode materials exhibiting tuneable electrochemical performance suitable for high-energy applications. The synthesis method results in an electrically interconnected 3D architecture from lignin 3D structure and the final CF acts as a current collector and Li insertion materials. This carbonized lignin has a disordered nanocrystalline microstructure.

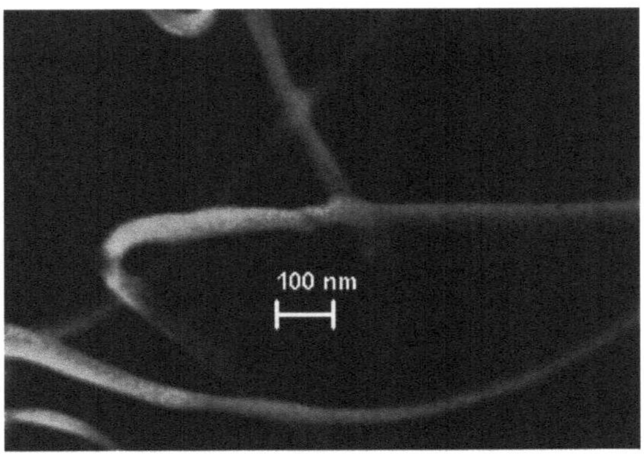

Figure 16

Scanning electron micrograph (SEM) of carbon nanofibres. Individual CNFs derived from lignin with diameters of <100 nm. *Reproduced with permission from Ref. [88], copyright American Chemical Society, 2012.*

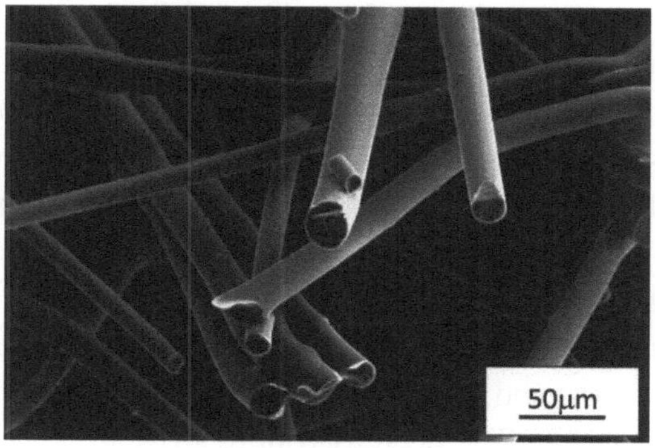

Figure 17
SEM image of varying levels of fibre–fibre fusion in LCF mats. *Reproduced with permission from Ref. [82], copyright Wiley-VCH, 2014.*

Structural modification of lignin with different ester groups results in a precursor highly compatible with the melt processing technique for fibre extrusion and further conversion into microstructured carbons by oxidative stabilization and subsequent carbonization [91]. In the acid anhydride modification of lignin hydroxyl groups (acyl substitution) under neutral conditions and under a long reaction time (6 h), most of the available (phenolic and aliphatic) hydroxyl groups of lignin would be esterified. However, this would not change the polymeric structural characteristics of lignin throughout the modification process. Interestingly, the surface areas of the resulting carbon materials fall in the range of those of commercial graphitic carbons. Phthalic anhydride-modified lignin contains a relatively higher microscale porosity as compared to that of the unmodified lignin-based CFs (Figure 18). This modification influenced the underlying mesostructure and microstructure development and pore creation, and the carbonization temperature can be adjusted to access carbons with either more graphitic or disordered structures.

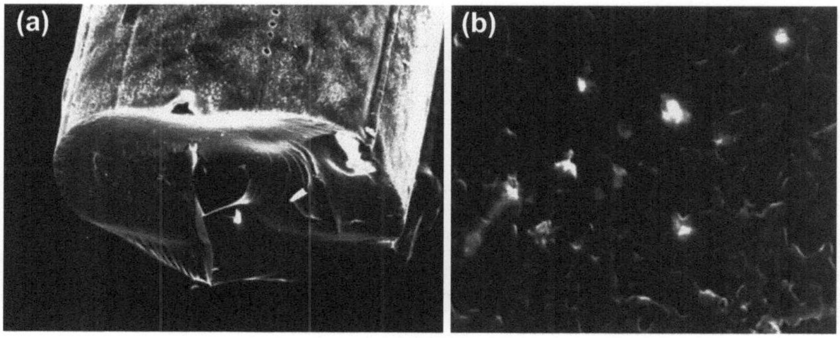

Figure 18
SEM of carbons from (a) phthalic anhydride-modified lignin and (b) microscale porosity of phthalic anhydride-modified lignin. *Reproduced with permission from Ref. [84], copyright RSC, 2013.*

Methods have been devised for ultrafine (100–500 nm) highly porous activated CFs (ACFs) by the electrospinning of aqueous solutions of predominantly alkaline lignin (low sulphone content) followed by simultaneous carbonization and activation [92]. The major advantage of this process is the simultaneous carbonization and activation in a single heating step that resulted a range of micropore- to mesopore-dominant ACFs with a superior specific surface and porosity obtained by varying the contents of alkali hydroxides. The use of the electrospinnable Poly(ethylene oxide) (PEO) carrier as one-ninth of the AL_{1s} made the mixture electrospinnable to fibres. Interestingly, size variation of fibres in the 100- to 300-nm and 300- to 500-nm range when using NaOH and KOH can also be achieved (Figure 19). The decreased fibre sizes are generally attributed to the increased charges of the solutions and improved charge dissipation with the added electrolytes in electrospinning.

This approach offers a highly porous and high specific surface of ACFs with both mesopores and micropores with different size proportions and with slit like micropores. This developed into a network of larger mesopores connected or embedded within the micropore matrix by narrow necks as the ratios increased to 0.3 and 0.5 along with an increase of mesopores. Production of carbon fibres by mixing carbon nanotubes with the carbonaceous precursor to generate carbon nanotube composite fibres. This depends on the chemical modification of

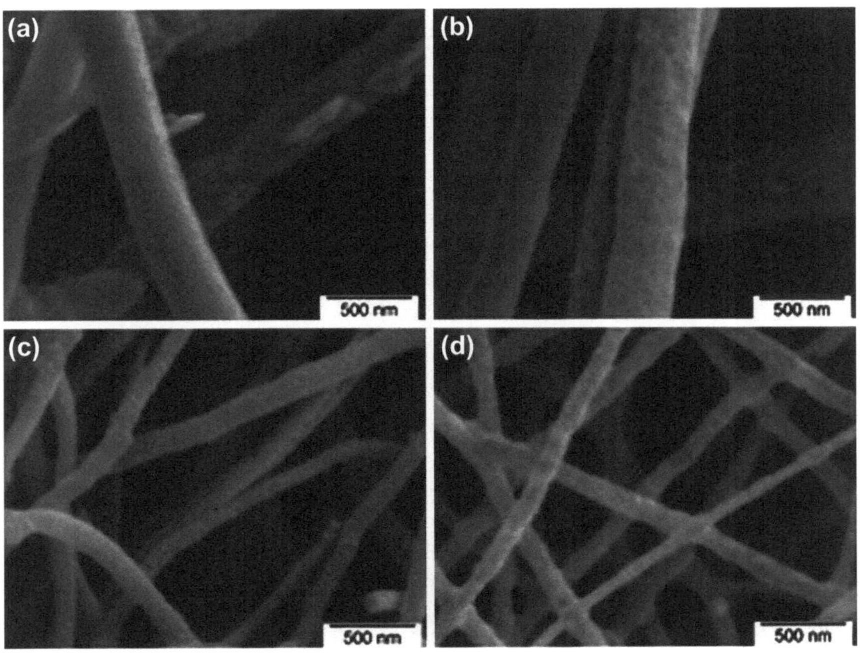

Figure 19

SEM images of AL1s–PEO hybrid fibres with NaOH (a, b) and KOH (c, d). *Reproduced with permission from Ref. [85], copyright RSC, 2014.*

the carbonaceous precursors [93]. Carbon fibres derived from lignin contains a controllable turbostatic disorder, which is controllable by the choice of functional agent and aspect ratio. In addition, fibres can be derived with desired transport and interfacial properties. These lignin-derived fibres found potential applications in Li-ion batteries.

Melt-spinning is already proved as a suitable technique to access CF when lignin was modified chemically and blended with poly(lactic) (PLA) before melt-spinning into lignin fibres [94]. This precarbonization functionalization involves butyration to install ester functional groups in place of polar hydroxyls (−OH), which enhances the miscibility of lignin with PLA. Microstructure analysis (scanning-electron microscopy (SEM) and transmission electron micrographs (TEM)) of CFs produced from lignin/PLA blends revealed the composition-dependent microporous structure inside the fine fibres (Figure 20). The microvoids in the cross-section of all fibres, which is absent in pure lignin-derived fibres, produced by PLA after carbonization and size of the microvoids decreased with the increase in lignin content.

A strategy of porous core construction was developed by controlled co-electrospinning and carbonization method of lignin and cellulose. This produces core−shell CFs with

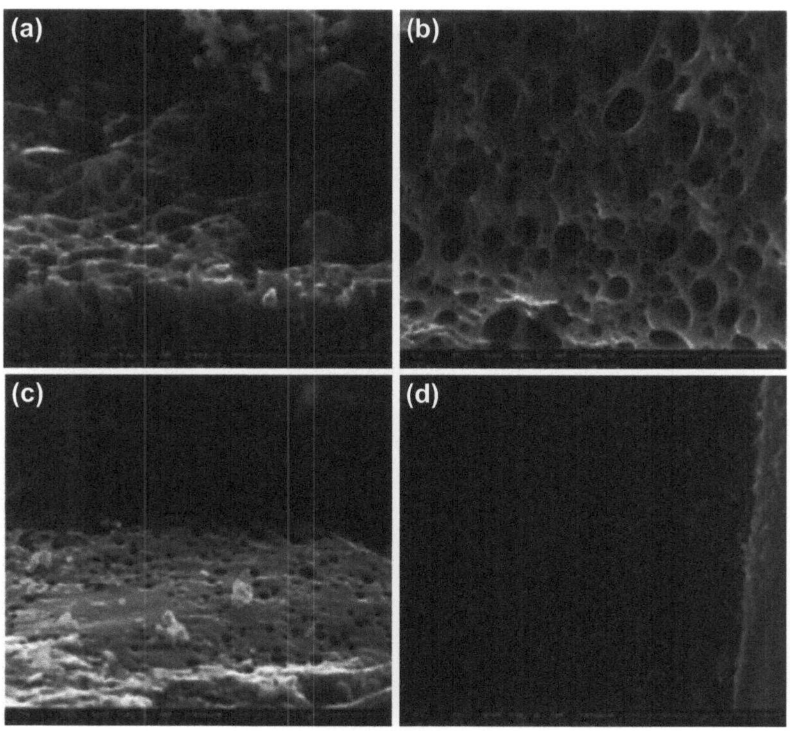

Figure 20
SEM images of the cross-section of CFs from B-lignin/PLA blends: (a) 50/50; (b) 75/25; (c) 90/10; (d) 100/0. *Reproduced with permission from Ref. [87], copyright Elsevier, 2014.*

cellulose nanofibrils (CeNF) with a porous core. Carbonization offers a hollow fibre structure with an increased surface and porosity [95]. It is advantageous that an increased surface area and porosity extended via co-electrospinning method provide an improved platform for various interactions between fibres and their surrounding environments, which potentially enhanced the performance of the fibres for advanced applications. Core-shell CNF–lignin/Pan fibres exhibit relatively smooth surfaces where the porous core consisted of a CNF network and the shell consists of lignin and PAN. The difference in lignin-based hollow fibres (lignin/PAN) and CNF lignin-based core–shell fibres (CNF-lignin/PAN) can be easily found while considering the cross-sectional surface as shown in Figure 21.

Phenolic resin has been broadly applied as a source of ACFs, and the lignin with phenyl propane structure similar to that of the phenol-formaldehyde resin was an ideal replacement for phenolic resin. Furthermore, the application of this resin to fabricate lignin-based carbon materials of a controllable surface and pore structure and physical properties is desired.

Nanostructured carbons with hierarchical pore arrangements are considered as one of the most suitable electrode materials for high-performance electrochemical capacitors. The ordered mesopore network of these materials provides shorter diffusion pathways along the carbon particles, which enhances the diffusivity of molecules and/or ions to make the inner surface more accessible than in microporous-activated carbons. The presence of certain oxygen surface groups on carbon materials can contribute significantly to the performance by improving the wettability of the surface and enhancing pseudocapacitance. Preparation of hierarchically porous carbon from biomass sources is currently an emerging area of extensive interest. The technique to access such novel nanostructured carbons majorly depends on a nanocasting technique using a hard or soft template. Microporous carbon with a 3D connected and ordered pore structure can be produced by liquid phase

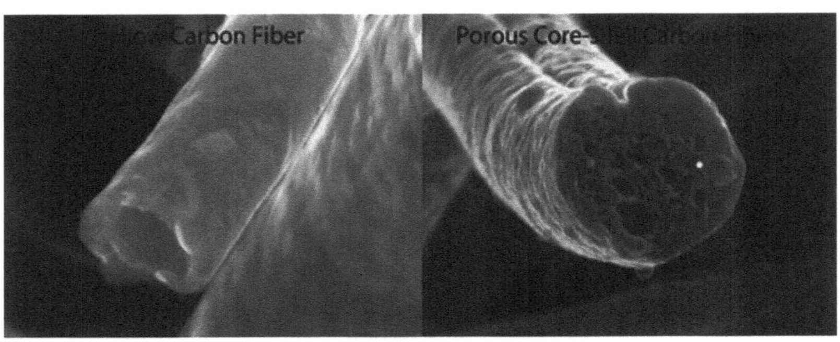

Figure 21
SEM images of hollow and porous core–shell CFs derived from lignin/PAN and CNF-lignin/PAN, respectively. *Reproduced with permission from Ref. [88], copyright Elsevier, 2013.*

impregnation with a carbon precursor over a porous zeolite and further template is traditionally removed by HF washing. To prepare ultraporous carbon, the use of a zeolite template and lignin as carbon precursor to obtain zeolite-template carbons (ZTCs) containing the fingerprint of zeolite was achieved recently [96]. This lignin-derived ZTCs contain a bimodal pore structure where microporosity results from both the removal of zeolite walls and the lignin carbonization, and a mesoporous network is originated as a result of the incomplete pore filling of the hard template. The interesting part is to analyse the origin of mesoporosity for the template carbon materials which replicate both the external and internal morphologies of the corresponding zeolite crystals.

5. Summary and Outlook

We have reviewed the recent work on a number emerging strategies of lignin valorization, including depolymerization into monomers, further upgradation of monomers, enzymatic oxidation and carbonization. Among these, depolymerization via hydrogenolysis of lignin into monomers has been the key strategies. However, with the development of a nanoscale catalyst, selective deoxygenation via the hydrogenolysis of the β–O–4 type C–O bond has emerged as an effective tool for further upgradation of monomers obtained via LDP. Emerging strategies of enzymatic oxidation of lignin are reviewed along with describing predictable potentials of lignin for the production of a structurally wide range of carbon materials.

As a resource of nanomaterials, biopolymers have many advantages and can be extracted industrially on a large scale, for example, lignin from kraft pulping [97]. Some of the current approaches showed that lignin can be a versatile source to obtain a high-quality wide range of nanostructured CFs that are emerging materials for electrochemical energy storage applications. The article also provides an analysis of the scope of the deoxygenation techniques for upgradation of lignin-derived aromatic functionalized monomers and its successful recent outcomes for accessing high-value deoxygenation products containing an intact aromatic ring. However, LDP from intact biomass currently produces a largely heterogeneous slate of molecules, which makes lignin valorization an enormously difficult problem. Such strategies of cleavage of polymeric units consists of aromatic fragments to demonstrate the careful selection of reactive molecules or nanoparticles capable of selective cleavage of biomass lignin. This is a successful route to disrupt the recalcitrance of lignin under controlled conditions to generate value-added products from lignin and to make the carbohydrate fraction more reactive for hydrolysis. Enzymatic strategies involve synthetic biological processes with appropriate enzymes, which are currently evolving and have a lot more to offer. In nature, lignin is slowly degraded by white-rot fungi such as *Phanerochaete chrysosporium*, which provides an extracellular lignin peroxidase enzyme to begin the degradation process. Manganese peroxidase and laccase enzymes are also capable of lignin breakdown to initial breakdown products such as phenylpropanoid dimers, which can be further degraded via aromatic *meta*-cleavage pathways. This is carried out in nature by soil bacteriae. Certain studies that contain a huge potential to change the face of lignin valorization are envisaged as follows:

1. Accessing non-pyrolytic lignin bio-oil by developing a new depolymerization strategy is essential since the lignin bio-oil is extremely amenable to HDO for further upgradation. Therefore, a catalytic biorefining method can be established. Transfer hydrogenation in alcohol media has found as a potential strategy as discussed earlier in this article. A new hydrogen transfer strategy under mild condition will be sorted for the future development of the depolymerization process to access lignin oil fraction. Efficient strategies would enable one to proceed by avoiding the use of recalcitrant solid polymers obtained from the organosolv process.
2. Establishing the role of a bimetallic alloy system for selective C—O bond hydrogenolysis is one of the fundamental interests related to lignin valorization and its chemical aspects. Development of a strong and effective homogeneous catalyst for HDO is something not often witnessed recently. This topic has significant scope for developing new lignin chemistry. Moreover, in the HDO of naturally occurring lignin species, what features of lignin surface and its role will be important, that are yet to be revealed, the HDO process is to be studied extensively in the future in addition to designing and improving conventional methods of LDP.
3. Identify new bacterial lignin degraders and test enzymes like manganese peroxidase and laccase in the solution phase and immobilized phase. The mechanism of lignin cleavage by enzymes is an interesting area to explore. Polysaccharide monooxygenase enzyme AA9 (GH61) is also another candidate for lignin valorization [98]. In this respect, bacterial peroxidase enzymes are potential bet which are capable of oxidizing lignin [65]. The article highlights this part to convey that experts in biology have a major role to play in the near future for the gene-specific or gene-altered enzyme-specific disruption of the biopolymer chains of lignin. Apart from exploring new oxygenase enzymes, novel nanomaterial—enzyme composites and enzyme-encapsulated nanoscale materials as biocatalysts should be explored. This holds a significant potential to further develop lignin chemistry.
4. The strategy of exemplifying the potential of different rigid templates such as hierarchical zeolites and large pore metal-organic frameworks can be extensively tested for the access of hierarchical porous carbon by infiltration of lignin as a carbon source. Novel strategies to obtain carbon by using lignin as a carbon precursor have plenty to offer for designing and developing energy storage devices. In order to develop strategies of chemical modification or chemical conversion of lignin, it is highly important to consider bonds that are to be cleaved.

References

[1] Mamman AS, Lee J-M, Kim Y-C, Hwang IT, Park N-J, Hwang YK, et al. Biofuels Bioprod Biorefin 2008;2:438–54.
[2] Hage RE, Brosse N, Chrusciel L, Sanchez C, Sannigrahi P, Ragauskas A, et al. Polym Degrad Stab 2009;94:1632–8.
[3] Nimz HH. Angew Chem Int Ed 1974;13:313–21.
[4] Adler E. Wood Sci Technol 1977;11:169–218.
[5] Chatel G, Rogers RD. ACS Sustain Chem Eng 2014;2:322–39.
[6] Zakzeski J, Bruijnincx PCA, Jongerius AL, Weckhuysen BM. Chem Rev 2010;110:3552.
[7] Holladay J, White JF, Bozell JJ, Johnson D. DOE report "Top value-added chemicals from biomass, volume II". 2007.
[8] Laskar DD, Yang B, Wang H, Lee J. Biofuels Bioprod Biorefin 2013;7:602–26.
[9] Yang H, Yan R, Chen H, Lee DH, Zheng C. Fuel 2007;86:1781–8.
[10] Regalbuto JR. Science 2009;325:822–4.
[11] Roberts VM, Stein V, Reiner T, Lemonidou A, Li X, Lercher JA. Chem Eur J 2011;17:5939–48.
[12] Janesko BG. Phy Chem Chem Phys 2014;16:5423–33.
[13] Sathitsuksanoh N, Holtman KM, Yelle DJ, Morgan T, Stavila V, Pelton J, et al. Green Chem 2014;16:1236–47.
[14] Ma Z, Troussard E, Van Bokhoven JA. Appl Catal A Gen 2012;423–424:130–6.
[15] Britt PF, Buchanan AC, Cooney MJ, Martineau DR. J Org Chem 2000;65:1376–89.
[16] Macala GS, Matson TD, Johnson CL, Lewis RS, Iretskii AV, Ford PC. ChemSusChem 2009;2:215–7.
[17] Li Z, Garedew M, Lam C-H, James JE, Miller DJ, Saffron CM. Green Chem 2012;14:2540–9.
[18] Thring RW, Breau J. Fuel 1996;75:795–800.
[19] Schinski WL, Kuperman AE, Han J and Naae DG. US 2009/0218061A0218061; 2009.
[20] Saisu M, Sato T, Watanabe M, Adschiri T, Arai K. Energy Fuels 2003;17:922–8.
[21] Osada M, Sato T, Watanabe M, Adschiri T, Arai K. Energy Fuels 2004;18:327–33.
[22] Osada M, Hiyoshi N, Sato O, Arai K, Shirai M. Energy Fuels 2007;21:1854–8.
[23] Huang J, Wang X, Wang ZL. Nano Lett 2006;6:2325–31.
[24] Lange H, Decina S, Crestini C. Eur Polym J 2013;49:1151–73.
[25] Gao Y, Zhang J, Chen X, Ma D, Yan N. ChemPlusChem 2014;79:825–34.
[26] Radoykova T, Nenkova S, Stanulov K. Chem Nat Compd 2010;46:807–8.
[27] Wahyudiono, Sasaki M, Goto M. Fuel 2009;88:1656–64.
[28] Karagöz S, Bhaskar T, Muto A, Sakata Y. Fuel 2004;83:2293–9.
[29] Voitl T, Rudolf von Rohr P. ChemSusChem 2008;1:763–9.
[30] Stärk K, Taccardi N, Bösmann A, Wasserscheid P. ChemSusChem 2010;3:719–23.
[31] Zakzeski J, Jongerius AL, Bruijnincx PCA, Weckhuysen BM. ChemSusChem 2012;5:1602–9.
[32] Coughlin RW, Davoudzadeh F. Nature 1983;303:789–91.
[33] Yan N, Zhao C, Dyson PJ, Wang C, Liu LT, Kou Y. ChemSusChem 2008;1:626–9.
[34] Li CZ, Zheng MY, Wang AQ, Zhang T. Energy Environ Sci 2012;5:6383–90.
[35] Harris EE, D'Ianni J, Adkins H. J Am Chem Soc 1938;60:1467–70.
[36] Jongerius AL, Jastrzebski R, Bruijnincx PCA, Weckhuysen BM. J Catal 2012;285:315–23.
[37] Nichols JM, Bishop LM, Bergman RG, Ellman JA. J Am Chem Soc 2010;132:12554–5.
[38] Runnebaum RC, Nimmanwudipong T, Block DE, Gates BC. Catal Sci Technol 2012;2:113–8.
[39] Song Q, Wang F, Xu J. Chem Commun 2012;48:7019–21.
[40] Song Q, Wang F, Cai J, Wang Y, Zhang J, Yu W, et al. Energy Environ Sci 2013;6:994–1007.
[41] Chan JMW, Bauer S, Sorek H, Sreekumar S, Wang K, Toste FD. ACS Catal 2013;3:1369–77.
[42] Ma X, Tian Y, Hao W, Ma R, Li Y. Appl Catal A Gen 2014;481:64–70.
[43] Yoshikawa T, Yagi T, Shinohara S, Fukunaga T, Nakasaka Y, Tago T, et al. Fuel Process Technol 2013;108:69–75.

[44] Yoshikawa T, Shinohara S, Yagi T, Ryumon N, Nakasaka Y, Tago T, et al. Appl Catal B Env 2014;146:289–97.
[45] Ma R, Hao W, Ma X, Tian Y, Li Y. Angew Chem Int Ed 2014;53:7310–5.
[46] Dutta S. ChemSusChem 2012;5:2125–7.
[47] Yi J, Liu S, Abu-Omar MM. ChemSusChem 2012;5:1401–4.
[48] Shiramizu M, Toste FD. Angew Chem 2012;124:8206–10. Angew. Chem. Int. Ed., 2012, 51, 8082–8086.
[49] Saidi M, Samimi F, Karimipourfard D, Niianwudipong T, Gates BC, Rahimpour MR. Energy Environ Sci 2014;7:103–29.
[50] Crossley S, Faria J, Shen M, Resasco DE. Science 2009;327:68–72.
[51] Wildschut J, Iqbal M, Mahfud FH, Melian-Cabrera I, Venderbosch RH, Heeres HJ. Energy Environ Sci 2010;3:962–70.
[52] Wildschut J, Melian-Cabrera I, Heeres HJ. Appl Catal B 2010;99:298–306.
[53] Mercader FD, Groeneveld MJ, Kersten SRA, Geantet C, Toussaint G, Way NWJ, et al. Energy Environ Sci 2011;4:985–97.
[54] Sergeev AG, Hartwig JF. Science 2011;332:439–43.
[55] Parsell TH, Owen BC, Klein I, Jarrell TM, Marcum CL, Haupert LJ, et al. Chem Sci 2013;4:806–13.
[56] Yan N, Dyson PJ. Curr Opin Chem Eng 2013;2:178–83.
[57] Zhang J, Teo J, Chen X, Asakura H, Tanaka T, Teramura K, et al. ACS Catal 2014;4:1574–83.
[58] Zhang J, Asakura H, van Rijn J, Yang J, Duchesne P, Zhang B, et al. Green Chem 2014;16:2432–7.
[59] Laskar DD, Tucker MP, Chen X, Helms GL, Yang B. Green Chem 2014;16:897–910.
[60] Zakzeski J, Weckhuysen BM. ChemSusChem 2011;4:369–78.
[61] Xu WY, Miller SJ, Agrawal PK, Jones CW. ChemSusChem 2012;5:667–75.
[62] Barta K, Matson TD, Fettig ML, Scott SL, Iretskii AV, Ford PC. Green Chem 2010;12:1640–7.
[63] Matson TD, Barta K, Iretskii AV, Ford PC. J Am Chem Soc 2011;133:14090–7.
[64] Jongerius AL, Bruijnincx PCA, Weckhuysen BM. Green Chem 2013;15:3049–56.
[65] Brown ME, Walker MC, Nakashige TG, Iavarone AT, Chang MCY. J Am Chem Soc 2011;133:18006–9.
[66] Wong DWS. Appl Biochem Biotechnol 2010;157:174–209.
[67] Bugg TDH, Ahmad M, Hardiman EM, Rahmanpour R. Nat Prod Rep 2011;28:1883–96.
[68] Ahmad M, Roberts JN, Hardiman EM, Singh R, Eltis LD, Bugg TDH. Biochemistry 2011;50:5096–107.
[69] Brown ME, Barros T, Chang MCY. ACS Chem Biol 2012;7:2074–81.
[70] Atsumi S, Hanai T, Liao JC. Nature 2008;451:86–90.
[71] Peng X, Masai E, Kitayama H, Harada K, Katayama Y, Fukuda M. Appl Environ Microbiol 2002;68:4407–15.
[72] Sainsbury PD, Hardiman EM, Ahmad M, Otani H, Seghezzi N, Eltis LD, et al. ACS Chem Biol 2013;8:2151–6.
[73] Lee JW, Na D, Park JM, Lee JM, Choi S, Lee SY. Nat Chem Biol 2012;8:536–46.
[74] Camarero S, Ibarra D, Martinez AT, Romero J, Gutierrez A, del Rio JC. Enzyme Microb Technol 2007;40:1264.
[75] Riva S. Trends Biotechnol 2006;24:219.
[76] Kamwilaisak K, Wright PC. Energy Fuels 2012;26:2400–6.
[77] Majumdar S, Lukk T, Solbiati JO, Bauer S, Nair SK, Cronan JE, et al. Biochemistry 2014;53:4047–58.
[78] Brown ME, Chang MC-Y. Curr Opin Chem Biol 2014;19:1–7.
[79] Ten HR, Teunissen PJ. Chem Rev 2001;101:3397–413.
[80] Masai E, Katayama Y, Nishikawa S, Yamasaki M, Morohoshi N, Haraguchi T. FEBS Lett 1989;249:348–52.
[81] Otsuka Y, Sonoki T, Ikeda S, Kajita S, Nakamura M, Katayama Y. Eur J Biochem 2003;270:2353–62.
[82] Hayashi J, Kazehaya A, Muroyama K, Watkinson AP. Carbon 2000;38:1873–8.
[83] Fierro V, Torne-Fernandez V, Celzard A. Microporous Mesoporous Mater 2007;101:419–31.
[84] Babel K, Jurewicz K. Carbon 2008;46:1948–56.
[85] Tenhaeff WE, Rios O, More K, McGuire MA. Adv Funct Mater 2013;24:86.
[86] Kubo S, Uraki Y, Sano Y. Carbon 1998;36:1119–24.

[87] Lallave M, Bedia J, Ruiz-Rosas R, Rodríguez-Mirasol J, Cordero T, Otero JC, et al. Adv Mater 2007;19:4292–6.
[88] Spender J, Demers AL, Xie X, Cline AE, Earle MA, Ellis LD, et al. Nano Lett 2012;12:3857–60.
[89] Tenhaeff WE, Rios O, More K, McGuire MA. Adv Funct Mater 2014;24:86–94.
[90] Park M, Zhang XC, Chung MD, Less GB, Sastry AM. J Power Sources 2010;195:7904–29.
[91] Chatterjee S, Clingenpeel A, McKenna A, Rios O, Johs A. RSC Adv 2014;4:4743–53.
[92] Hu S, Hsieh Y-L. J Mat Chem A 2013;1:11279–88.
[93] Rios O, Tenhaeff WE, Daniel C, Dudney NJ, Johns A, Alexander G, Baker FS. US Patent 2014/0038034 A1.
[94] Thunga M, Chen K, Gewell D, Kessler MR. Carbon 2014;68:159–66.
[95] Xu X, Zhou J, Jiang L, Lubineau G, Chen Y, Wu X-F, et al. Mater Lett 2013;109:175–8.
[96] Ruiz-Rosas R, Valero-Romero MJ, Salinas-Torres D, Rodriguez-Mirasol J, Cordero T, Morallon E, et al. ChemSusChem 2014;7:1458–67.
[97] Schnepp Z. Angew Chem Int Ed 2012;52:1096–108.
[98] Hu J, Arantes V, Pribowo A, Gourlay K, Saddler JN. Energy Environ Sci 2014;7:2308–15.

Integrated Bio- and Chemocatalytic Processing for Biorenewable Chemicals and Fuels

Md Imteyaz Alam, Shelaka Gupta, Ejaz Ahmad, M. Ali Haider
Renewable Energy and Chemicals Laboratory, Department of Chemical Engineering, Indian Institute of Technology, Delhi, India

1. Introduction

Depleting fossil fuel reserves and increase in the demand of fossil fuel-based products have necessitated efforts in the search for an alternative source of feedstock. In this regard, biomass, with its abundant availability, represents an attractive renewable option to explore. At present, fossil fuels contribute to nearly 92% of the commodity chemicals produced, which is expected to change, and a large part of it can be generated from biomass in the near future [1]. For a proposed biorefinery, apart from fuels, high-value chemicals constitute an important component of the product portfolio [2]. Current efforts to develop a biomass-based technology for producing fuels and chemicals are directed towards the utilization of non-edible lignocellulosic biomass [3,4]. In general, lignocellulosic biomass contains 35–50% of cellulose, 20–35% of hemicellulose and 10–25% lignin [5]. Lignin provides rigidity to the network of cellulose and hemicelluloses by filling the spaces in between. It is made of aromatic monomers and contains no sugars. On the contrary, celluloses and hemicelluloses are linear chain and branched polymers consisting of six-member and five-member cyclic sugar monomers such as glucose and xylose, respectively. In order to exploit biomass as a feedstock for making a desired product, it should first be broken down into constituent sugar monomers that can be selectively upgraded. While biomass can be directly converted into chemicals via pyrolysis, or gasification, in general these methods require high temperature and pressure conditions [6]. In contrast, aqueous phase enzymatic, acid or alkaline hydrolysis can be achieved at relatively low temperatures ($T < 200$ °C) and ambient pressure [7], resulting in the depolymerization of structurally complex lignocelluloses, yielding to a range of sugar molecules similar in structure to xylose, glucose, galactose or mannose. The aqueous

Sustainable Catalytic Processes. http://dx.doi.org/10.1016/B978-0-444-59567-6.00006-6

sugars thus obtained can be converted directly into a product or a platform molecule, which may act as a precursor for producing a diverse array of commodity chemicals, fuels or fuel additives.

The derived sugars can be selectively processed either by a biological fermentation method or by chemical catalytic processing to a desired chemical or a precursor molecule [8]. Several biomass-derived molecules having carboxylic, cyclic ester and furan functionalities are proposed as platform molecules, which can either be obtained directly from aqueous phase hydrolysis or via a biocatalytic processing [9,10]. While some of the molecules from these groups (e.g. succinic acid) can be derived by both biological and chemical routes, some (e.g. lactic acid) are solely obtained from a biological route. New opportunities are arising in the preparation of higher-value chemicals and biofuels from renewable feedstocks via biocatalytic routes [11] owing to the high regio- and chemoselectivity of a biocatalyst, which can be achieved under ambient temperature and pressure conditions [12,13]. In general, chemical catalytic routes are observed to be superior to the biological routes, in terms of achieving better economics, yield, scale-up and purification of the desired product. Natural biocatalysts are not favoured for industrial applications, due to their high price, low stability and operations in a limited temperature range. Additionally, the products of fermentation are highly functionalized and of shorter carbon chain as compared to the petroleum-derived hydrocarbon fuels and chemicals. However, applying a carbon—carbon coupling reaction on these products combined with a defunctionalizing reaction may produce long chain hydrocarbon fuels and chemicals. The platform molecule thus obtained from a biocatalytic route can be further upgraded to a commodity chemical or fuel by utilizing a chemical catalytic reaction. It has been recently observed that an integrated strategy of combining bio- and chemocatalytic processing could provide a wider range of products with higher selectivity and yield [14]. Such an integrated strategy, as proposed in Figure 1, will directly utilize biomass or biomass-derived aqueous sugars in a fermenter. The products as obtained from the fermenter will be fed to a separator and the purified platform molecules can be stored, or directly fed as a reactant to a suitable catalytic reactor for the production of fuels and chemicals.

This chapter is focused on studying the selected platform molecules, which are obtained from bioprocessing and can be catalytically upgraded into a higher-value chemical or fuel. Fermentation of biomass, or derived aqueous sugars, produces biogenic impurities such as protein fragments, residual nutrients from fermentation and byproducts, which tend to interfere in catalytic processing, leading to catalyst deactivation and thereby reduced stability [15,16]. Since the product of the fermentation process acts as a reactant for catalytic processing, it is important to develop a suitable purification method. The strategy for catalytic transformations primarily constitute a selective oxygen removal process in the aqueous sugars, which can be carried out by a deoxygenation reaction, composed of a single or multiple reaction steps of decarboxylation, decarbonylation, dehydration and

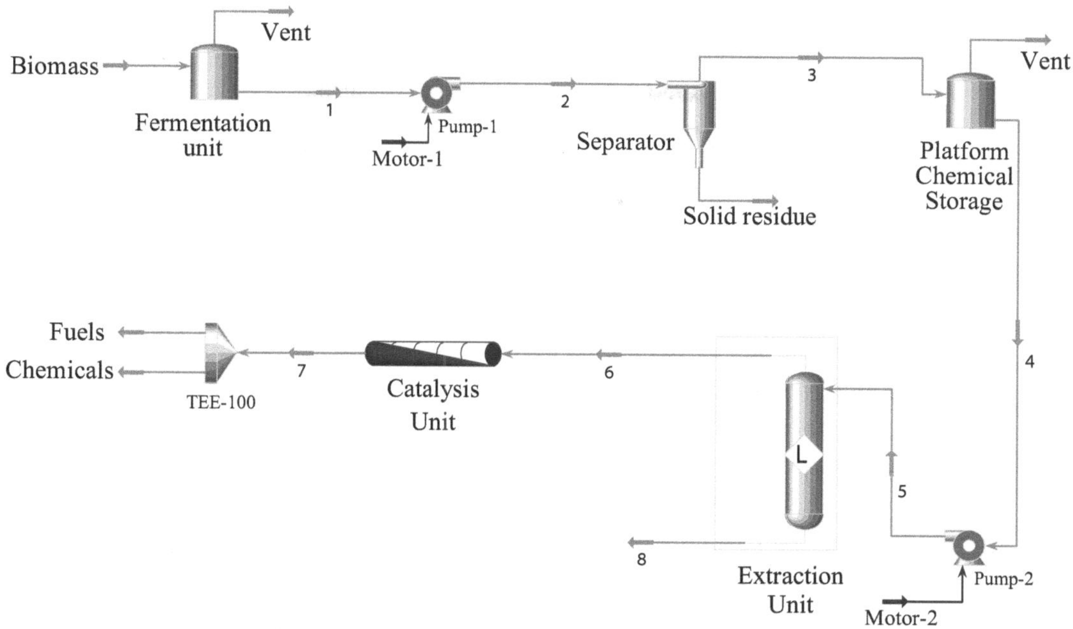

Figure 1
Representative process flow diagram for integrated processing of biomass into fuels and chemicals.

hydrogenolysis. The impurities present in biomass may interfere with some or all of these steps. Therefore, the nature of catalysis in integrated processing needs to be explored. Following up on this discussion, this study is broadly categorized into the following:

- Biocatalytic transformation to platform molecules
- Challenges related to the nature of biogenic impurities
- Strategies for catalytic transformation of platform molecules

2. Biocatalytic Transformation to Platform Molecules

The biochemical production of platform molecules is carried out by enzymatic materials, which are produced extracellularly or intracellularly from a microorganism. The three-dimensional nature of the active site in the enzyme acts with high specificity towards a substrate molecule, which helps in selective transformation [17]. For example, cellulytic and hemicellulytic enzymes are specifically employed for the deconstruction of cellulose and hemicellulose into their respective sugar monomers. Enzyme-assisted conversion of lignocellulosic biomass can potentially reduce the cost and unwanted byproducts. Under optimum conditions, the reaction efficiency may improve multifold while keeping the regio-, diastereo-, enantio- or chemoselectivity of the desired product [12]. Enzymatic reactions can be carried out at ambient temperature and pressure conditions, without effecting the rate of isomerization, racemization, epimerization and rearrangement of the compound, which are generally altered in relatively harsh conditions. The hexoses and pentoses can be selectively converted into a desired platform molecule by utilizing a range of available biological routes. Several metabolic pathways act selectively on a substrate and the choice of pathway dictates the kind of microorganism employed. In addition, genetic engineering techniques may be utilized to shut off the side reaction in a metabolic pathway and thus improve the overall yield and selectivity. One such example is the polyketide biosynthetic pathway, which has traditionally been exploited to produce a diverse variety of products such as fatty acids, chalcones, stilbenes and antibiotics [18]. Recently, the polyketide biosynthetic pathway has been utilized to produce triacetic acid lactone (TAL) as a precursor to make biorenewable chemicals [19,20]. Polyketide biosynthesis offers a unique ability to selectively produce alkyl chains with varying length (C_4 to C_{30} and more) [21,22]. The structural diversity of polyketide depends on the starter and extender units, steps involved in condensation reactions, and the mode of offloading or ring closure of the product of polyketide chains [21].

Compared to a chemical catalytic conversion, biocatalytic conversion suffers serious drawbacks as it offers limited ability on the desired functional group modification, which makes it less applicable. In addition, since enzymes are functional only in a limited range of operating temperatures, biological transformation cannot be carried out at high temperatures to derive higher conversions or selectivity [23]. As an example, we consider here the case of isoprene, which is industrially produced as a byproduct of ethylene via steam cracking of crude oil or naphtha. Isoprene is widely used in the synthesis of natural rubber and is applicable in the manufacturing of materials such as baby toys, medical equipment, tyres, shoe soles, etc. Current production levels of isoprene are estimated to be about 1 million tonnes, 95% of which is directly utilized in producing polyisoprene [24]. In an alternative biocatalytic route, isoprene is naturally synthesized by biological organisms such as algae, plants, bacteria and animals. Interestingly, natural production of isoprene by plants is estimated to be around 500−600 million tonnes, which is emitted into the atmosphere. However, it is extremely difficult to recover isoprene from the atmosphere, making it less viable for commercial use. In the polyketide biosynthetic production of isoprene, two pathways are distinctly identified: mevalonate (MVA) and non-mevalonate. Microorganisms such as bacteria or algae can utilize either one or both metabolic pathways to synthesize isoprene. While the presence of the *isoprenesynthase* (*IspS*) gene is necessary to produce isoprene as the final product, it can also be expressed heterologously in the fermentation process. Lindberg et al. have shown the production of isoprene through heterologous expression of the gene encoding for the *IspS* in a photosynthetic *Cyanobacteria* [25]. Several other genetic engineering techniques can be employed to increase the yield of the final product [26]. For example, in a recent study by

Yang et al. on optimizing the MVA pathway in *Enterococcus faecalis* by introducing mutation in the *mvaS* gene, isoprene production was observed to be enhanced to a level of 6.3 g/L after 40 h of fed-batch fermentation [27]. While these results are encouraging, they may not prove to be economical from a commercial perspective where higher production levels are expected for a process to become attractive for commercialization. In conclusion, even after employing various genetic engineering techniques, the biocatalytic production of isoprene remains a challenge, owing to the limitations offered by the enzymatic reactions. Therefore, it is desirous to develop an integrated processing strategy, where platform molecules as obtained from fermentation can be catalytically upgraded into a desired product.

Recent reports by the Department of Energy have identified the top 10 building block chemicals, out of which five are primarily obtained from biocatalytic routes as shown in Table 1 [28]. These molecules show significant potential to be upgraded into high-value chemicals or fuels by catalytic conversion. For example, bacterial species such as *Carnobacterium*, *Lactobacillus*, *Leuconostoc*, *Pediococcus*, *Tetragenococcus*, *Vagococcus* and *Streptococcus* produces lactic acid [29]. Over the years, lactic acid has found applications in the food, textile, pharmaceutical and chemical industries with a potential market value of about US$552–636 million [30]. Interestingly, the derived products from lactic acid have a much higher market value of about US$21,900 million. This is primarily due to the high demand of polylactic acid in the polymer industry for the manufacturing of fibres, packaging and chemical products. In addition, more value is added to lactic acid as a precursor molecule by making products such as propylene glycol and acrylic acid [31]. On successful commercialization, several derived products from lactic acid such as polylactide, 1,2-propandiol, 2-chloropropionic acid, ethyl lactate, oxalic acid, propionic acid, pyruvic acid, etc. may further escalate the market value.

Compared to lactic acid, succinic acid is produced from petrochemicals, primarily from the hydrogenation of maleic acid or oxidation of 1,4-butanediol [32]. Recently, a biological route for the production of succinic acid has been set up, which relies on fermentation by bacterial strains such as *Actinobacillus*, *Succiniciproducens* and *Mannheimia* [33]. Based on 100,000 TPA (tonnes per annum) of production, the market price of succinic acid is estimated to be US$2400–3000 million/tonne [34]. The market share of succinic acid is expected to grow sevenfold by 2020 [35]. On subsequent hydrogenation, 1,4-butanediol is synthesized, which alone has a potential market of about US$3000 million [36]. Other important derivatives of succinic acid are tetrahydrofuran and γ-butyrolactone, which are used as cleaning solvents for electronics and pharmaceuticals.

Chemically, 3-hydroxy propionic acid can be produced from acrylic acid, allyl alcohol, vinyl acetate or 1,3-propanediol. However, biological routes via the fermentation of glucose using strains of *Escherichia coli* and *Klebsiella pneumoniae* are preferred [37].

Table 1: Top Six Platform Molecules Derived from Biomass, which Are Potential Candidates for Integrated Processing

S. No.	Platform Chemical	Production Route[a]	Market Value[b]	Upgraded Products[c]	Market Value of the Product[d]	References
1	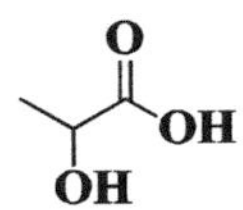 Lactic acid	Fermentation (400,000)	552–636	Propylene glycol, acrylic acid, lactate esters, chiral synthons, food additives, polylactic acid, polylactide, 1,2-propandiol, ethyl lactate, oxalic acid, pyruvic acid,	>21,900	[31,35,43]
2	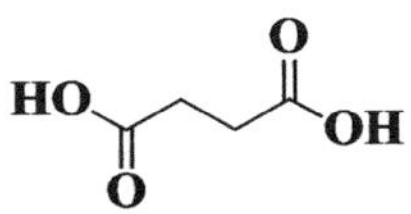 Succinic acid	Chemical/Fermentation (700,000)	1680–2100	2-pyrrolidones, γ-butyrolactone, succinate ester, 1,4-butandiol	>4240	[35,36,44]
3	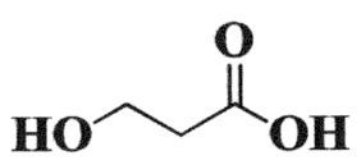 3-Hydroxypropionic acid	Fermentation (3,600,000)	NA	1,3-Propandiol, malonic acid, acrylic acid, methyl acrylate, acryl amide	>3020	[39,45,46]
4	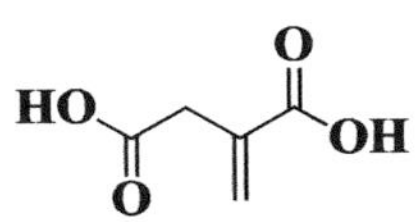 Itaconic acid	Fermentation (410,000)	984	Itaconic diamide, *N*-substituted pyrrolidones, methyl tetrahydrofuran, methyl γ-butyrolactone, 2-methyl-1,4-butandiol	>1200–1600	[35,41]

5	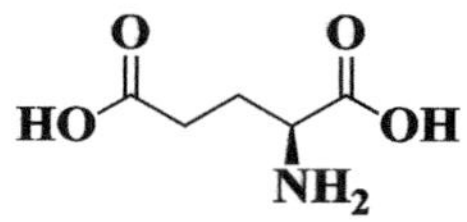 Glutamic acid	Fermentation (2,500,000)		Pyroglutaminol, pyroglutamic acid, glucaric acid, 1,5-pentandiol, glutaminol, polyglutamic acid, proline, prolinol,	NA	[35,47]
6	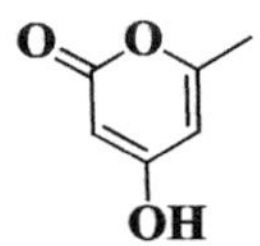 Triacetic acid lactone	Chemical/Fermentation (NA)	NA	2,4-Pentanedione, 1,3-pentadiene, 3-Penetene-2-one, 4-hydroxypentanone, Parasorbic acid, Sorbic acid, Resin precursor	>1000	[19,48]

[a]Projected production in TPA (tonnes per year).
[b]Market value based on projected production in million US dollars.
[c]Up-graded products are obtained through their respective platform chemical only.
[d]Projected market value of product derived economy in million US dollars.

The derivatives of 3-hydroxy propionic acid such as 1,3-propanediol, malonic acid, acrylic acid, methyl acrylate and acryl amide have found applications in the making of speciality chemicals and polymers [38]. 1,3-Propanediol was reported to have a market value of US$80 million in the year 2009, whereas malonic acid has an estimated market value of US$14 million [39] and acrylic acid between US$1296 and US$2908 million. Similarly, itaconic acid is commercially produced from the fermentation of sugars such as sucrose and glucose using the fungus *Aspergillus terreus* [40]. The market value of itaconic acid was estimated to be US$160 million in 2009 [41]. Itaconic acid is an important monomer in the production of a variety of polymers for applications in resins, fibres and paints. Glutamic acid is another important biomass-derived molecule, produced on fermentation by bacterial strains such as *Cephalosporium*, *Arthrobacter* and *Corynebacterium* [42]. It has an estimated market potential of US$16,250 million. The important derivatives of glutamic acid include γ-aminobutyric acid (GABA), *N*-methylpyrrolidone (NMP), *N*-vinylpyrrolidone (NVP), succinonitrile and acrylonitrile.

All of these platform molecules can be purified from fermentation media and directly utilized in a catalytic reactor in integrated processing. Recent work by Dumesic and co-workers has shown the potential of 2-pyrones such as TAL in producing biorenewable chemicals by applying integrated processing [14]. On catalytic hydrogenation, dehydration, ring-opening and decarboxylation reaction, TAL yields to 1,3-pentadiene, 3-pentene-2-one, 4-hydroxypentanone, sorbic acid, hexenoic acid, hexalactone and capralactone, Table 1 [20]. The potential market value of TAL is yet to be calculated; however, a higher market value for the derived products of TAL is expected.

In order to develop an integrated process, it is essential to understand the nature of impurities present in a biocatalytic process. For example, in lactic acid fermentation, reaction media are comprised of nutrients and minerals. Some inorganic bases, such as $Ca(OH)_2$, are added during fermentation to neutralize the acid produced. Subsequently, the resultant fermented broth is treated with H_2SO_4 several times to obtain pure lactic acid, followed by a series of concentration and purification steps. The choice of the purification step depends on the economics, recyclability and efficiency. Purification cost is almost half of the overall production cost of the pure molecule [16]. Consequently, the production of these platform molecules could become highly expensive. As an alternative, lesser pure molecules, derived from fermentation media, can be directly utilized. However, impurities present in the fermentation media may inhibit the catalytic reaction, which necessitates further exploration for a better solution to remove deactivation. Consequently, it is desirable to study the deactivation mechanism to maintain the stability and activity of the heterogeneous catalyst in biogenic aqueous environments.

3. Challenges Related to the Nature of Biogenic Impurities

During the heterogeneous catalytic reactions, the fermented product along with residual impurities, diffuses into the catalyst pores and interacts with the catalyst surface. Model biogenic impurities such as amino acids, proteins and vitamins and their mechanism of deactivation of the metal catalyst (Ru, Ni, Pt, Pd) for a hydrogenation reaction are listed in Table 2. As reported by Miller and co-workers, for lactic acid hydrogenation to propylene glycol by an Ru/C catalyst, sulphur-containing amino acids such as cysteine and methionine can cause irreversible catalyst deactivation. In contrast, amino acids having no sulphur lead to partial and reversible reduction in catalyst activity [49]. Dumesic and co-workers have further asserted to this observation for the hydrogenation reaction of cyclohexene on Pt, Pd and Ni catalyst, by correlating reaction rate in the presence of biogenic impurities to the measurements of respective adsorption isotherms of representative impurities such as thiamine, cysteine, methionine, biotin, tryptophan, niacin, threonine and *p*-aminobenzoic acid. Not only amino acids but sulphur-containing vitamins such as thiamine and biotin show strong inhibition of the catalytic hydrogenation reaction. Catalyst activity in the presence of the biogenic impurity can therefore be attributed to the adsorption strength of the catalyst surface. In addition, geometry of the adsorbed structure is important in determining the precise nature of the deactivation mechanism. For example, thiamine and cysteine are expected to bind with the catalyst surface by forming a strong metal sulphur bond. However, thiamine is predicted to dissociate on the metal surface leading to the formation of atomic sulphur, while cysteine remains intact on the metal surface [49]. Interestingly, biotin requires a relatively larger ensemble of surface metal atoms for adsorption from its S-atom. Therefore it shows lesser deactivation as compared to thiamine.

Similarly, threonine and alanine, having a free amino group, are likely to bind with their N-atom. While alanine has shown a competitive and irreversible inhibition of the lactic acid hydrogenation, theronine has shown a negligible effect on the three catalysts for cyclohexene hydrogenation [49]. Compared to a constituent amino acid, a full protein molecule like albumin is expected to block the pores leading to deactivation. In addition, the amino acid units (e.g. cysteine in albumin) of the protein molecule may adsorb on the surface of the metal catalyst by their respective sulphur or non-sulphur components resulting in an irreversible or reversible deactivation. A detailed study is thus required of the nature of protein folding and denaturation in the catalyst pores at the reaction conditions. Recently, an interesting therapy has been introduced in which a coating of poly(vinyl alcohol) (PVA) has shown significant improvement in the catalyst stability for hydrogenation reaction with fermented TAL and lactic acid [15]. The discovery, further

Table 2: Structure and Mechanism of Inhibition of the Catalyst Surface by Representative Biogenic Impurities

Impurities	Structure	Mechanism of Inhibition
Methionine (Met) (amino acid)	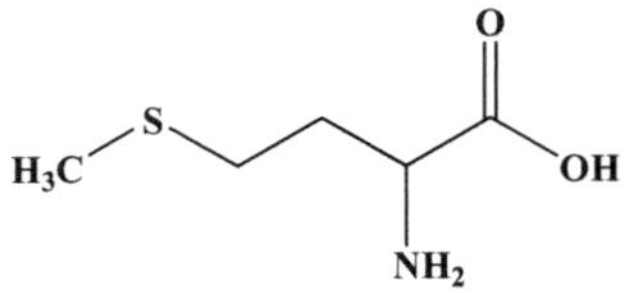	The sulphur atom strongly binds at the metal surface. The deactivation is observed to be irreversible on Ru catalyst. On Pd, the molecule decomposes to form SCH_3 species. On Ni and Pt, complete decomposition to atomic sulphur may occur [49].
Cysteine (Cys) (amino acid)	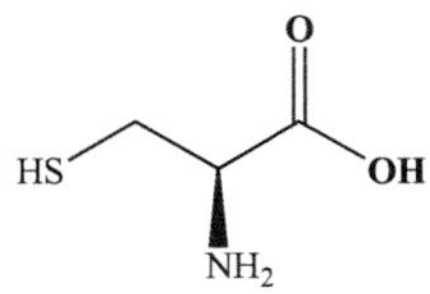	Cysteine binds to the metal surface by forming a strong metal—sulphur bond, leading to complete and irreversible deactivation. Further decomposition to elemental sulphur is not expected [49].
Alanine (Ala) (amino acid)	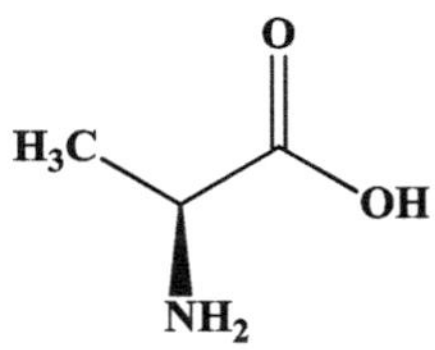	Alanine, having a free amino group, is expected to bind to metal catalyst by nitrogen, leading to a competitive and reversible deactivation, specifically in the case of the hydrogenation of the lactic acid [16].
Tryptophan (Trp) (amino acid)	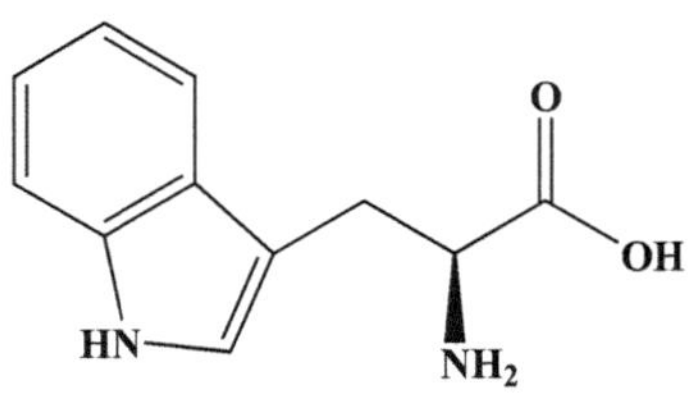	The nitrogen-containing heterocyclic compound binds strongly to the metal surface through the nitrogen atom in a flat or a tilted orientation leading to strong deactivation [50].
Glutathione (GSH) (amino acid)	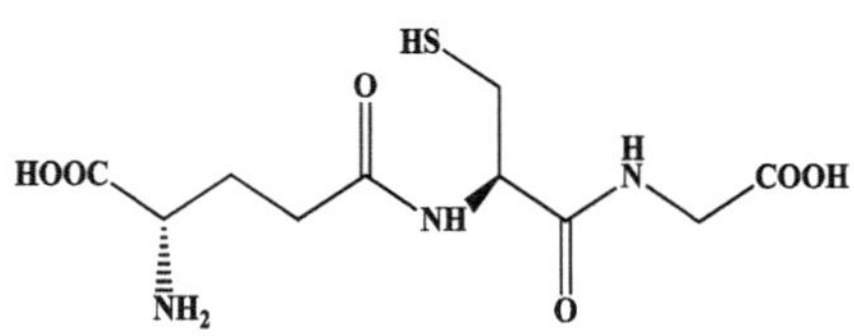	The thiol group has shown an inhibitory effect towards homogeneous catalysts by poisoning the site of the transition metals [51].
Hystidine (His) (amino acid)	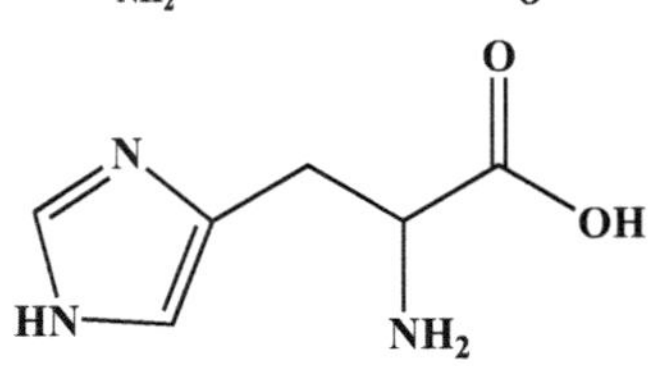	The N-atom of the heterocycle forms a stable bond with the Rh-central atom of pentamethylcyclopentadienyl rhodium bipyridine ($[Cp^*Rh(bpy)(H_2O)]^{2+}$) [52].

Table 2: Structure and Mechanism of Inhibition of the Catalyst Surface by Representative Biogenic Impurities—cont'd

Impurities	Structure	Mechanism of Inhibition
Albumin (protein)		Proteins like albumin are expected to adsorb into the pores of the supported metal catalyst leading to pore blockage. Sulphur present in the amino acid units irreversibly poisons the metal surface [16].
Thiamine (vitamin)	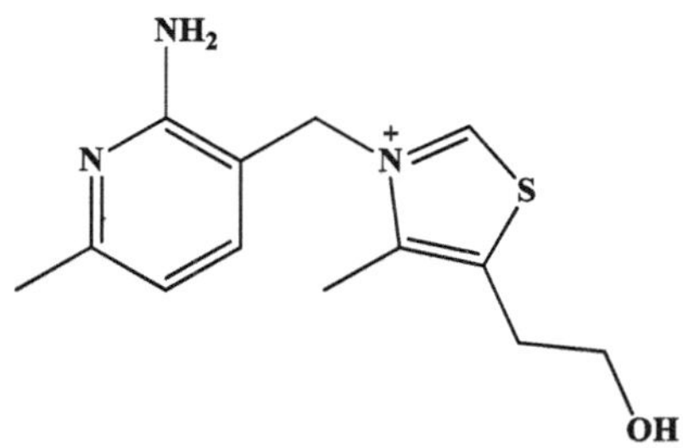	Thiamine strongly inhibits metals such as Pd, Pt and Ni. The inhibition is due to the presence of Cl^- and S atoms. It is suggested that the thiamine may decompose to elemental sulphur [49].
Niacin (vitamin)	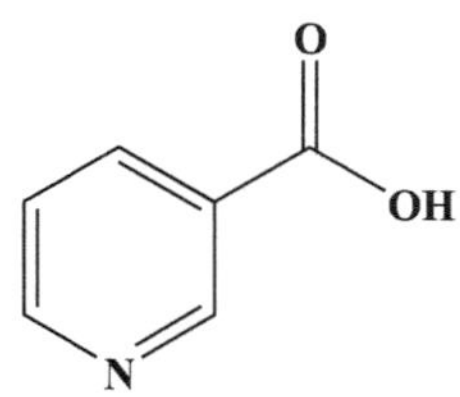	Niacin adsorbs on the surface of a metal through nitrogen atoms in a tilted configuration leading to moderate inhibition [50].
Biotin (vitamin)	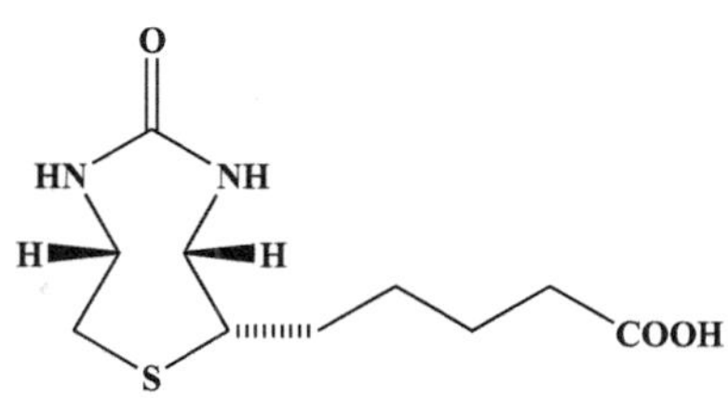	Biotin adsorbs only on large ensembles and as a result shows a less inhibitory effect as compared to other sulphur-containing amino acid or vitamin impurities.
Threonine (Thr) (Amino Acid)	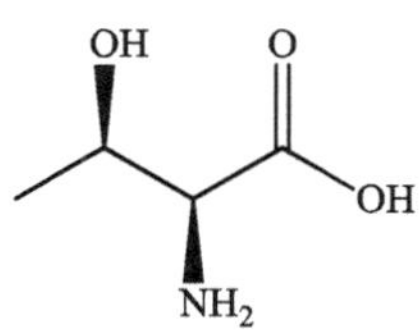	Threonine may bind either through the amine or the hydroxyl group to the catalyst surface. However, the inhibition is less significant as compared to S-containing amino acids [49].
p-Aminobenzoic acid	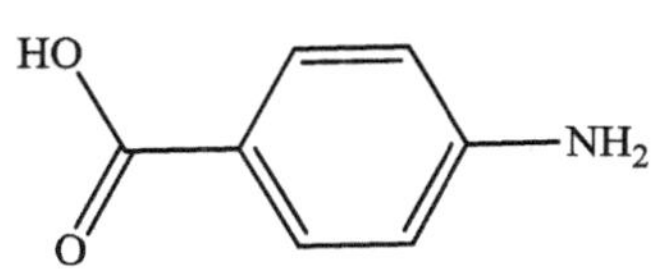	*p*-Aminobenzoic acid may bind through amine or carboxylic acid moiety to the catalyst surface [49].

pressures on the importance of developing a comprehensive understanding of the nature of biogenic impurities, its interactions with the metal surface and the mechanism of inhibition may provide insights for a better design of the heterogeneous catalyst for integrated processing.

In summary, it is essential to develop an efficient and economic method for purification of microbially synthesized platform molecules, which will yield a product of desirable purity and is deemed suitable for catalytic processing. In this respect, a number of approaches including solvent extraction, direct distillation, and adsorption can be utilized with an objective to reduce expenses, minimize waste and simplify the product recovery process. For all such processes, several limitations are observed. For example, in solvent extraction, high solvent loss will lead to additional cost and damage to the environment. In addition, a solvent extraction may result in the formation of high boiling internal esters as dimers and polymers. For resin-based extraction, regeneration of ion exchange resin and adjustment of feed pH to increase sorption efficiency remain the main concerns.

An interesting and alternative approach could be electrodialysis (ED), which has been successfully applied at a commercial scale for the purification of lactic acid from fermentation media. ED has been proven to be relatively fast and effective in terms of product recovery with no byproducts. In an ED process, ion (cation and anion) exchange membranes are used alternatively to separate ions from an aqueous solution under electrical potential. The maximum transport rate of ions is achieved at the limiting current density [53]. Integrated with the fermentation and catalytic processing, the ED technique could help in a unique process integration method, which will serve as a setup for future applications in the commercial production of higher value chemicals and biopharmaceuticals. Membrane separation helps in enhancing product recovery [53] and reducing energy consumption, where no solvent or heating is required. With this approach, one such process has been successfully implemented at the industrial scale by Eurodia Industrie (Rue Saint Martin, Pertuis, France) for the purification of fermentation media. Through this separation method, similar biomass-based platform chemicals can be purified. Figure 2 shows a schematic of a batch electrodialysis cell comprising four chambers having a cathode, anode, feed and recovery cell. The cells are designed in such a way that by using an appropriate membrane, pure fermented product can be selectively transported from feed to the recovery cell.

4. Strategies for Catalytic Transformation of Platform Molecules

The deactivation studies on heterogeneous catalysts by biogenic impurities have been limited in their scope as the studies are conducted only for the hydrogenation reactions on metal catalysts [49,54]. Catalytic transformation of platform molecules in integrated

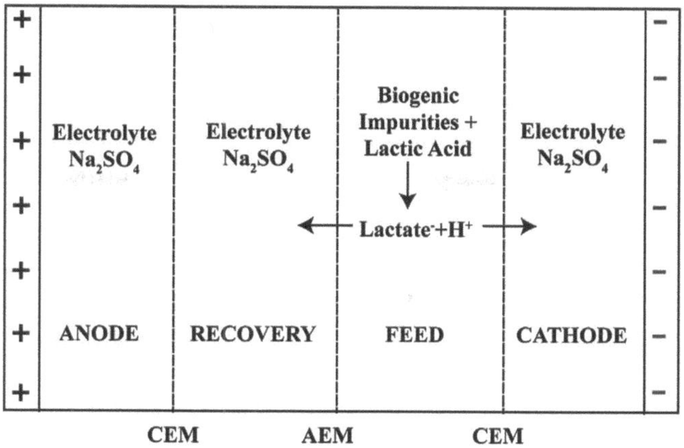

Figure 2

Schematic diagram for a batch electrodialysis process for the recovery of lactic acid from fermentation broth. CEM and AEM are cation exchange membrane and anion exchange membrane, respectively.

processing will require a comprehensive understanding of a pool of reactions that can be applied for making a desired product on an oxide, metallic, bimetallic or a solid acid catalyst, where we expect similar interactions of biogenic impurities on the catalyst surface.

In general, biomass-based molecules have relatively higher oxygen content in their structure (e.g. −OH, −C=O and COOH functionalities) as compared to the petrochemical-derived hydrocarbon fragments. The challenge, therefore, in making chemicals and fuels from biomass lies is the selective removal of oxygen. In addition, multiple polyol (−OH) groups in sugar molecules can cause low solubility (in organic solvents) and thermal instability limiting its scope for valorization into intermediate and speciality chemicals. The bond dissociation energies for the oxygen-containing functional groups (−OH, −COOH and −C=O) are relatively higher [55] and hence require appropriate catalytic processing to cleave a C−O bond. Deoxygenation is generally achieved in one of the following steps:

1. Dehydration of vicinal diols followed by hydrogenation of carbonyl groups
2. Dehydration of alcohols and hydrogenation of C−C double bonds
3. Hydrogenolysis of cyclic ethers
4. Decarbonylation of keto and carboxyl functionalities
5. Decarboxylation of carboxyl and cyclic ester functionalities

Dehydration of biomass-derived molecules involves acid-assisted elimination of water from reacting oxygenates. From this reaction, a number of useful molecules including furfural, 5-hydroxymethyl furfural (HMF), acrolein and acrylic acid can be synthesized from xylose, glucose, glycerol and lactic acid, respectively. Dehydration has been extensively used in chemical catalytic reactions, e.g. to convert sorbitol to sorbitan and then to isosorbide. A summary of products obtained on various platform molecules is shown in Table 3. Dehydration could be potentially important in integrated processing where glutamic acid can be upgraded to succinimide on dehydration [54]. Similarly,

Table 3: Biorenewable Platform Molecules and Products Obtained on Their Dehydration

Entry	Reactant	Reactant Structure	Dehydration Product	References
1	Lactic acid	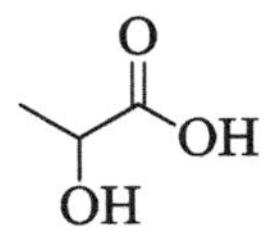	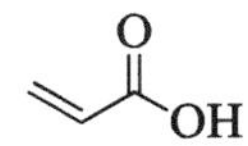 acrylic acid	[56]
2	Glutamic acid	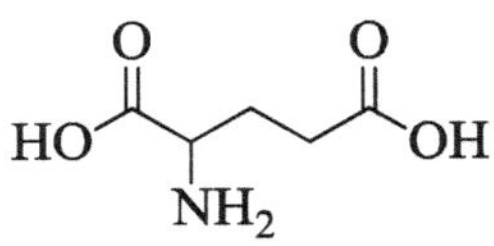	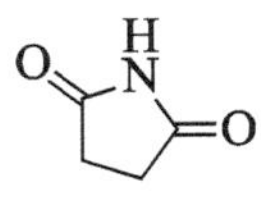 succinimide	[54]
3	4-Hydroxy-6-methyltetrahydro-2-pyrone	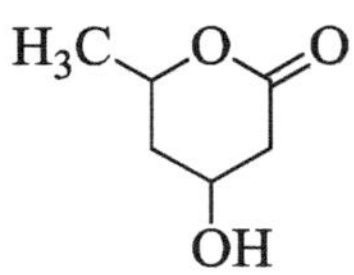	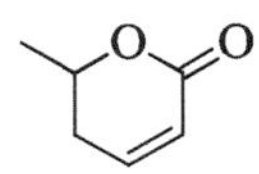 parasorbic acid	[19]
4	Glycerol	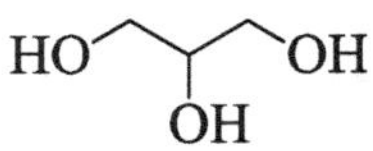	1 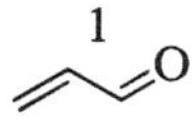 acrolein 2 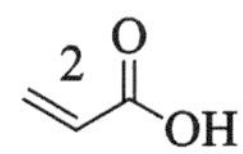 acrylic acid	[57]
5	Sorbitol	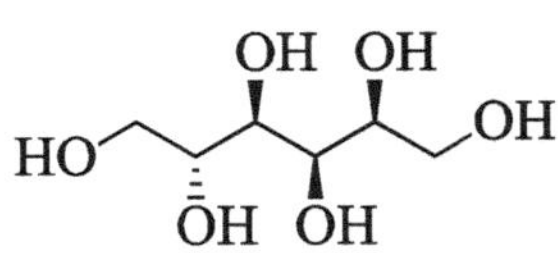	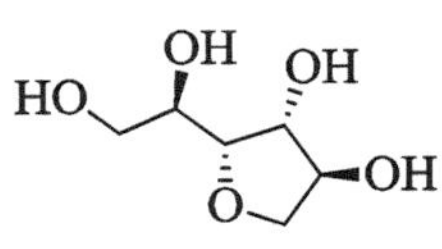 sorbitan	[57]

Table 3: Biorenewable Platform Molecules and Products Obtained on Their Dehydration—cont'd

Entry	Reactant	Reactant Structure	Dehydration Product	References
6	Sorbitan	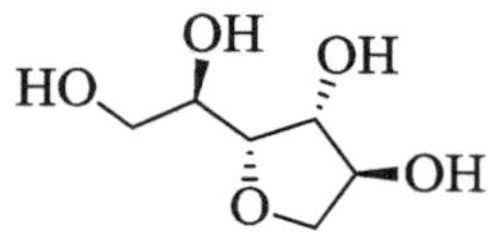	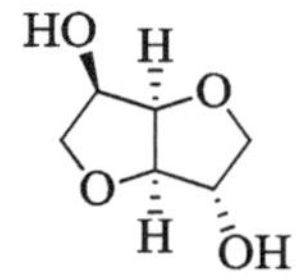 isosorbide	[57]
7	Fructose	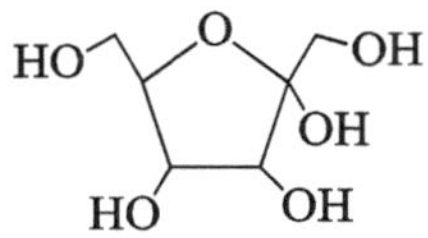	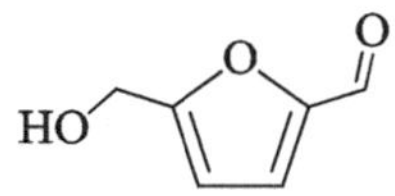 HMF	[58]
8	Xylose	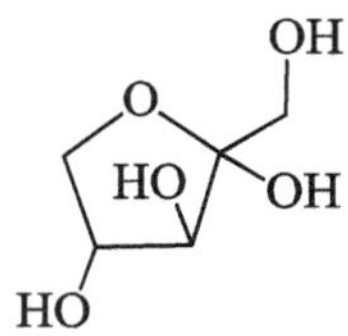	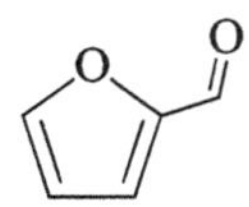 furfural	[59]

glycerol produced from microbial fermentation can be upgraded to acrolein or hydroxyacetone by catalytic dehydration. 4-Hydroxy-6-methyl-tetrahydro-2-pyrone is obtained on hydrogenation of TAL, which on catalytic dehydration yields parasorbic acid [20]. The biogenic impurities present in the fermentation media, which deactivate the hydrogenation reaction, are therefore likely to inhibit their subsequent dehydration.

Hydrogenolysis comprises a C—X (X can be C, O, N, etc.) bond cleavage reaction. Selective C—O bond cleavage helps in oxygen removal from highly oxygenated species into monofunctional species such as carboxylic acids and alcohols. The C—O bond cleavage can be assisted in the presence of hydrogen in a hydrodeoxygenation reaction where oxygen is eliminated from the reacting molecule in the form of water by the consumption of one molecule of hydrogen for each oxygen atom. Hydrodeoxygenation has shown wide applications in chemical catalysis, whenever several hydroxyl groups are present in the platform molecule, which needs to be removed. Furfural and HMF can undergo hydrogenolysis into methyl furan and 2,5-dimethyl furan [60,61], which can be used as fuel additives owing to their high energy density (28.5 and 29.3 MJ/L, respectively) and high octane rating as shown in Figure 3 [62]. However, it remains unclear as to how and where hydrogenolysis may be applied to integrated processing. Nevertheless, hydrogenolysis is carried out on supported metal catalysts (Pt, Pd, Re, Rh, Zn, Fe, Ni, Cu, etc.) [60], some of which have been proven to show adverse effects with biogenic impurities, which need to be further explored to design a novel reaction.

Catalytic decarbonylation and decarboxylation are widely used for selective removal of one or two oxygen atoms; however, one carbon atom is lost. Decarbonylation of several bio-derived carboxylic acids leads to the formation of alkyl acrylates, styrene and acrylonitrile from mono-alkyl succinates, hydrocinnamic acid and 3-cyanopropanoic acid, respectively, as shown in Figure 4(i—iii) [63]. HMF has recently been decarbonylated to

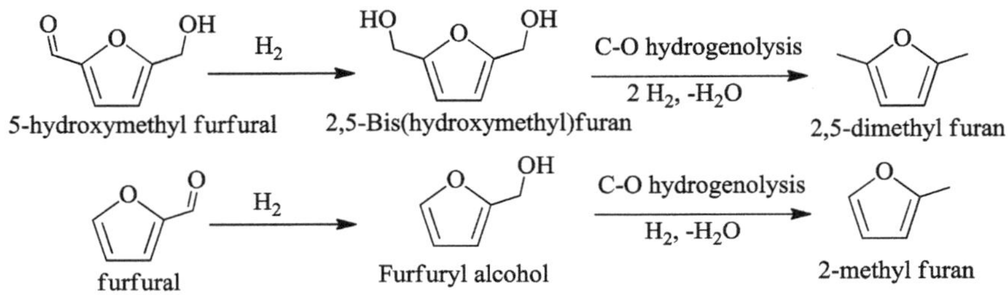

Figure 3
Products of hydrogenolysis reaction.

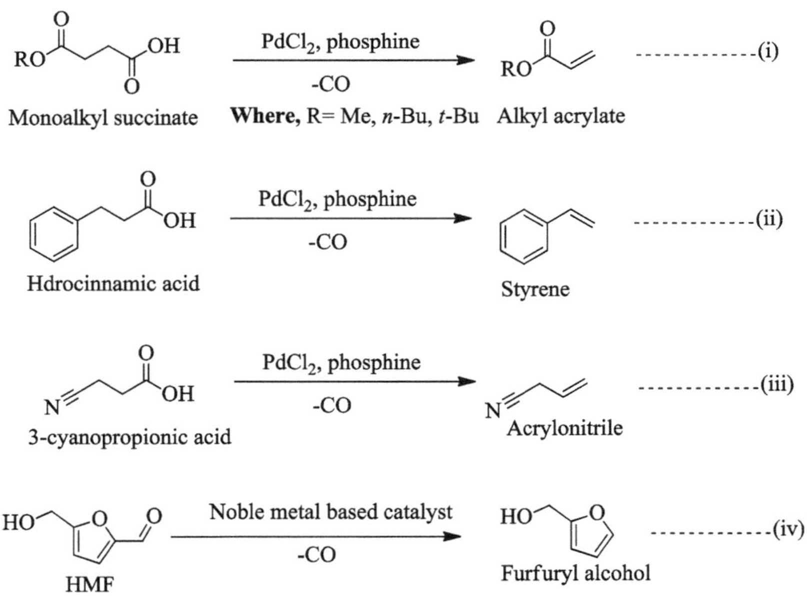

Figure 4
Reactions involving decarbonylation of biomass-derived molecules.

produce furfuryl alcohol on a Pd catalyst (Figure 4(iv)) [64]. We expect similar interactions with biogenic impurities on the Pd catalyst as observed in the case of hydrogenation.

Decarboxylation reactions have been more direct and important in upgrading bio-based chemicals. For example, glutamic acid is decarboxylated into acrylonitrile and succinonitrile via the formation of an intermediate (3-cyanopropionic acid) (Figure 5(i)) [65]. In saturated or unsaturated lactones, decarboxylation is combined with ring-opening, which yields the production of linear chain hydrocarbons. For example, on catalytic ring-opening and decarboxylation, γ-valerolactone yields to butene (Figure 5(ii)), which can further be converted into fuels and chemicals [66]. The reaction is carried out in the presence of a solid acid catalyst at elevated temperatures (>350 °C) and pressure (~36 atm) conditions [66]. Interestingly, for unsaturated lactones such as TAL, the ring-opening and decarboxylation reaction needed to produce 2,4-pentanedione proceeds at low temperatures (<100 °C) and under autogenous pressure conditions. The reaction may not even require an acid catalyst and the aqueous environment is favourable in yielding high conversions (>90%) at relatively low temperatures (<100 °C). Such reactions are expected to proceed in the presence of biogenic impurities; however, the interaction may lead to several unknown products, which remain to be studied. The ring-opening and decarboxylation reaction can be combined with a hydrogenation step to yield a larger array of products. For example, TAL on partial hydrogenation yields to 5,6-dihydro-4-hydroxy-6-methyl-2-pyrone (DHHMP) as shown in Figure 5(iii). DHHMP ring-opens and decarboxylates readily in water without the requirement of a catalyst at temperatures as low as 50 °C. However, addition of an acid catalyst such as Amberlyst may help the reaction to derive higher conversions [67,68]. Biogenic impurities, therefore, will have a similar role to play in these reactions.

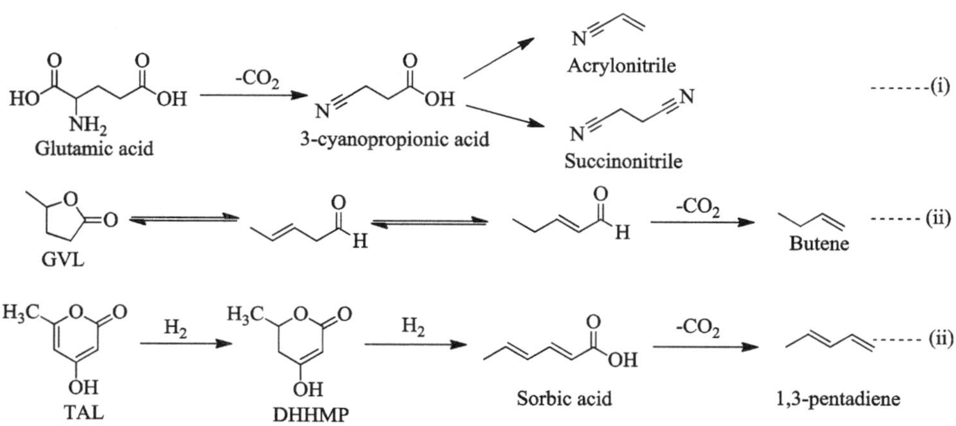

Figure 5
Reactions involving decarboxylation of biomass-derived molecules.

5. Conclusions

Biomass or derived aqueous sugars can be converted into useful products by using a biocatalytic fermentation process. Recent advances in the development of molecular biology tools for metabolic pathway engineering has raised the prospects of achieving higher yield of a desired product for economic commercial-scale production. The products derived from biomass via a biocatalytic process can be further valorized into chemicals and fuels by chemocatalytic processing. This is termed integrated processing of biomass, which has shown significant potential for the overall development of a future biorefinery. For successful integrated processing, a fundamental problem needs to be addressed, which is related to the impurities present in fermentation media, observed to interact with the catalyst surface and resulting in catalyst deactivation. While impurities containing sulphur molecules are expected to interact with the metal catalyst surface with strong irreversible binding and complete inhibition, non-sulphur impurities are expected to show moderate and reversible inhibition. Protein molecules may clog the catalyst pores leading to total deactivation by disallowing mass transfer. Novel strategies, such as PVA coating on the metal catalyst, are likely to improve catalyst activity, which needs to be explored in detail. Electrodialytic purification of the fermented product could form an interesting option owing to its low cost, efficient transport and relatively easier integration to the fermentation and catalytic processes. The concepts introduced in the literature on the inhibition of metal catalyzed hydrogenation reactions by biogenic impurities can be extended to other catalytic reactions such as dehydration, hydrogenolysis, decarbonylation and decarboxylation. These reactions are commonly applied with an overall goal of deoxygenation of platform molecules to commodity chemicals and fuels. Depending upon the catalyst used and the combination of these catalytic reactions, biogenic impurities are expected to play an important role in determining the catalyst stability, activity and overall yield of the product.

References

[1] Reed D. Impact of climate change on biodiversity. In: Chen W-Y, Seiner J, Suzuki T, Lackner M, editors. Handbook of climate change mitigation SE - 15. Springer US; 2012. p. 505–30.
[2] Chheda JN, Huber GW, Dumesic JA. Liquid-phase catalytic processing of biomass-derived oxygenated hydrocarbons to fuels and chemicals. Angew Chem January 2007;46(38):7164–83.
[3] Dutta S, De S, Alam MI, Abu-Omar MM, Saha B. Direct conversion of cellulose and lignocellulosic biomass into chemicals and biofuel with metal chloride catalysts. J Catal 2012;288:8–15.
[4] Imteyaz Alam M, De S, Dutta S, Saha B. Solid-acid and ionic-liquid catalyzed one-pot transformation of biorenewable substrates into a platform chemical and a promising biofuel. RSC Adv 2012;2(17):6890.
[5] Sun Y, Cheng J. Hydrolysis of lignocellulosic materials for ethanol production: a review. Bioresour Technol May 2002;83(1):1–11.
[6] sun J, Changiun Liu YW, Wang H, Karim AM. Catalytic fast pyrolysis of lignocellulosic biomass. Chem Soc Rev 2014;43(22):7594–623.

[7] Kumar P, Barrett DM, Delwiche MJ, Stroeve P. Methods for pretreatment of lignocellulosic biomass for efficient hydrolysis and biofuel production. Ind Eng Chem Res April 2009;48(8):3713–29.
[8] Murat Sen S, Henao CA, Braden DJ, Dumesic JA, Maravelias CT. Catalytic conversion of lignocellulosic biomass to fuels: process development and technoeconomic evaluation. Chem Eng Sci January 2012;67(1):57–67.
[9] Li GC, Anastas P, Gallezot P. Conversion of biomass to selected chemical products w. Chem Soc Rev 2012;41:1538–58.
[10] Straathof AJJ. Transformation of biomass into commodity chemicals using enzymes or cells. Chem Rev 2014;114:1871–908.
[11] Lee Y-C, Dutta S, Wu KC-W. Integrated, cascading enzyme-/chemocatalytic cellulose conversion using catalysts based on mesoporous silica nanoparticles. ChemSusChem 2014;7(1):3241–6.
[12] Leresche JE, Meyer H-P. Chemocatalysis and biocatalysis (biotransformation): some thoughts of a chemist and of a biotechnologist. Org Process Res Dev May 2006;10(3):572–80.
[13] Woodley JM. New opportunities for biocatalysis: making pharmaceutical processes greener. Trends Biotechnol June 2008;26(6):321–7.
[14] Chia M, Schwartz TJ, Shanks BH, Dumesic JA. Triacetic acid lactone as a potential biorenewable platform chemical. Green Chem 2012;14(7):1850–3.
[15] Schwartz TJ, Johnson RL, Cardenas J, Okerlund A, Da Silva NA, Schmidt-Rohr K, et al. Engineering catalyst microenvironments for metal-catalyzed hydrogenation of biologically derived platform chemicals. Angew Chem Int Ed Engl November 2014;53(47):12718–22.
[16] Zhang Z, Jackson JE, Miller DJ. Effect of biogenic fermentation impurities on lactic acid hydrogenation to propylene glycol. Bioresour Technol 2008;99:5873–80.
[17] Neurock M. Engineering molecular transformations for sustainable energy conversion. Ind Eng Chem Res 2010;49:10183–99.
[18] Pfeifer BA, Khosla C. Biosynthesis of polyketides in heterologous hosts. Microbiol Mol Biol Rev 2001;65(1):106–18.
[19] Chia M, Haider MA, Pollock G, Kraus GA, Neurock M, Dumesic JA. Mechanistic insights into ring-opening and decarboxylation of 2-pyrones in liquid water and tetrahydrofuran. J Am Chem Soc March 2013;135(15):5699–708.
[20] Chia M, Schwartz TJ, Shanks BH, Dumesic JA. Triacetic acid lactone as a potential biorenewable platform chemical. Green Chem 2012;14:1850.
[21] Yu D, Xu F, Zeng J, Zhan J. Type III polyketide synthases in natural product biosynthesis. IUBMB Life April 2012;64(4):285–95.
[22] Nikolau BJ, Perera MADN, Brachova L, Shanks B. Platform biochemicals for a biorenewable chemical industry. Plant J 2008;54:536–45.
[23] Daniel RM, Danson MJ. Temperature and the catalytic activity of enzymes: a fresh understanding. FEBS Lett 2013;587(17):2738–43.
[24] Greev H-H. Ullmann's encyclopedia of industrial chemistry. Wiley-VCH Verlag GmbH & Co. KGaA; 2000. p. 1–25.
[25] Lindberg P, Park S, Melis A. Engineering a platform for photosynthetic isoprene production in cyanobacteria, using synechocystis as the model organism. Metab Eng January 2010;12(1):70–9.
[26] Yang J, Zhao G, Sun Y, Zheng Y, Jiang X, Liu W, et al. Bio-isoprene production using exogenous MVA pathway and isoprene synthase in *Escherichia coli*. Bioresour Technol January 2012;104:642–7.
[27] Yang J, Xian M, Su S, Zhao G, Nie Q, Jiang X, et al. Enhancing production of bio-isoprene using hybrid MVA pathway and isoprene synthase in *E. coli*. PLoS One April 2012;7(4):e33509.
[28] Bozell JJ, Petersen GR. Technology development for the production of biobased products from biorefinery carbohydrates – the US Department of Energy's 'Top 10' revisited. Green Chem 2010;12(4):539.
[29] Hofvendahl K, Hahn-Hägerdal B. Factors affecting the fermentative lactic acid production from renewable resources. Enzyme Microb Technol 2000;26(2):87–107.

[30] Wee Y, Kim J, Ryu H. Biotechnological production of lactic acid and its recent applications. Food Technol Biotechnol 2006;44(2):163–72.
[31] Gruber PR. Commodity polymers from renewable resources: polylactic acid. In: Carbon management: implications for R & D in the chemical sciences and technology. Washington D.C: National Academy Press; 2001. p. 166–84.
[32] Dutta S, Wu L, Mascal M. Efficient, metal-free production of succinic acid by oxidation of biomass-derived levulinic acid with hydrogen peroxide. Green Chem 2015;17(4):2335–8.
[33] Delhomme C, Weuster-Botz D, Kühn FE. Succinic acid from renewable resources as a C4 building-block chemical – a review of the catalytic possibilities in aqueous media. Green Chem 2009;11:13.
[34] Cheng K-K, Zhao X-B, Zeng J, Zhang J-A. Biotechnological production of succinic acid: current state and perspectives. Biofuels Bioprod Biorefin 2012;6:302–18.
[35] Choi S, Woo C, Ho J, Yup S. Biorefineries for the production of top building block chemicals and their derivatives. Metab Eng 2015;28:223–39.
[36] Erickson B, Nelson, Winters P. Perspective on opportunities in industrial biotechnology in renewable chemicals. Biotechnol J 2012;7:176–85.
[37] Kumar V, Ashok S, Park S. Recent advances in biological production of 3-hydroxypropionic acid. Biotechnol Adv 2013;31(6):945–61.
[38] Fernando S, Adhikari S, Chandrapal C, Murali N. Biorefineries: current status, challenges, and future direction. Energy Fuels 2006;20:1727–37.
[39] Johson DT, Taconi KA. The glycerin glut: options for the value -added conversion of crude glycerol resulting from biodiesel production. Environ Prog 2007;26(4):338–48.
[40] Okamoto S, Chin T, Nagata K, Takahashi T, Ohara H, Aso Y. Production of itaconic acid in *Escherichia coli* expressing recombinant α-amylase using starch as substrate. J Biosci Bioeng November 2014;119(5):548–53.
[41] Okabe M, Lies D, Kanamasa S, Park EY. Biotechnological production of itaconic acid and its biosynthesis in *Aspergillus terreus*. Appl Microbiol Biotechnol 2009;84:597–606.
[42] Kumar R, Vikramachakravarthi D, Pal P. Production and purification of glutamic acid: a critical review towards process intensification. Chem Eng Process Process Intensif July 2014;81:59–71.
[43] Datta R, Henry M. Lactic acid: recent advances in products, processes and technologies – a review. J Chem Technol Biotechnol 2006;81:1119–29.
[44] Bechthold I, Bretz K, Kabasci S, Kopitzky R, Springer A. Succinic acid: a new platform chemical for biobased polymers from renewable resources. Chem Eng Technol 2008;31(5):647–54.
[45] Sauer M, Porro D, Mattanovich D, Branduardi P. Microbial production of organic acids: expanding the markets. Trends Biotechnol 2008;26(2):100–8.
[46] Rathnasingh C, Raj SM, Jo JE, Park S. Development and evaluation of efficient recombinant *Escherichia coli* strains for the production of 3-hydroxypropionic acid from glycerol. Biotechnol Bioeng 2009;104(4):729–39.
[47] Tuck CO, Pérez E, Horváth IT, Sheldon RA, Poliakoff M. Valorization of biomass: deriving more value from waste. Science August 2012;337(6095):695–9.
[48] Xie D, Shao Z, Achkar J, Zha W, Frost JW, Zhao H. Microbial synthesis of triacetic acid lactone. Biotechnol Bioeng 2006;93(4):727–36.
[49] Schwartz TJ, Brentzel ZJ, Dumesic JA. Inhibition of metal hydrogenation catalysts by biogenic impurities. Catal Lett 2015;145:15–22.
[50] Ferri D, Burgi T. An in situ attenuated total reflection infrared study of a chiral catalytic solid-liquid interface: cinchonidine adsorption on Pt. J Am Chem Soc 2001;123(25):12074–84.
[51] Wilson YM, Durrenberger M, Nogueira ES, Ward TR. Neutralizing the detrimental effect of glutathione on precious metal catalysts. J Am Chem Soc 2014;136(136):8928–32.
[52] Poizat M, Arends IWCE, Hollmann F. On the nature of mutual inactivation between $[Cp^*Rh(bpy)(H_2O)]^{2+}$ and enzymes – analysis and potential remedies. J Mol Catal B Enzyme May 2010;63(3–4):149–56.

[53] Gyo Lee E, Moon SH, Keun Chang Y, Yoo IK, Nam Chang H. Lactic acid recovery using two-stage electrodialysis and its modelling. J Memb Sci 1998;145:53–66.

[54] Eng J, Zhang Q, Pan T, Xu Q, Guo Q, Fu Y. Synthesis of biobased succinimide from glutamic acid via silver catalyzed decarboxylation. RSC Adv 2014;4:27541–4.

[55] Yan N, Dyson PJ. Transformation of biomass via the selective hydrogenolysis of CO bonds by nanoscale metal catalysts. Curr Opin Chem Eng 2013;2(2):178–83.

[56] Carlos Serrano-Ruiz J, Dumesic JA. Catalytic upgrading of lactic acid to fuels and chemicals by dehydration/hydrogenation and C-C coupling reactions. Green Chem 2009;11(8):1101–4.

[57] Sheldon RA. Green and sustainable manufacture of chemicals from biomass: state of the art. Green Chem 2014;16(3):950–63.

[58] van Putten R-J, van der Waal JC, de Jong E, Rasrendra CB, Heeres HJ, de Vries JG. Hydroxymethylfurfural, a versatile platform chemical made from renewable resources. Chem Rev February 2013;113(3):1499–597.

[59] Dutta S, De S, Saha B, Alam MI. Advances in conversion of hemicellulosic biomass to furfural and upgrading to biofuels. Catal Sci Technol 2012;2(10):2025.

[60] Scholz D, Aellig C, Hermans I. Catalytic transfer hydrogenation/hydrogenolysis for reductive upgrading of furfural and 5-(hydroxymethyl)furfural. ChemSusChem 2014;7:268–75.

[61] Saha B, Bohn CM, Abu-Omar MM. Zinc-assisted hydrodeoxygenation of biomass-derived 5-hydroxymethylfurfural to 2,5-dimethylfuran. ChemSusChem 2014;7(ii):3095–101.

[62] Saha B, Bohre A, Dutta S, Abu-Omar MM. Upgrading furfurals to drop-in biofuels: an overview. ACS Sustainable Chem Eng. Just accepted manuscript. http://dx.doi.org/10.1021/acssuschemeng.5b00271. <http://pubs.acs.org/doi/abs/10.1021/acssuschemeng.5b00271>.

[63] Miranda MO, Pietrangelo A, Hillmyer MA, Tolman WB. Catalytic decarbonylation of biomass-derived carboxylic acids as efficient route to commodity monomers. Green Chem 2012;14:490–4.

[64] Mitra J, Zhou X, Rauchfuss T. Pd/C-catalyzed reactions of HMF: decarbonylation, hydrogenation, and hydrogenolysis. Green Chem 2015;17:307–13.

[65] But A, Nôtre L, Scott EL, Wever R, Sanders JPM. Selective oxidative decarboxylation of amino acids to produce industrially relevant nitriles by vanadium chloroperoxidase. ChemSusChem 2012;5:1199–202.

[66] Bond JQ, Martin Alonso D, West RM, Dumesic JA. Y-valerolactone ring-opening and decarboxylation over SiO_2/Al_2O_3 in the presence of water. Langmuir 2010;26(7):16291–8.

[67] Wang D, Hakim SH, Alonso DM, Dumesic JA. A highly selective route to linear alpha olefins from biomass-derived lactones and unsaturated acids. Chem Commun (Camb) 2013;49:7040–2.

[68] Schwartz TJ, Johnson RL, Cardenas J, Okerlund A, Da Silva NA, Schmidt-Rohr K, et al. Engineering catalyst microenvironments for metal-catalyzed hydrogenation of biologically derived platform chemicals. Angew Chem Int Ed 2014;53:12718–22.

Catalytic Coal Gasification

Xin Huang[1,2], Fan Zhang[1,2], Maohong Fan[2], Yonggang Wang[1]
[1]School of Chemical and Environmental Engineering, China University of Mining and Technology, Beijing, PR China; [2]Department of Chemical and Petroleum Engineering, University of Wyoming, Laramie, WY, USA

1. Introduction

Coal is one of the most important energy resources in human history due to its abundance, wide distribution and low cost (Figure 1). With a higher ratio of carbon to hydrogen than for petroleum, coal utilization today is facing greater challenges because of its great carbon dioxide emissions [1−3]. During the Asia-Pacific Economic Cooperation Summit in 2014, United States and China − the two largest coal consumers and carbon dioxide emitters around the globe, signed the US−China Joint Announcement on Climate Change, in which the US government promised that greenhouse gas emissions will be one-quarter lower in

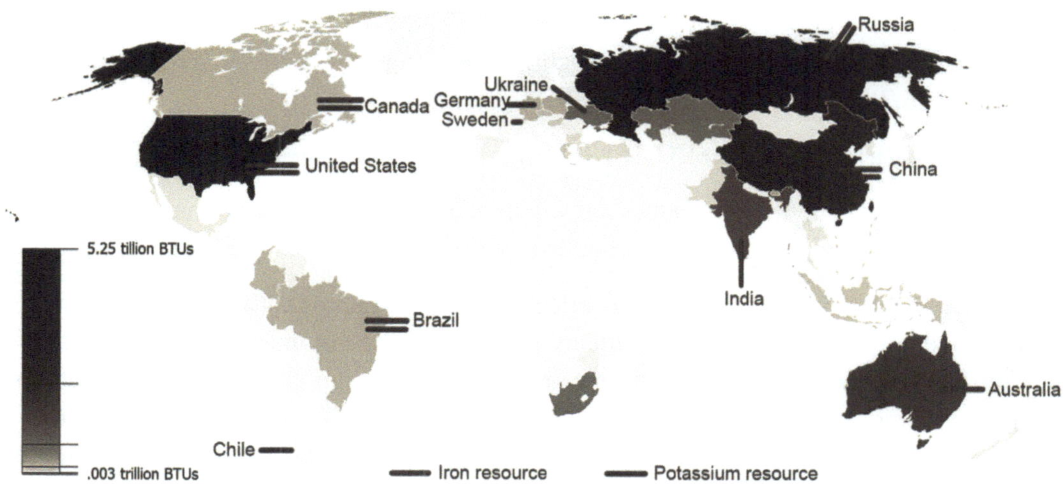

Figure 1
Areas of major coal, iron and potassium resources in the world. *Mineral Commodity Summaries 2014, Coal Reserves in BTUs 2007.*

Sustainable Catalytic Processes. http://dx.doi.org/10.1016/B978-0-444-59567-6.00007-8

2025 than they were back in 2005, while the Chinese government will try to reduce CO_2 emissions per unit of the gross domestic product by 40–45% by 2020, compared with the level of 2005. Likewise, other countries, including developed and developing countries, are also under the pressure of United Nations Framework Convention on Climate Change which aims to tackle the challenge of greenhouse gases by intergovernmental cooperation. Therefore, the utilization of fossil fuels especially coal more efficiently and cleanly is now a difficult task for human beings in the twenty-first century.

Integrated gasification combined cycle (IGCC) with carbon capture and storage (CCS) is a promising technology to deal with the main environmental challenge of coal utilization, as CCS could potentially realize a carbon capture rate of at least 90%. Compared to 740–800 kg/MWh without CCS, IGCC power generation technology with CCS has CO_2 emissions specified in the range of 76–87 kg/MWh [4]. At the heart of IGCC are gasification-based systems, including gasifiers, gasifying agents, and carbon-based feedstocks such as coal, petroleum coke, biomass and municipal hazardous wastes [5]. Coal gasifiers are typically fed with coal, steam, CO_2 and/or oxygen or air, in which coal undergoes a pyrolysis process and then precipitates gasification reactions with steam or CO_2 in the form of

$$C(s) + H_2O(g) \rightarrow CO(g) + H_2(g) \quad \Delta H^\circ 298K = +131.46\ kJ/mol \tag{R1}$$

$$C(s) + CO_2(g) \rightarrow 2CO(g) \quad \Delta H^\circ 298K = +172.67\ kJ/mol \tag{R2}$$

The heat requirements for R1 and R2 are provided mainly by the exothermic combustion reaction occurring in the gasifier, as shown in R3:

$$C(s) + O_2(g) \rightarrow CO_2(g) \quad \Delta H^\circ 298K = -393.98\ kJ/mol \tag{R3}$$

In some cases, a water-gas shift (WGS) reaction will also occur in the coal gasifier, as shown in R4 [6,7]:

$$CO(g) + H_2O(g) \rightarrow CO_2(g) + H_2(g) \quad \Delta H^\circ 298K = -41.2\ kJ/mol. \tag{R4}$$

The WGS reaction benefits H_2 production, which is an important part of coal utilization. About 30% of industrial hydrogen is produced by the gasification of coal [8].

Due to its high gasification rates, coal gasification technologies adopting entrained flow gasifiers have attracted the attention of many researchers and commercial entities such as Shell and Texaco. However, entrained flow gasifiers require high gasification temperatures provided by combustion reactions (e.g. 1500–1900 °C for Shell and 1350–1500 °C for Texaco) (R3). In other words, more carbon is consumed and more CO_2 is produced in the combustion reaction when gasifying at a higher temperature. A further disadvantage of high gasification temperatures is that, as an exothermic reaction, WGS is restrained with increased gasification temperatures, leading to a reduction in H_2 production. Additionally, high-temperature gasification requires high construction and maintenance costs. As such,

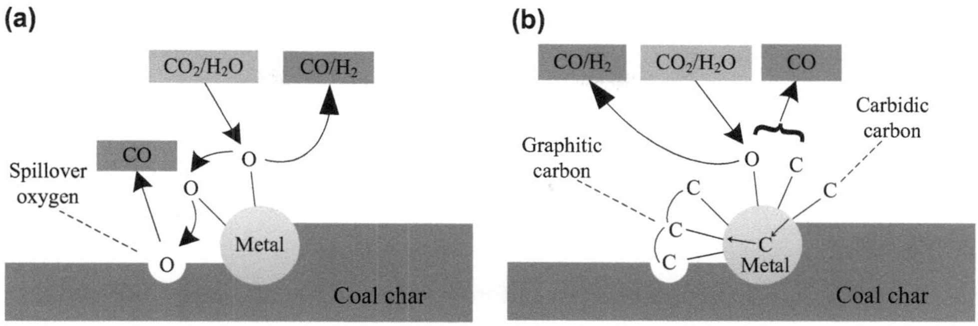

Figure 2
Mechanism of metal-catalysed carbon gasification: (a) spillover mechanism and (b) carbon dissolution mechanism [9].

the challenge in coal gasification lies in reducing gasification temperature while maintaining gasification rates at high levels.

Catalytic coal gasification is an important technique used to improve the gasification process. Many research groups have investigated the catalytic gasification of coal, focussing their attention on the salts and oxides of alkali and alkaline earth metals (AAEM), as well as transitional metals, due to their efficiency, availability and low cost. Figure 1 shows the wide distribution of iron and potassium resources in the world, not to mention sodium resources; these metals are widely available to coal producers to realize catalytic coal gasification. The efficiency of AAEM and transitional metals on coal gasification has been widely proved, with two mechanisms of metal-catalysed carbon gasification proposed [9], namely, the spillover mechanism and the carbon dissolution mechanism (Figure 2).

Many researchers have proposed specific catalytic mechanisms for different kinds of catalysts. However, metal catalysts with different properties have varying effects on the kinetics and gas products of coal gasification. Therefore, the objective of this chapter is to review the performance of different types of catalysts on coal gasification. The compounds of alkali metals, alkali earth metals, transitional metals and their composites used as catalysts in coal gasification are reported, respectively, and for sustainable processes, catalyst recovery is also addressed in this chapter.

2. Catalysts for Coal Gasification

2.1 Alkali Metal Compounds

The alkali metal series are among the most active and inexpensive catalysts in coal gasification. Results from Quyn et al. [10] showed that sodium chloride, a naturally occurring material, was an effective catalyst for the gasification of Loy Yang brown coal.

The data collected by Shinya Yoshida et al. [11] demonstrated that the rate of CO_2 gasification using Na_2CO_3 was 3.3 times higher than without the use of Na_2CO_3. Ye et al. [12] found that Na and K could considerably accelerate CO_2 gasification of a South Australian low-rank coal at atmospheric pressure and temperatures between 714 °C and 892 °C. Popa et al. [13] found that Na_2CO_3 was active during both the pyrolysis and gasification steps, and could considerably decrease the activation energy of the gasification step.

Several catalytic cycles have been proposed for steam and CO_2 gasification of carbon, represented by the following processes [14]:

$$M + H_2O \leftrightarrow M(O) + H_2 \tag{R5}$$
$$M(O) + C \leftrightarrow M + C(O) \tag{R6}$$
$$M(O) + C(O) \leftrightarrow M + CO_2 \tag{R7}$$
$$C(O) \leftrightarrow CO \tag{R8}$$

Suzuki et al. [15] studied the CO_2-pulsed gasification of Yallourn coal char loaded with Na_2CO_3, and proposed the following redox cycles as the CO_2 gasification mechanism:

$$2Na + CO_2 \rightarrow Na_2O + CO \tag{R9}$$
$$Na_2O + C \rightarrow 2Na + CO \tag{R10}$$

However, Alam et al. [16] suggested that the effects of K_2CO_3 and Na_2CO_3 may be explained on the basis of the vapour cycle mechanism, represented by the following reactions:

$$\frac{1}{2}M_2CO_3 + C \rightarrow M(g) + \frac{3}{2}CO \tag{R11}$$
$$M(g) + CO_2 \rightarrow \frac{1}{2}M_2CO_3 + \frac{1}{2}CO \tag{R12}$$

where M is either K or Na. Bulk phase cycles were also proposed by some researchers [17]:

$$K_2CO_3 + C \leftrightarrow 2K + CO + CO_2 \tag{R13}$$
$$2K + CO_2 \rightarrow K_2O + CO \tag{R14}$$
$$K_2O + CO_2 \rightarrow K_2CO_3 \tag{R15}$$

The catalyst takes the oxygen from the gas (i.e. CO_2, see R15) and transfers it to the surface where it reacts with the carbon to form CO. Namkung [7] used X-ray diffraction (XRD) to detect the change in the crystalline structure of K_2CO_3 during coal-steam gasification at 800 °C and found that, after 20 min of reaction, K_2CO_3 was transformed to $KHCO_3$ after reacting with CO_2 and H_2O; K_2SO_4 was detected at 60 min, due to the chemical reaction between potassium salts and the sulphur compounds in coal.

Others have embodied the redox cycle in localized sites where the catalyst and carbon interact [18]:

$$KC(site) + CO_2 \leftrightarrow KC - O + CO \quad (R16)$$

$$KC - O \rightarrow CO + K(s) \quad (R17)$$

$$K(s) + C \rightarrow KC \quad (R18)$$

In this scheme, the KC grouping represents a generalized site with proper potassium–carbon contact. The third step (R18) symbolizes site regeneration, which requires mobility on the part of the potassium. Despite numerous studies in this area, the interaction between catalyst, carbon and a given type of reactive surface intermediate is still debatable. Sam and Shadman [19,20] suggested the formation of possible carboxylic ($-CO_2K$), phenolic ($-COK$) or even completely reduced potassium–carbon ($-CK$) groups. Based on their results, Kopyscinski et al. [6] proposed a reaction scheme including $-COK$ and $-CK$ surface intermediates, as shown in Figure 3.

In most research, the order of catalytic effects of alkali metals for coal gasification is Li > Na > K [21,22], Rb and Cs are not often considered, as they are too costly to be used commercially [23]. Sun et al. [24] conducted catalytic gasification with CO_2 on maceral chars at 1173 K and 0.1 MPa, result showed that the catalytic effects of the loading catalysts on carbon conversion were in the order of Na > K. On the other hand, the catalytic effects of alkali salts also depend on the anion. Lang [25] investigated the anion effect on the alkali-catalysed steam gasification of coal chars, and found that the alkali salts of weak acids were good gasification catalysts, while those of strong acids were poor. Ye et al. [12] also found that NaAc, Na_2CO_3 (the weak acid salt) and NaOH

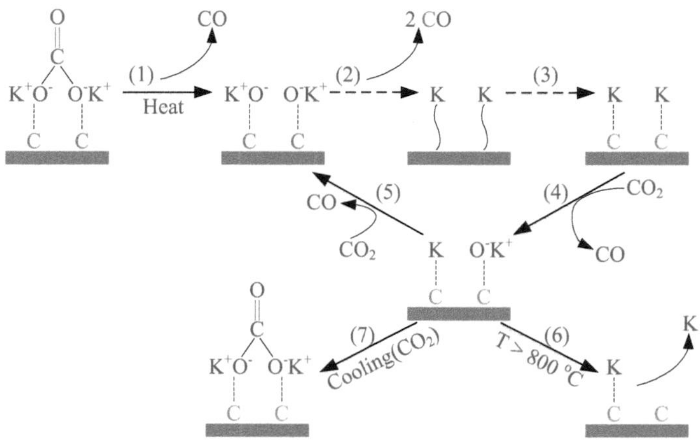

Figure 3

Reaction scheme of K_2CO_3 on ash-free coal heated under N_2 or CO_2 atmosphere [6].

Table 1: Characteristic Parameters of Petroleum Coke Steam Gasification [26].

Catalytic and Non-catalytic	T_{max} (°C)	R_{max} (min^{-1})	C_{max}
Original sample	1000	0.0248	0.314
KCl	882	0.0966	0.745
K_2SO_4	932	0.0867	0.462
K_2CO_3	892	0.0594	0.551
KNO_3	890	0.0572	0.429
KAC	921	0.05	0.463
$FeCl_3$	997	0.0293	0.399
$CaCl_2$	998	0.0284	0.465

T_{max}: The temperature corresponding to the maximum reaction rate.
R_{max}: The maximum reaction rate.
C_{max}: Char conversion corresponding to the maximum reaction rate.

exhibit strong catalytic effects with minor differences among them, while NaCl (the strong acid salt) showed very little catalytic effect. However, Li [26] compared the performance of KCl, K_2CO_3, K_2SO_4, KAC and KNO_3 on petroleum coke steam gasification, and found that the catalytic effects of K-based materials on char gasification could be ordered as KAC < K_2SO_4 < K_2CO_3 < KNO_3 < KCl. The results are listed in Table 1. The possible reason for this is that some anions could compete with carbon materials for alkali cations, thus inhibiting the formation of an alkali–carbon complex (M–O–C).

Variation in the composition of gas products based on the added catalyst is also a method of characterizing catalytic gasification. In steam gasification, part of the CO would participate in the WGS reaction (R4); thus, the composition of the gas product does not obey R1 strictly, as H_2 yield is usually greater than the CO yield. If the added catalyst has an effect on the WGS reaction, the composition of gas products would be changed. Popa et al. [13]. studied the catalytic steam gasification of a Powder River Basin coal using Na_2CO_3, and the results demonstrated that the total amounts of H_2, CO, CO_2 and CH_4 generated during the entire gasification with 5 g of dry ash-free coal were 0.32, 0.11, 0.10 and 0.01 mol, respectively, while correspondingly, those with catalyst were 0.37, 0.06, 0.16 and 0.02 mol. An Exxon catalytic coal gasification plant used 20 wt% K_2CO_3 to catalyse the reaction of coal-steam and methanation for methane at roughly 700 °C and 3.5 Mpa [27]. Wang studied the reaction selectivity of catalytic steam-coal gasification by potassium carbonate and found that hydrogen-rich gas was produced in a temperature range of 700–750 °C. The results are shown in Figure 4.

It has been proved that the alkali metal is easy to evaporate during the gasification process above 800 °C due to its low melting and boiling point, which makes the catalyst more fluid and active among the carbon molecules [6]. Li [29] found that Na was always

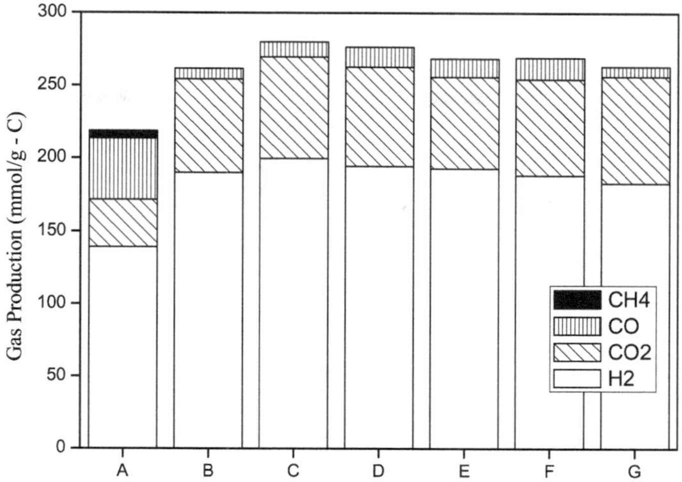

Figure 4
Total production of major gases from typical gasification runs. A: non-catalytic gasification of the 700 °C char at 1000 °C; B: gasification of the 700 °C char with 17.5% K_2CO_3 at 700 °C; C: gasification of the 700 °C char with 17.5% K_2CO_3 at 725 °C; D: gasification of the 700 °C char with 17.5% K_2CO_3 at 750 °C; E: gasification of 500 °C char with 10% K_2CO_3 at 750 °C; F: gasification of the 700 °C char with 10% K_2CO_3 at 750 °C; G: gasification of the 900 °C char with 10% K_2CO_3 at 750 °C [28].

volatilized to a much larger extent than the Mg and Ca under similar pyrolysis conditions. At high temperatures (900–1200 °C), the dramatic volatilization of Na (up to 80%) was accompanied by increases in tar yield during the pyrolysis of the Na-form samples [30], as shown in Figure 5. Kitsuka [31] also proved that forced gas flow through the bed of char particles could diminish the amount of alkali metal during steam gasification at 900 °C, thereby causing completely different changes in the char conversion over time. Quyn et al. [10] suggested that the high mobility of Na allows it to migrate onto the char surface to catalytically active species such as –O–Na and/or Na clusters, thus improving char reactivity. Zhang et al. [32] studied the CO_2 gasification rate for 5 wt% Na-doped coal samples between 750 and 950 °C. The results (Figure 6) may be explained by the high mobility of the sodium, which was believed to rapidly redistribute and form new active sites in the coal matrix during gasification. As the char gasified, the sodium concentration in char increased, thus leading to an increase in active sites and, therefore, the rate of gasification.

2.2 Alkaline Earth Metal Compounds

Alkaline earth elements are also wildly researched for coal catalytic gasification. McKee [33] found that the kinetics of steam graphite gasification was strongly influenced by the

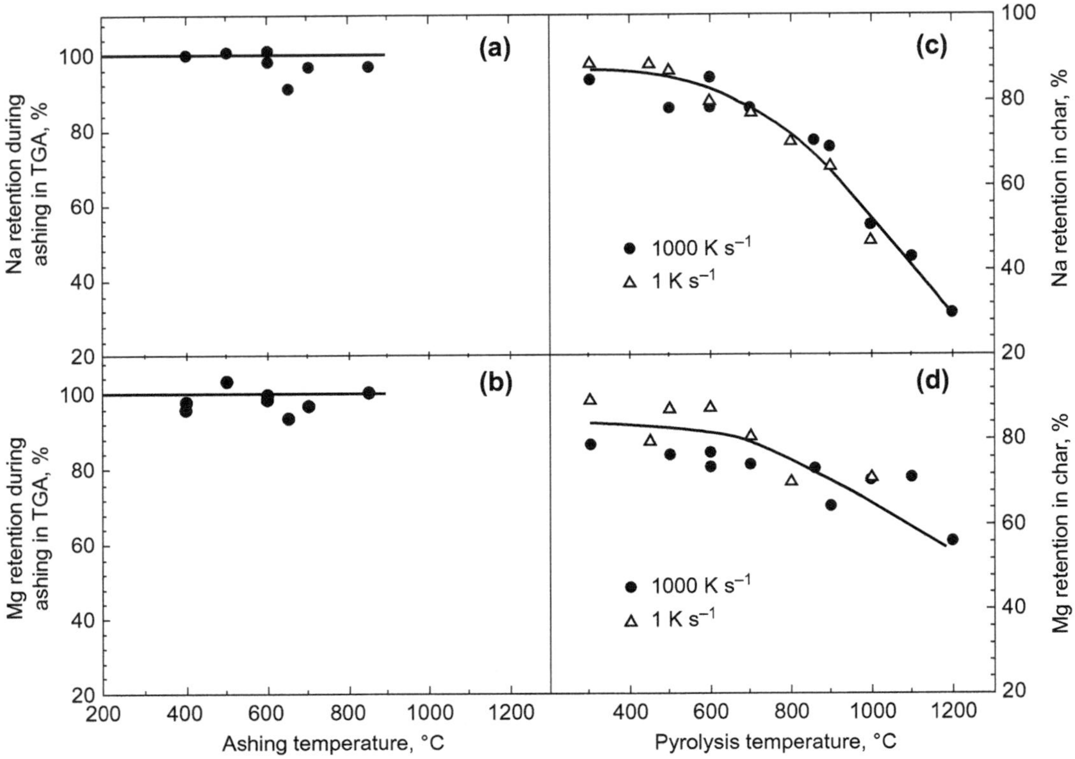

Figure 5

Effects of ashing temperature (a) and (b) in O_2 in a thermogravimetric analyser, and pyrolysis temperature (c) and (d) in He in a wire-mesh reactor on the volatilization of AAEM species from Loy Yang raw coal [30].

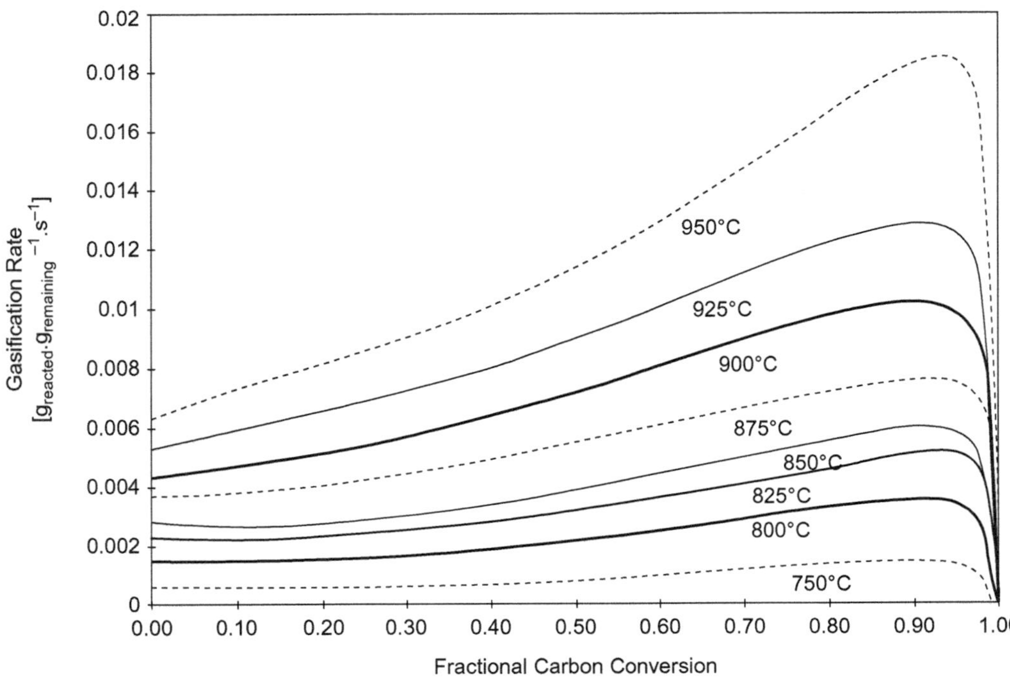

Figure 6

Variation of gasification rate with fractional carbon conversion for the 5 wt% Na-doped coal sample in pure CO_2 at different temperatures between 750 and 950 °C [32].

presence of salts of the Group IIA alkaline earth metals, in which Ba and Cr were the most active catalysts, whereas Ca and Mg were less effective. Kapteijn et al. [34] concluded that, for both steam and CO_2 gasification, the activity of the alkaline earth metal elements increased in the following order: activated carbon $\approx$ Be $\approx$ Mg $<$ Ba $\approx$ Sr $<$ Ca; and the activity of Ca turned out to be equal to that of K at temperatures around 1000 K. Their results are shown in Figure 7. Chen [35] studied five AAEM catalysts in coal gasification with H_2O and CO_2 by measuring the turnover frequencies on the edge carbon sites; their catalytic activities followed the order K $>$ Ba $>$ Ca $>$ Na $>$ Li. Hengel et al. [36] studied catalytic lignite char gasification by exchangeable calcium and magnesium, and found that the gasification of exchanged lignite was strongly catalysed by Ca, while its activity was not affected by the presence of Mg on the char.

Because of its wide availability and good efficiency, only calcium is extensively employed in catalytic coal gasification among the alkaline earth metals. Hattingh [37] tested the catalytic effect of inorganic species within coal ash on the CO_2 gasification of three South African coals at temperatures of 900 and 1000 °C. The results showed that coal reactivity increased with increasing calcite, dolomite and CaO contents more than with other inorganic salts (e.g. iron or titanium salt). Ohtsuka et al. [38] investigated the steam gasification of calcium-loaded Yallourn coal. Their results showed that calcium hydroxide, carbonate, acetate, nitrate and chloride exhibited a similar catalytic effectiveness and could lower reaction temperature by 150 K [38]. The same group also carried out the steam

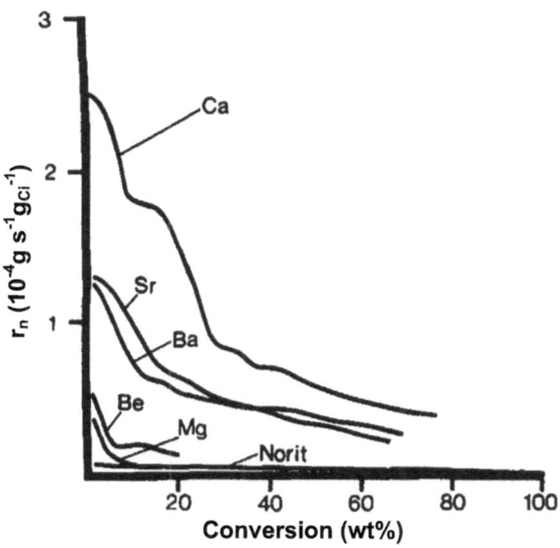

Figure 7
Normalized reaction rate r_n as a function of conversion for isothermal CO_2 gasification of alkaline earth nitrate/Norit RX extra samples [34].

gasification of 16 coals with different ranks and sulphur contents in the presence of $Ca(OH)_2$, and found that the catalyst at 5 wt% Ca promoted the gasification of all the coals and lowered gasification temperatures by 110−150 K.

The reaction mechanism of alkaline metal species in carbon gasification has been widely discussed. Mckee [39] suggested that the catalytic mechanism appears to involve a carbonate−oxide oxidation−reduction cycle, and inferred that the carbonate was the active species that decomposed by the reaction with carbon:

$$MCO_3 + C \rightarrow MO + 2CO \tag{R19}$$

$$MO + CO_2 \rightarrow MCO_3 \tag{R20}$$

Sear et al. [40] explained the catalytic activity of Ca with the formation of a surface peroxide:

$$CaO + CO_2 \rightarrow CaO \cdot O + CO \tag{R21}$$

$$CaO \cdot O + C \rightarrow CaO + C[O] \tag{R22}$$

$$C[O] \rightarrow CO \tag{R23}$$

Using high-temperature X-ray diffraction analysis, Ohtsuka et al. [38] found that interconversion between calcium carbonate and oxide takes place readily in the gasification temperature region, and that calcium carbonate was the predominant species during gasification at 923 K. Chen et al. [35] proposed that the active intermediates are catalysts in the form of particles and clusters, while the C−O−M (where M denotes metal) phenolate groups, although present in abundant amount, are not the active intermediates.

It has been reported that catalytic activity strongly depends on the preparation method of the samples. When ion exchange or impregnation is applied, gasification activity depends upon the catalyst concentration [34]. $Ca(OH)_2$ was proven to be an effective catalytic raw material in the gasification of brown coal, as Ca^{2+} ions in an aqueous solution of $Ca(OH)_2$ can be easily exchanged with the protons of the carboxyl groups present in brown coal, as shown in Figure 8 [38,41]. Ohtsuka [42] studied the difference between the two forms of calcium in affecting the behaviour of coal gasification. Results indicated that the exchanged calcium showed significant catalytic activity on coal gasification, especially for low-rank coals, due to the higher content of oxygen-functional groups, including carbonyl and phenolic hydroxyl as ion-exchangeable sites [42].

Compared with alkali metals, alkaline earth metals showed a much lower mobility during the pyrolysis and gasification of coal. Karimi [43] investigated the distribution of K, Na, Ca and Mg by scanning electron microscopy coupled with energy dispersive X-ray spectroscopy (SEM/EDX) and found that Ca and Mg were present in the coke particles only as relatively large distinct crystals either with little contact with the coke or trapped

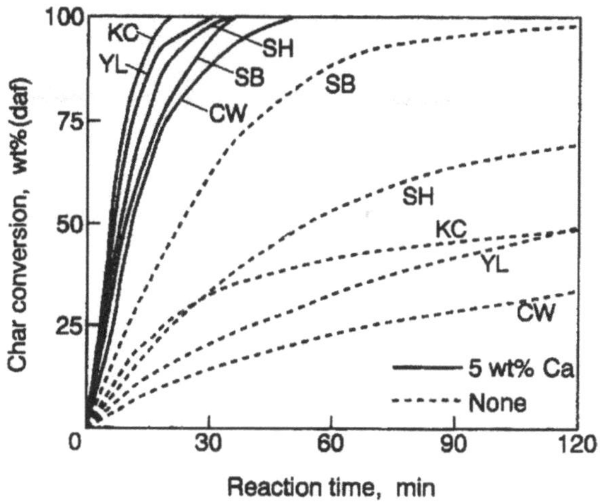

Figure 8
Steam gasification profiles at 973 K for low-rank coals with and without $Ca(OH)_2$ [41].

inside it, while the K and Na were spread relatively uniformly over the coke particles. Due to their different properties, deactivation of the catalytic alkali metals was attributed to interaction with inherent aluminosilicates, while the deactivation of calcium was related to its sintering via crystallite growth [44]. Additionally, high-sulphur coals would result in the deactivation of the calcium catalyst, as the sulphur that evolved during pyrolysis or gasification would react with the calcium [41].

2.3 Transition Metal Compounds

In addition to AAEMs, transition metals are also effective catalysts in coal gasification. Holstein et al. [45] studied carbon gasification with steam and CO_2, and found that the platinum-catalysed reactions were about 100 times faster than noncatalysed reactions. The same group also confirmed the order of catalytic activity for the $C-H_2O$ reaction was Ru > Rh > Ir > Pd > Co > Ni > Fe [46]. Kodama et al. [47] studied the metal oxide-catalysed CO_2 gasification of coal in a fluidized-bed reactor and found that the activity of the metal oxides used followed the order of $In_2O_3 > ZnO > Fe_2O_3$ at high temperatures above 1123 K, while the order changed to $In_2O_3 > Fe_2O_3 > ZnO$ at lower temperatures (Figure 9).

Due to its abundance, low cost and environmental friendliness, iron has been regarded by some researchers as the most practical catalyst among the transition metals in coal gasification [48–51]. Popa et al. [50] studied the steam gasification of a Powder River Basin coal and found that, by using iron catalyst, the apparent activation energy decreased from 89.5 kJ/mol (0 wt% Fe) to 50.6 kJ/mol (3 wt% Fe). Asami [48] studied the effects of

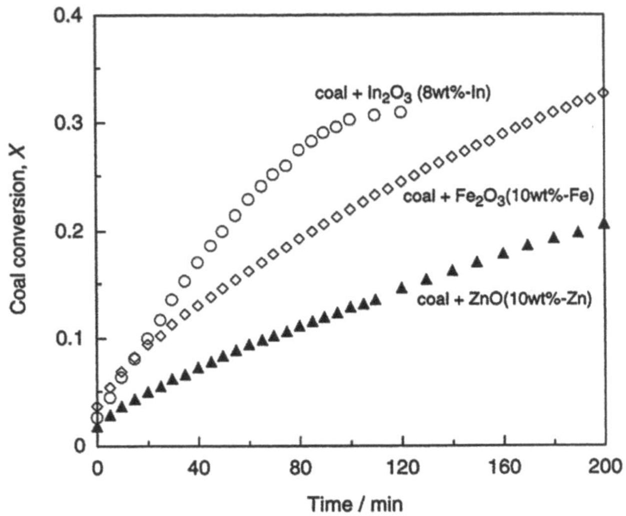

Figure 9
Time variation of coal conversion during coal gasification. The coal bed with a metal oxide catalyst of In_2O_3, ZnO or Fe_2O_3 in an infrared furnace at lower temperatures (1098 K) [47].

iron catalysts on brown coal gasification. The results showed that loading the iron catalyst could lower gasification temperatures by >100 K and increase the gasification rate by a factor of 4.

The mechanism of gasification reaction for particular transition metals has been proposed as [52–54]:

$$Me + H_2O \rightarrow Me(O) + H_2 \quad (R24)$$
$$Me(O) + C \rightarrow C(O) + Me \quad (R25)$$
$$C(O) \rightarrow CO \quad (R26)$$

Me refers to elemental transition metals.

Suzuki et al. [15] studied the temperature-programmed desorption and pulsed-CO_2 gasification of Yallourn coal char loaded with $Fe(NO)_3$, and confirmed the following redox cycles using a $^{13}CO_2$ pulse:

$$Fe_nO_m + CO_2 \rightarrow Fe_nO_{m+1} + CO \quad (R27)$$
$$Fe_nO_{m+1} + C \rightarrow Fe_nO_m + CO \quad (R28)$$

Furimsky et al. [53] depicted the iron-catalysed redox cycle as

$$xFe + CO_2/H_2O \rightarrow Fe_xO + CO/H_2 \quad (R29)$$
$$Fe_xO + C \rightarrow xFe + CO \quad (R30)$$

The chemical form of iron depends upon the gasification temperature and reaction atmosphere. Ohtsuka et al. [55] examined the iron catalyst in situ during carbon gasification by a controlled atmosphere, with high-temperature X-ray diffraction method; Fe_3O_4, FeO, Fe_3C, α-Fe and γ-Fe were identified between 700 and 1000 °C. At 750 °C, the reduction of magnetite by carbon was observed to be a rate-determining step [53]. At low oxygen levels, stable iron forms at 900 °C were metallic iron and wustite (FeO) [51]. At 950 °C, both magnetite and wustite were present, and the amount of wustite increased with decreasing CO_2 concentration. Wustite decomposed into metallic iron and magnetite (Fe_3O_4) below 570 °C on slow cooling [56]. Using temperature programmed desorption spectra and XRD, Ohme et al. [57] determined that, in iron-catalysed gasification, the active species was the highly dispersed iron metal, while the deactivated species were sintered iron and highly oxidized iron.

Popa et al. [50] reported that iron could improve H_2 generation by WGS reaction during steam-coal gasification, as summarized in the following reactions:

$$3FeO + H_2O \rightarrow Fe_3O_4 + H_2 \qquad (R31)$$

$$Fe_3O_4 + CO \rightarrow 3FeO + CO_2 \qquad (R32)$$

$$CO_2 + C \leftrightarrow 2CO \qquad (R33)$$

Their results show that the improvement of H_2 generation was obvious at low gasification temperatures, while it was little pronounced at high temperatures, as listed in Table 2.

2.4 Composite Catalysts

In order to accomplish the synergistic effects and overcome the shortcomings of individual metal catalysts, several gasification studies with favourable results have been performed

Table 2: The Quantities of Gasification Products During Coal Gasification Normalized for the Quantities of Carbon for Samples Treated and Untreated with $FeCO_3$ (3 wt% Fe) [50].

Fe (wt%)	Temperature (°C)	H_2 (mol/mol C)	CH_4 (mol($\times 10^{-2}$)/mol C)	CO (mol($\times 10^{-1}$)/mol C)	CO_2 (mol($\times 10^{-1}$)/mol C)
0	700	1.558	3.739	3.285	6.424
0	750	1.386	3.086	5.040	4.651
0	800	1.336	2.830	6.410	3.305
0	850	1.182	2.286	7.308	2.024
0	900	1.119	1.720	8.746	1.081
3	700	1.623	3.813	3.007	6.584
3	750	1.464	3.307	4.702	4.967
3	800	1.315	2.776	6.145	3.577
3	850	1.191	2.287	7.536	2.230
3	900	1.117	1.329	8.73	1.120

using composite catalysts [28,58–63]. Meng et al. [21] studied the catalytic effect of mixed carbonates containing $Li_2CO_3 + Na_2CO_3$, or $Li_2CO_3 + K_2CO_3$. The results showed that mixed carbonates with a mole ratio of 9:1 (Li:Na/K) had the largest catalytic effect for Pingshuo coal, and a synergistic effect between Li and Na/K appeared at temperatures >1300 K. Jiang et al. [61] investigated the effect of calcium additive on potassium deactivation, as well as the interactive effect between potassium, calcium and char in gasification. It was found that each calcium additive acted as a deterrent to potassium's deactivation, thus promoting catalytic gasification. Hu et al. [62] investigated the synergistic catalysis between calcium species and K_2CO_3 for char gasification and found that each of three calcium species ($Ca(OH)_2$, $Ca(Ac)_2$ or $CaCO_3$) synergistically promoted the catalytic activity of K_2CO_3 for the gasification of all samples, and the degree of promotion was dependent on calcium species and carbon or char samples. Zhang et al. [63] studied the catalytic effects of Na_2CO_3, $Ca(OH)_2$ and their mixture on the gasification of a sub-bituminous coal. Results showed that Ca/Na-loaded coal exhibited the highest reactivity and promoted the WGS reaction, leading to lower yields of heavy tar and soot, and higher H_2 production, as shown in Figure 10.

To some degree, coal ash or biomass ash can be regarded as composite catalysts in coal gasification, as these are mixtures of metal oxides and salts with silicon compounds. Demineralized coal or high-rank coal has lower gasification reactivity than raw coal or brown coal, as they contain fewer primary minerals [12,64]. Brown [65] proved that the addition of switchgrass ash had an eightfold increase on the coal char gasification rate with CO_2 at 895 °C and achieved a 160 °C decrease in gasification temperature, as shown in Figure 11.

The interaction of the added metals plays an important role in the synergistic effect occurring in catalytic coal gasification. Jiang et al. [61] suggested that the formation of $K_2Ca(CO_3)_2$, in which the potassium atoms form a visualization in molecular size, can be more effective for dispersion on char due to the greater parting effect of calcium atoms than in K_2CO_3. Pereira et al. [66] confirmed that bimetallic oxide ($CaKO_x$) had a high catalytic activity for graphite or ash-free carbon than monoxide alone. Monterroso et al. [60] studied the iron–sodium catalyst for coal gasification, and suggested a set of reactions to form the potential catalyst intermediate, namely,

$$Na_2CO_3 + \alpha - Fe_2O_3 \rightarrow 2NaFeO_2 + CO_2 \qquad \text{(R34)}$$

$$Na_2O_2 + \alpha - Fe_2O_3 \rightarrow 2\alpha - NaFeO_2 + \frac{1}{2}O_2 \qquad \text{(R35)}$$

It is possible that the catalytic effect of iron on coal gasification is enhanced by its coexistence with sodium due to the function of the latter in improving the mobility of iron.

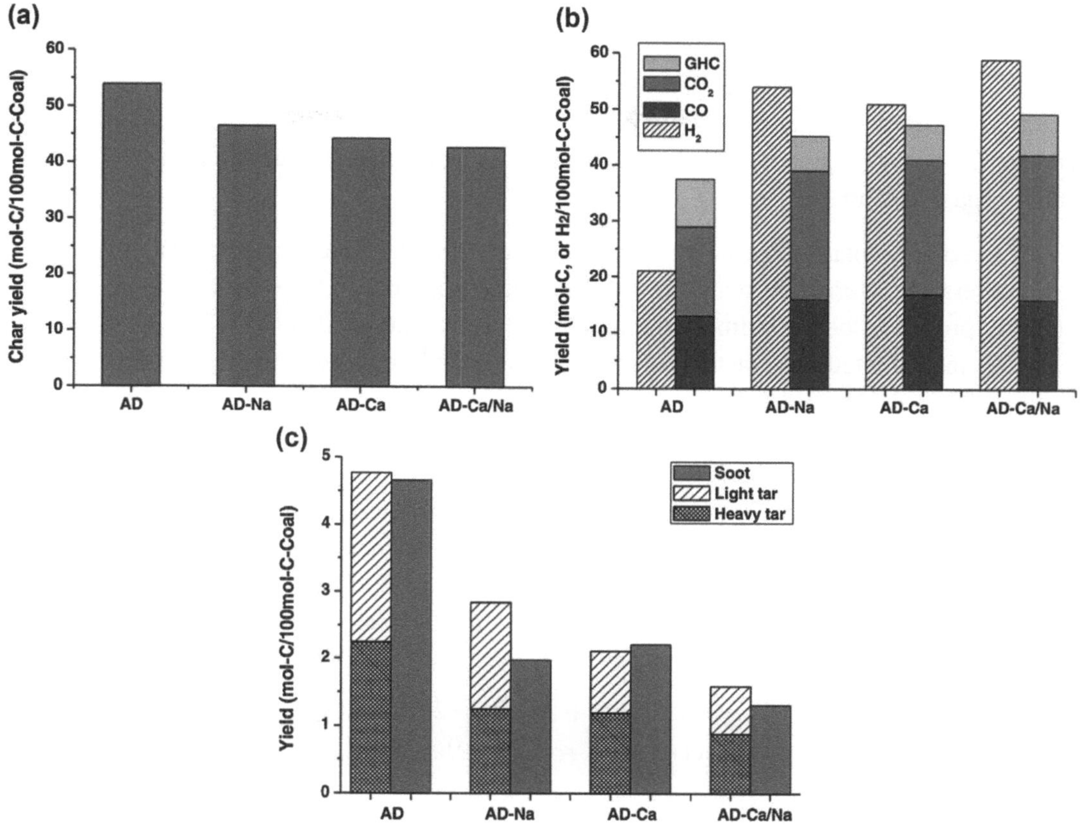

Figure 10

Product distribution of AD (raw coal), AD-Na, AD-Ca and AD-Ca/Na gasified at 900 °C in the drop-tube reactor. (a) Char yields; (b) gaseous products yields; (c) light tar, heavy tar and soot yields [63].

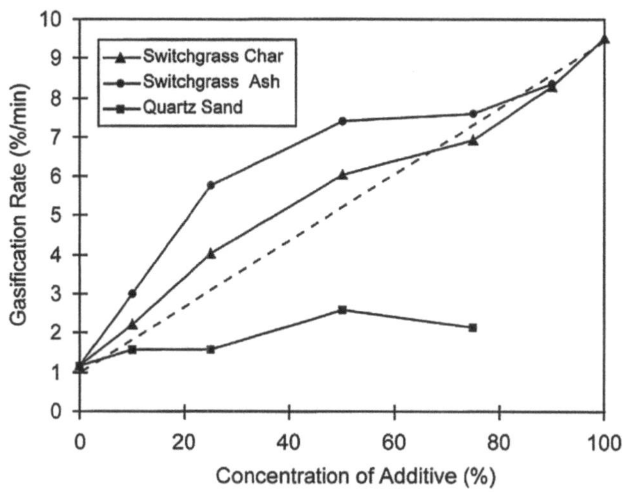

Figure 11

Coal char gasification rate as a function of the amount of additive in the form of switchgrass char, switchgrass ash, or quartz sand at 895 °C [65].

2.5 Invalidation and Recovery of Catalyst

The aforementioned metal catalysts – especially the iron and sodium species are basically low cost and environmentally friendly. However, catalyst recovery is still necessary to further decrease operation costs, thus promoting the application of catalytic gasification.

One of the disadvantages in catalytic coal gasification is the deactivation of catalysts via their irreversible interaction with the minerals present in coal [6,67–69]. Si and Al are generally proved to be the prime retardants in catalytic coal gasification [70]. Kühn et al. [69] investigated the inactivation reaction of added potassium, and found that Osumilit ($K_2O \cdot 4.4MgO \cdot 4.6Al_2O_3 \cdot 20.4SiO_2$), Sanidine ($K_2O \cdot Al_2O_3 \cdot 6SiO_2$), Leucite ($K_2O \cdot Al_2O_3 \cdot 4SiO_2$) and Kaliophilite ($K_2O \cdot Al_2O_3 \cdot 2SiO_2$) were four possible potassium deactivation products, and the last product counted for 29.8%, which was the highest possible deactivation reaction at 900 °C for German coal catalytic gasification. Zhang et al. [71] studied the interaction of added Na_2CO_3 and primary minerals in coal gasification, and found that the combination of sodium with either silicon or aluminium changed with carbon conversion and gasification temperatures. The combination was suggested as the following reactions:

$$Na_2O + 2SiO_2 + Al_2O_3 \rightarrow 2NaAlSiO_4 \quad \text{(R36)}$$

$$Na_2O + SiO_2 + CaO \rightarrow Na_2CaSiO_4 \quad \text{(R37)}$$

Ash-free coal is used to deal with the problem in laboratory experiments. HyperCoal, the hydrocarbon extracted from raw coal by 1-methylnaphthalene at 360 °C containing <0.05% ash, was widely used for catalytic coal gasification/liquefaction and relevant catalyst recycling experiments [8,72–74]. The results of Wang's experiment showed that 84–85% of the potassium in the HyperCoal chars was recovered by water leaching compared to 15–34% of raw coal, meanwhile the recycled catalysts were nearly lossless after three gasifications of HyperCoal (Figure 12). Sherma et al. [73] recycled the K_2CO_3 catalyst four times, and only insignificant conversion rates or gas composition changes were observed during HyperCoal gasification. The profile is shown in Figure 13:

Zhang et al. [75] used biomass-derived crude vinegars to recover the calcium catalyst. The primary and recycled calcium catalysts had nearly identical catalytic effects, which showed dramatically increased coal gasification reactivity. Additionally, the biomass-derived crude vinegars could promote biomass waste utilization. This technology signals a promising future for calcium-catalysed coal gasification. Compared to calcium, the sodium and potassium in coal gasification slag are easier to recycle due to their high solubility in aqueous solutions. Chen et al. [76] studied the effect of the water scrubbing method on the recovery of potassium in KOH catalytic coal gasification slag, and

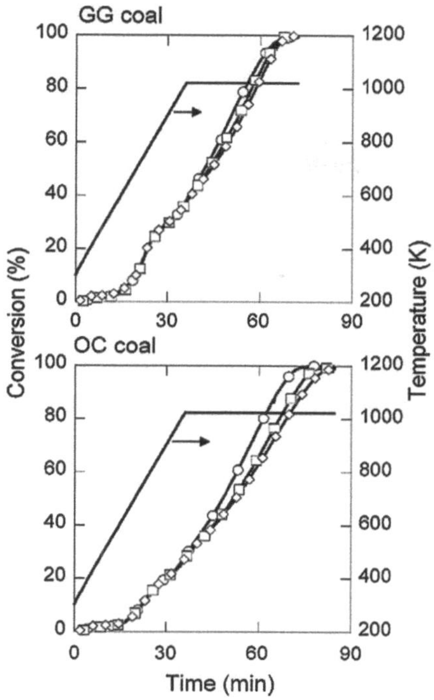

Figure 12
Coal conversion by steam gasification of HyperCoal with the recycled catalyst: (-○-) 5.8% fresh K_2CO_3, (-□-) first recycled catalyst and (-◇-) second recycled catalyst [72].

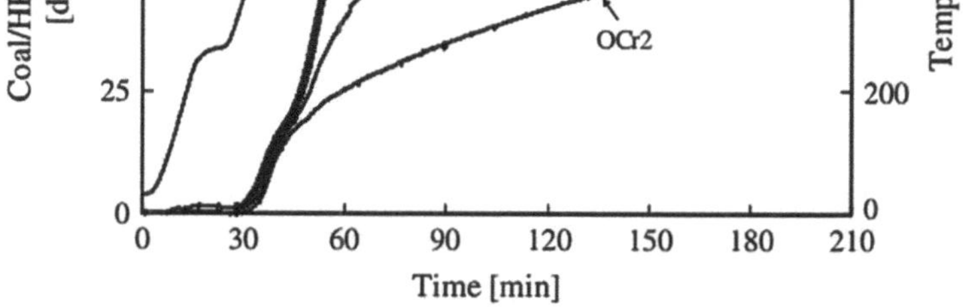

Figure 13
Gasification profiles with recycled catalyst at 775 °C. OCr1 and OCr2: OC coal recycle 1 and 2; OCHPCr1, OCHPCr2, OCHPCr3 and OCHPCr4: OC HyperCoal recycle 1, 2, 3 and 4 [73].

Table 3: Change in the Zinc Amount by ZnO Separation Process[a] [47].

Step	Zinc Content[b] (wt%-Zn)	Zinc Content[b] (g)	Recovery Ratio (%)
Before separation	10	0.1422	97.7
After separation	0.02	0.0027	

[a] The mixture of coal (1.0 g) and ZnO (10 wt%-Zn) in the packed-bed reactor was heated in an infrared furnace to 1173 K for 6 h while passing CO_2 ($pCO_2 = 1.0$) at a flow rate of 15 Ndm^3/min. After the CO_2 gasification, a CH_4/CO_2 mixture (CH_4:CO_2 ratio = 1:9) was introduced into the reactor at a flow rate of 10 Ncm^3/min and the remaining coal ash in the reactor was heated to 1423 K for 2 h.
[b] The zinc content in the starting coal–metal oxide mixture and the remaining coal ash [47].

confirmed that not only can this method recover up to 90% of total potassium, but that the recovered potassium still had nearly the same catalytic potency. Kodama et al. [47] demonstrated the high-temperature separation of used ZnO catalyst from coal ash at 1423 K. As shown in Table 3, 97.7% of zinc was evaporated and removed from the coal ash by the zinc separation process.

3. Conclusion and Outlook

Catalysts offer a promising way for industrializing coal gasification process under mild conditions. The alkali, alkaline earth metals, transition metals and their mixtures used as catalysts in coal gasification have been extensively studied, especially sodium, potassium, calcium and iron due to their efficiency, availability and environmental friendliness. The addition of a catalyst can effectively improve the carbon reaction rate, the H_2 production via the WGS reaction and can also decrease the gasification temperature. However, their disadvantages as catalysts in coal gasification cannot be ignored, for example, their corrosion effect to the gasifier and difficulty to recycle. Therefore, a lot of work still needs to be done to give an impetus to the industrialization of the coal catalytic gasification.

References

[1] Kwon S, Fan M, Dacosta HFM, Russell AG, Tsouris C. Reaction kinetics of CO_2 carbonation with Mg-rich minerals. J Phys Chem A 2011;115:7638–44.
[2] Zhang B, Fan M, Bland AE. CO_2 separation by a new solid K–Fe sorbent. Energy & Fuels 2011;25:1919–25.
[3] He L, Fan M, Dutcher B, Cui S, Shen X, Kong Y, et al. Dynamic separation of ultradilute CO_2 with a nanoporous amine-based sorbent. Chem Eng J 2012;189–190:13–23.
[4] Cormos CC. Integrated assessment of IGCC power generation technology with carbon capture and storage (CCS). Energy 2012;42:434–45.
[5] Stiegel GJ, Ramezan M. Hydrogen from coal gasification: an economical pathway to a sustainable energy future. Int J Coal Geol 2006;65:173–90.
[6] Kopyscinski J, Lam J, Mims CA, Hill JM. K_2CO_3 catalyzed steam gasification of ash-free coal. Studying the effect of temperature on carbon conversion and gas production rate using a drop-down reactor. Fuel 2014;128:210–9.

[7] Namkung H, Yuan X, Lee G, Kim D, Kang T, Kim H. Reaction characteristics through catalytic steam gasification with ultra clean coal char and coal. J Energy Inst 2014;87:253–62.
[8] Kim J, Choi H, Lim J, Rhim Y, Chun D, Kim S, et al. Hydrogen production via steam gasification of ash free coals. Int J Hydrog Energy 2013;38:6014–20.
[9] Tomita A. Catalysis of carbon–gas reactions. Catal Surv Jpn 2001;5:17–24.
[10] Quyn DM, Wu H, Hayashi J, Li C. Volatilisation and catalytic effects of alkali and alkaline earth metallic species during the pyrolysis and gasification of Victorian brown coal. Part IV. Catalytic effects of NaCl and ion-exchangeable Na in coal on char reactivity. Fuel 2003;82:587–93.
[11] Yoshida S, Matsunami J, Hosokawa Y, Yokota O, Tamaura Y, Kitamura M. Coal/CO_2 gasification system using molten carbonate salt for solar/fossil energy hybridization. Energy & Fuels 1999;13:961–4.
[12] Ye DP, Agnew JB, Zhang DK. Gasification of a South Australian low-rank coal with carbon dioxide and steam: kinetics and reactivity studies. Fuel 1998;77:1209–19.
[13] Popa T, Fan M, Argyle MD, Slimane RB, Bell DA, Towler BF. Catalytic gasification of a Powder River Basin coal. Fuel 2013;103:161–70.
[14] Hashimoto K, Miura K, Xu J, Watanabe A, Masukami H. Relation between the gasification rate of carbons supporting alkali metal salts and the amount of oxygen trapped by the metal. Fuel 1986;65:489–94.
[15] Suzuki T, Inoue K, Watanabe Y. Temperature-programmed desorption and carbon dioxide-pulsed gasification of sodium- or iron-loaded Yallourn coal char. Energy & Fuels 1988;2:673–9.
[16] Alam M, DebRoy T. Reaction between CO_2 and coke doped with NaCN. Carbon 1987;25:279–88.
[17] McKee DW. Mechanisms of the alkali metal catalysed gasification of carbon. Fuel 1983;62:170–5.
[18] McKee DW. The catalyzed gasification reactions of carbon. Chem Phys Carbon 1981;16:118.
[19] Sams DA, Shadman F. Mechanism of potassium-catalyzed carbon/CO_2 reaction. AIChE J 1986:1132–7.
[20] Shadman F, Sams DA, Punjak WA. Significance of the reduction of alkali carbonates in catalytic carbon gasification. Fuel 1987;66:1658–63.
[21] Meng L, Wang M, Yang H, Ying H, Chang L. Catalytic effect of alkali carbonates on CO_2 gasification of Pingshuo coal. Min Sci Technol (China) 2011;21:587–90.
[22] Lahijani P, Zainal ZA, Mohamed AR, Mohammadi M. CO_2 gasification reactivity of biomass char: catalytic influence of alkali, alkaline earth and transition metal salts. Bioresour Technol 2013;144:288–95.
[23] Nahas NC. Exxon catalytic coal gasification process: fundamentals to flowsheets. Fuel 1983;62:239–41.
[24] Sun Q, Li W, Chen H, Li B. The CO_2-gasification and kinetics of Shenmu maceral chars with and without catalyst. Fuel 2004;83:1787–93.
[25] Lang R. Anion effects in alkali-catalysed steam gasification. Fuel 1986;65:1324–9.
[26] Li Y, Yang H, Hu J, Wang X, Chen H. Effect of catalysts on the reactivity and structure evolution of char in petroleum coke steam gasification. Fuel 2014;117(Part B):1174–80.
[27] Siefert N, Shekhawat D, Litster S, Berry D. Molten catalytic coal gasification with in situ carbon and sulphur capture. Energy & Environ Sci 2012;9:8660–72.
[28] Wang J, Jiang M, Yao Y, Zhang Y, Cao J. Steam gasification of coal char catalyzed by K_2CO_3 for enhanced production of hydrogen without formation of methane. Fuel 2009;88:1572–9.
[29] Li CZ. Some recent advances in the understanding of the pyrolysis and gasification behaviour of Victorian brown coal. Fuel 2007;86:1664–83.
[30] Li C, Sathe C, Kershaw J, Pang Y. Fates and roles of alkali and alkaline earth metals during the pyrolysis of a Victorian brown coal. Fuel 2000;79:427–38.
[31] Kitsuka T, Bayarsaikhan B, Sonoyama N, Hosokai S, Li CZ, Norinaga K, et al. Behavior of inherent metallic species as a crucial factor for kinetics of steam gasification of char from coal pyrolysis. Energy & Fuels 2007:387–94.
[32] Zhang D, Poeze A. Variation of sodium forms and char reactivity during gasification of a South Australian low-rank coal. Proc Combust Inst 2000;28:2337–44.
[33] McKee DW. Catalysis of the graphite-water vapor reaction by alkaline earth salts. Carbon 1979;17:419–25.

[34] Kapteijn F, Porre H, Moulijn JA. CO_2 gasification of activated carbon catalyzed by earth alkaline elements. AIChE J 1986;32:691–5.

[35] Chen SG, Yang RT. Mechanism of alkali and alkaline earth catalyzed gasification of graphite by CO_2 and H_2O studied by electron microscopy. J Catal 1992;138:12–23.

[36] Hengel TD, Walker PL. Catalysis of lignite char gasification by exchangeable calcium and magnesium. Fuel 1984;63:1214–20.

[37] Hattingh BB, Everson RC, Neomagus HWJP, Bunt JR. Assessing the catalytic effect of coal ash constituents on the CO_2 gasification rate of high ash, South African coal. Fuel Process Technol 2011;92:2048–54.

[38] Ohtsuka Y, Tomita A. Calcium catalysed steam gasification of Yallourn brown coal. Fuel 1986;65:1653–7.

[39] McKee DW. Catalytic effects of alkaline earth carbonates in the carbon-carbon dioxide reaction. Fuel 1980;59:308–14.

[40] Sears JT, Muraldihara HS, Wen CY. Reactivity correlation for the coal char CO_2 reaction. Ind Eng Chem Process Des Dev 1980;19:358–64.

[41] Ohtsuka Y, Asami K. Steam gasification of coals with calcium hydroxide. Energy & Fuels 1995;9:1038–42.

[42] Ohtsuka Y, Asami K. Highly active catalysts from inexpensive raw materials for coal gasification. Catal Today 1997;39:111–25.

[43] Karimi A, Gray MR. Effectiveness and mobility of catalysts for gasification of bitumen coke. Fuel 2011;90:120–5.

[44] Radovic LR, Walker PL, Jenkins RG. Catalytic coal gasification: use of calcium versus potassium. Fuel 1984;63:1028–30.

[45] Holstein WL, Boudart M. Uncatalyzed and platinum-catalyzed gasification of carbon by water and carbon dioxide. J Catal 1982;75:337–53.

[46] Holstein WL, Boudart M. Transition metal and metal oxide catalysed gasification of carbon by oxygen, water, and carbon dioxide. Fuel 1983;62:162–5.

[47] Kodama T, Funatoh A, Shimizu K, Kitayama Y. Kinetics of metal oxide-catalyzed CO_2 gasification of coal in a fluidized-bed reactor for solar thermochemical process. Energy & Fuels 2001;15:1200–6.

[48] Asami K, Sears P, Furimsky E, Ohtsuka Y. Gasification of brown coal and char with carbon dioxide in the presence of finely dispersed iron catalysts. Fuel Process Technol 1996;47:139–51.

[49] Tanaka S, U-emura T, Ishizaki K, Nagayoshi K, Ikenaga N, Ohme H, et al. CO_2 gasification of iron-loaded carbons: activation of the iron catalyst with CO. Fuel Energy Abstr 1995;36:187.

[50] Popa T, Fan M, Argyle MD, Dyar MD, Gao Y, Tang J, et al. H_2 and CO_x generation from coal gasification catalyzed by a cost-effective iron catalyst. Appl Catal A: General 2013;464–465:207–17.

[51] Irfan MF, Usman MR, Kusakabe K. Coal gasification in CO_2 atmosphere and its kinetics since 1948: a brief review. Energy 2011;36:12–40.

[52] Hermann G, Hüttinger KJ. Mechanism of iron-catalyzed water vapour gasification of carbon. Carbon 1986;24:429–35.

[53] Furimsky E, Sears P, Suzuki T. Iron-catalyzed gasification of char in carbon dioxide. Energy & Fuels 1988;2:634–9.

[54] Yu J, Tian F, Chow M, McKenzie LJ, Li C. Effect of iron on the gasification of Victorian brown coal with steam: enhancement of hydrogen production. Fuel 2006;85:127–33.

[55] Ohtsuka Y, Kuroda Y, Tamai Y, Tomita A. Chemical form of iron catalysts during the CO_2 -gasification of carbon. Fuel 1986;65:1476–8.

[56] Biswas AK. Principles of blast furnace iron making. Brisbane: Cootha Publishing House; 1981.

[57] Ohme H, Suzuki T. Mechanisms of CO_2 gasification of carbon catalyzed with group VIII metals. 1. Iron-Catalyzed CO_2 gasification. Energy & Fuels 1996;10:980–7.

[58] Byung HS, Sang DK. Catalytic activity of alkali and iron salt mixtures for steam-char gasification. Fuel 1993;72:797–803.

[59] Matsumoto S, Philip L, Walker J. Char gasification in steam at 1123 K catalyzed by K, Na, and Fe – effect of H_2, H_2S and COS. Carbon 1986:277–85.

[60] Monterroso R, Fan M, Zhang F, Gao Y, Popa T, Argyle MD, et al. Effects of an environmentally-friendly, inexpensive composite iron−sodium catalyst on coal gasification. Fuel 2014;116:341−9.
[61] Jiang M, Zhou R, Hu J, Wang F, Wang J. Calcium-promoted catalytic activity of potassium carbonate for steam gasification of coal char: Influences of calcium species. Fuel 2012;99:64−71.
[62] Hu J, Liu L, Cui M, Wang J. Calcium-promoted catalytic activity of potassium carbonate for gasification of coal char: the synergistic effect unrelated to mineral matter in coal. Fuel 2013;111:628−35.
[63] Zhang L, Kudo S, Tsubouchi N, Hayashi JI, Ohtsuka Y, Norinaga K. Catalytic effects of Na and Ca from inexpensive materials on in-situ steam gasification of char from rapid pyrolysis of low rank coal in a drop-tube reactor. Fuel Process Technol 2013;113:1−7.
[64] Spiro CL, McKee DW, Kosky PG, Lamby EJ. Catalytic CO_2-gasification of graphite versus coal char. Fuel 1983;62:180−4.
[65] Brown RC, Liu Q, Norton G. Catalytic effects observed during the co-gasification of coal and switchgrass. Biomass Bioenergy 2000;18:499−506.
[66] Pereira P, Csencsits R, Somorjai GA, Heinemann H. Steam gasification of graphite and chars at temperatures <1000 K over potassium-calcium-oxide catalysts. J Catal 1990;123:463−76.
[67] Kong Y, Kim J, Chun D, Lee S, Rhim Y, Lim J, et al. Comparative studies on steam gasification of ash-free coals and their original raw coals. Int J Hydrog Energy 2014;39:9212−20.
[68] Alauddin ZABZ, Lahijani P, Mohammadi M, Mohamed AR. Gasification of lignocellulosic biomass in fluidized beds for renewable energy development: a review. Renew Sustain Energy Rev 2010;14:2852−62.
[69] Kühn L, Plogmann H. Reaction of catalysts with mineral matter during coal gasification. Fuel 1983;62:205−8.
[70] Zhang Y, Hara S, Kajitani S, Ashizawa M. Modeling of catalytic gasification kinetics of coal char and carbon. Fuel 2010;89:152−7.
[71] Zhang F, Xu D, Wang Y, Wang Y, Gao Y, Popa T, et al. Catalytic CO_2 gasification of a Powder River Basin coal. Fuel Process Technol 2015;130:107−16.
[72] Wang J, Sakanishi K, Saito I, Takarada T, Morishita K. High-yield hydrogen production by steam gasification of HyperCoal (ash-Free coal extract) with potassium carbonate: comparison with raw coal. Energy & Fuels 2005;19:2114−20.
[73] Sharma A, Takanohashi T, Morishita K, Takarada T, Saito I. Low temperature catalytic steam gasification of HyperCoal to produce H_2 and synthesis gas. Fuel 2008;87:491−7.
[74] Koyano K, Takanohashi T, Saito I. Catalytic hydrogenation of HyperCoal (ashless coal) and reusability of catalyst. Energy & Fuels 2009;23:3652−7.
[75] Zhang Y, Ashizawa M, Kajitani S, Hara S. A new approach to catalytic coal gasification: the recovery and reuse of calcium using biomass derived crude vinegars. Fuel 2010;89:417−22.
[76] Chen S, Yang R. Unified mechanism of alkali and alkaline earth catalyzed gasification reactions of carbon by CO_2 and H_2O. Energy & Fuels 1997;11:421−7.

Index

Note: Page numbers followed by "f" or "t" indicates figures and tables respectively.

A

A^3 coupling. *See* Aldehyde, alkyne and amine coupling (A^3 coupling)
AAEM. *See* Alkali and alkaline earth metals (AAEM)
ACFs. *See* Activated CFs (ACFs)
Acid catalysis, 84–85, 101
Acid–base catalysed reactions, 29. *See also* Coupling reactions; Polymerization reactions; Redox reactions
  aldol condensation reactions, 31–32
  Baeyer–Villiger oxidation, 33
  FC reaction, 30–31
  Fischer esterification and transesterification reactions, 33–35
  Knoevenagel condensation reaction, 32–33
Activated CFs (ACFs), 148, 150
Adipic acid, 69
Alcell lignin electrospun nanofibres, 145–146
Alcohol(s), 13
  alkylation reactions using, 30
  benzyl alcohol, 69–70
  dehydrodimerization, 86–87
  oxidation, 39–40, 39f
  transesterification and, 33–34
Aldehyde, alkyne and amine coupling (A^3 coupling), 50
Aldol
  adduct, 116–118
  condensation product, 12
  condensation reactions, 31–32
Alkali and alkaline earth metals (AAEM), 181, 185–187
Alkali metal compounds, 181–185
Alkaline earth metal compounds, 185–189
AlPO-n. *See* Aluminophosphate molecular sieves (AlPO-n)
Aluminium and zirconium substituted meso–microcomposite (AlZrMMSH), 9–10
Aluminophosphate molecular sieves (AlPO-n), 14–15
AlZrMMSH. *See* Aluminium and zirconium substituted meso–microcomposite (AlZrMMSH)
Ammoximation reactions, 36, 36f
Amycol, 139–140, 140f
Aqueous-phase dehydration/hydrogenation (APD/H), 116–118

B

Baeyer–Villiger oxidation, 33
Benzothiophene (BT), 13
BET. *See* Brunauer–Emmett–Teller (BET)
β-etherases, 144
β-hydroxyketone, 31–32
β–aryl ether bond, 144
β–O–4 ether bond, 133–134
Biodiesel, 33–35
Biofuels, 108, 109f
Biogenic impurities, 165–168, 166t–167t
Biomass. *See also* 5-hydroxymethyl furfural (HMF)
  acid catalysts, 106t
  biomass conversion
    catalyst for, 105–106
    strategy for, 100–101
  biomass-based molecules, 169
  biomass-derived molecules, 158
    decarbonylation reactions, 172f
    decarboxylation reactions, 173f
  and compositions, 99–100
  fermentation, 158–159
  lignocellulose composition, 100f
  platform chemicals from, 102f
  platform molecules derived from, 162t–163t
  second-generation biofuels from, 108–119
  to value-added chemicals, 101
Biorenewable chemicals, 159–160, 164. *See also* Integrated bio-and chemocatalytic processing
Boiling point (BP), 11–12
Bottom-up method, 6–7
BP. *See* Boiling point (BP)

Brunauer–Emmett–Teller (BET), 28–29
BT. *See* Benzothiophene (BT)

C

Candida antarctica lipase B (CALB), 72
Carbohydrate ($C_nH_{2n}O_n$), 99
Carbon capture and storage (CCS), 180
Carbon fibres (CFs), 144–151
Carbon–carbon bond (C–C bond), 44–45
Carbonization, 144–151
Cascade process for jet fuels, 116–118
Catalysts, 2, 6
  composite, 191–192
  invalidation and recovery, 194–196
Catalytic coal gasification, 181
  alkali metal compounds, 181–185
  alkaline earth metal compounds, 185–189
  coal, 179–180, 179f
    gasifiers, 180
    resource in world, 179f
  composite catalysts, 191–192
  invalidation and recovery of catalyst, 194–196
  iron, 179f
  metal-catalysed, 181f
  potassium, 179f
  reaction scheme of K_2CO_3, 183f
  transition metal compounds, 189–191
Catalytic/catalysis, 23–24
  acid catalysis, 84–85, 101
  decarbonylation and decarboxylation, 172–173
  hydroxyalkylation/alkylation, 115–116
C–C bond. *See* Carbon–carbon bond (C–C bond)
C–C coupling reactions, 43–49
CCS. *See* Carbon capture and storage (CCS)
Cellulose, 99–100
  conversion to HMF, 78–80
  depolymerization, 77–78
  dissolution, 74–77
  ethanol, 109
Cellulose nanofibils (CNFs), 138, 149–150
CFs. *See* Carbon fibres (CFs)
Chemical marriage broker, 23–24
Chemical techniques, 128–138
5-chloromethylfurfural (CMF), 111
Choline (H_2PO_4), 71
CMF. *See* 5-chloromethylfurfural (CMF)
CNFs. *See* Cellulose nanofibils (CNFs)
Co-condensation route, 25–27
CO_2 transformation to cyclocarbonates, 80–83
Coal, 179–180. *See also* Catalytic coal gasification
  gasifiers, 180
  resource in world, 179f
Composite catalysts, 191–192
Cooperative self-assembly mechanism (CSA mechanism), 5, 5f
Coupling reactions, 43. *See also* Acid–base catalysed reactions; Polymerization reactions; Redox reactions
  C–C coupling reactions, 43–49
  C–S coupling reactions, 49–50
Cr-based silica, 38
Cracking. *See* Hydrogenolysis
C–S coupling reactions, 49–50
CSA mechanism. *See* Cooperative self-assembly mechanism (CSA mechanism)

D

DBT. *See* Dibenzothiophene (DBT)
Decarboxylation reactions, 173
Deep eutectic mixtures (DES), 81–82
Degree of polymerization (DP), 74–75
Dehydration, 31–32, 101, 103–105, 169–171
Density, 62, 64
Density functional theory (DFT), 78–79
DES. *See* Deep eutectic mixtures (DES)
DFF. *See* 2,5-diformylfuran (DFF)
DFT. *See* Density functional theory (DFT)
DHHMP. *See* 5,6-dihydro-4-hydroxy-6-methyl-2-pyrone (DHHMP)
Dibenzothiophene (DBT), 11–12
Dielectric constant, 65
2,5-diformylfuran (DFF), 107
5,6-dihydro-4-hydroxy-6-methyl-2-pyrone (DHHMP), 173
3,4-dihydropyrimidin-2(1H)-ones, 35
Dimethyl formamide (DMF), 44, 47–48
Dimethyl sulphoxide (DMSO), 44–45, 104–105
1-(N,N-dimethylaminoethyl)-2,3-dimethylimidazolium trifluoromethane-sulphonate, 88
2,5-dimethylfuran (DMF), 102–103, 111–114
  conversion of fructose to, 113f
  one-pot DMF production, 114f
  production from HMF, 112f
DMA–LiCl. *See* *N*, *N*-dimethyl acetamide with lithium chloride (DMA–LiCl)
DMF. *See* Dimethyl formamide (DMF); 2,5-dimethylfuran (DMF)
DMSO. *See* Dimethyl sulphoxide (DMSO)
DP. *See* Degree of polymerization (DP)

E

Electrodialysis (ED), 168
EMF. *See* 5-ethoxymethyl furfural (EMF)

Enzymatic reactions in ionic liquids, 70–74
biofuels production, 73–74
H-bond basicity effects, 72–73
Hofmeister effect, 71
hydrophobicity effect of ionic liquids, 72
nucleophilicity effects, 72–73
Enzymatic techniques, 138–144
Enzyme-catalysed transesterification, 73
Ethanol, 109
5-ethoxymethyl furfural (EMF), 102–103, 110–111
Extended X-ray absorption fine structure (EXAFS), 133–134

F

FC reaction. *See* Friedel–Crafts reaction (FC reaction)
FCC. *See* Fluid catalytic cracking (FCC)
FDCA. *See* 2,5-furandicarboxylic acid (FDCA)
Fischer esterification and transesterification reactions, 33–35
Fluid catalytic cracking (FCC), 10–11
Fossil fuel, 99, 157–158
Fourier transform infrared measurements (FT-IR measurements), 80–81
Friedel–Crafts reaction (FC reaction), 30–31
FT-IR measurements. *See* Fourier transform infrared measurements (FT-IR measurements)
Functionalized mesoporous materials
organic transformations, 28–29
acid–base catalysed reactions, 29–35
coupling reactions, 43–50
polymerization reactions, 52
redox reactions, 36–42
role of functionalization over ordered mesoporous silicas, 52–53
synthesis and types of, 25–28
2,5-furandicarboxylic acid (FDCA), 101
HMF oxidation to, 107, 107f

G

γ-aminobutyric acid (GABA), 161–164
GHG. *See* Green house gas (GHG)
Glutamic acid, 161–164, 162t–163t, 173
Graphene oxide (GO), 111
Green house gas (GHG), 99

H

H-bond basicity effects, 72–73
HDN. *See* Hydrodenitrogenation (HDN)
HDO. *See* Hydrodeoxygenation (HDO)
HDS. *See* Hydrodesulphurization (HDS)
Hiyama coupling, 44–45, 46f
HMF. *See* 5-hydroxymethyl furfural (HMF)
Hofmeister effect, 71
Hybrid zeolitic-ordered mesoporous materials (HZM), 12
Hydrocarbons oxidation, 36–38, 38t
Hydrodenitrogenation (HDN), 11–12
Hydrodeoxygenation (HDO), 127–128, 131–133, 134f, 137–138
Hydrodesulphurization (HDS), 10–11
Hydrogenation reactions, 40–42
Hydrogenolysis, 128–131, 134–135
Hydroisomerization, 15
Hydrophobicity effect of ionic liquids, 72
3-hydroxy propionic acid, 161–164
5-hydroxymethyl furfural (HMF), 74, 99–100, 102, 169–171
aspects of HMF synthesis, 103–105
by-products in HMF production, 105
to LA, 107–108
oxidation
to DFF, 107
to FDCA, 107
HyperCoal gasification, 194
HZM. *See* Hybrid zeolitic-ordered mesoporous materials (HZM)

I

IGCC. *See* Integrated gasification combined cycle (IGCC)
ILs. *See* Ionic liquids (ILs)
Integrated bio-and chemocatalytic processing
biogenic impurities nature, 165–168, 166t–167t
biomass integrated processing, 159f
platform molecules
biocatalytic transformation to, 159–164
derived from biomass, 162t–163t
strategies for catalytic transformation, 168–173
Integrated gasification combined cycle (IGCC), 180
Ionic liquids (ILs), 30–31, 61, 101
CO_2 transformation to cyclocarbonates, 80–83
dissolution and conversion of biomass in, 74
cellulose conversion to HMF, 78–80
cellulose depolymerization, 77–78
cellulose dissolution, 74–77
enzymatic reactions in, 70–74
biofuels production, 73–74
H-bond basicity effects, 72–73

Ionic liquids (ILs) (*Continued*)
Hofmeister effect, 71
hydrophobicity effect of ionic liquids, 72
nucleophilicity effects, 72–73
transition metal catalysis reactions, 66–70
hydrogenation in, 67–68
oxidation in, 68–70
Iron and wustite (FeO), 191
Isoprenesynthase gene (IspS gene), 160–161
Itaconic acid, 161–164, 162t–163t

J

Jet fuels, cascade process for, 116–118

K

Knoevenagel condensation reaction, 32–33
Kosmotropes, 71

L

LA. *See* Levulinic acid (LA)
Laccases from *S. viridosporus* T7A (SVLAC), 142–143
LCFs. *See* Lignin CFs (LCFs)
LCT mechanism. *See* Liquid crystal templating mechanism (LCT mechanism)
LDP. *See* Lignin depolymerization (LDP)
Levulinic acid (LA), 99–100
to transportation fuels, 118–119
Lewis acidic property, 29
Light gas oil (LGO), 11–12
Lignin, 99–100, 144–145
deconstruction, 125–126
carbonization, 144–151
enzymatic techniques, 138–144
linkages and lignin model compounds, 126f
softwood and hardwood lignin, 126t
monomers upgradation, 131–138
valorization techniques
catalytic valorization, 127
chemical techniques, 128–138
reactions at water/oil interface, 132f
Lignin CFs (LCFs), 146
Lignin depolymerization (LDP), 129–130
into monomers, 129–131
Lignocellulosic biomass, 101
Liquid
phase catalytic reactions, 23
range, 62–64
Liquid crystal templating mechanism (LCT mechanism), 4–5, 5f
LogP scale, 72
Long chain hydrocarbons, 114–119

M

MAPO-*n*. *See* Metal aluminophosphate (MAPO-*n*)
MAPSO-*n*. *See* Metal aluminophosphate (MAPO-*n*)
MAS. *See* Meso-structure aluminosilicate (MAS)
MAS-5. *See* Mesoporous aluminosilicate (MAS-5)
MCM-41 mesophases, 4–5, 4f
MCM-41. *See* Mobil Composition of Matter-41 (MCM-41)
MCM-48 mesophases, 4–5, 4f
MCM-50 mesophases, 4–5, 4f
Melt-spinning, 149
Meso-SAPO-34, 15, 15t
Meso-structure aluminosilicate (MAS), 11–12
Meso/microaluminosilicate composite materials, 8–12
Meso/microcomposite of aluminophosphate-based materials, 14–18
Mesoporous aluminosilicate (MAS-5), 8
Mesoporous molecular sieves, 4–7
Mesoporous organic–inorganic hybrid polymers, 27–28
Metal aluminophosphate (MAPO-*n*), 14–15
2-methylfuran diesel process. *See* Sylvan diesel process
Mevalonate (MVA), 160–161
Micellar liquid crystalline phase, 5
Microporous titanosilicate (Mime-TS-1), 13
Mizoroki–Heck reaction, 47–48, 48f
Mobil Composition of Matter-41 (MCM-41), 23
Molecular sieve, 2f
MVA. *See* Mevalonate (MVA)

N

N,N-dimethyl acetamide with lithium chloride (DMA–LiCl), 104–105
N-methylpyrrolidone (NMP), 47–48, 161–164
N-vinylpyrrolidone (NVP), 161–164
Negishi coupling, 47–48
NMP. *See* *N*-methylpyrrolidone (NMP)
Non-radical ligninolytic enzymes, 144
Nucleophilicity effects, 72–73
NVP. *See* *N*-vinylpyrrolidone (NVP)

O

Oxygen, 116–118

P

p-MAP. *See* *p*-methoxyacetophenone (*p*-MAP)
p-methoxyacetophenone (*p*-MAP), 12
p-toluenesulphonic acid (*p*-TosOH), 115–116

p-TosOH. *See* *p*-toluenesulphonic acid (*p*-TosOH)
PAMAM. *See* Polyaminoamide (PAMAM)
PAN. *See* Polyacrylonitrile (PAN)
PEF. *See* Poly(ethylene furancarboxylate) (PEF)
Periodic mesoporous organosilica (PMO), 25–27, 39–40
PET. *See* Poly(ethylene terephthalate) (PET)
Petrochemical process, 1
4-PGol. *See* 4-propylguaiacol (4-PGol)
PLA. *See* Poly(lactic) acid (PLA)
Platform molecules
  biocatalytic transformation to, 159–164
  biorenewable platform molecules and products, 170t–171t
  derived from biomass, 162t–163t
  strategies for catalytic transformation, 168–173
PMO. *See* Periodic mesoporous organosilica (PMO)
Polarity, 64–65
Poly(ethylene furancarboxylate) (PEF), 107
Poly(ethylene terephthalate) (PET), 107
Poly(lactic) acid (PLA), 149
Poly(vinyl alcohol) (PVA), 165–168
Polyacrylonitrile (PAN), 144–145
Polyaminoamide (PAMAM), 41–42
Polymerization reactions, 52. *See also* Acid–base catalysed reactions; Coupling reactions; Redox reactions
Post-grafting technique, 25
Propargylamine, 50
4-propylguaiacol (4-PGol), 129–130
PVA. *See* Poly(vinyl alcohol) (PVA)

R

Redox reactions, 36. *See also* Acid–base catalysed reactions; Coupling reactions; Polymerization reactions
  alcohols oxidation, 39–40
  ammoximation reactions, 36
  hydrocarbons oxidation, 36–38, 38t
  hydrogenation reactions, 40–42
  sulphides oxidation, 39–40
Research octane number (RON), 111–112

S

SAPO-*n*. *See* Silicoaluminophosphate (SAPO-*n*)
SBUs. *See* Secondary building units (SBUs)
SCC medium. *See* Supercritical carbon dioxide medium (SCC medium)
SCWT. *See* Wild-type *Streptomyces coelicolor* (SCWT)
SDAs. *See* Structure-directing agents (SDAs)
Second-generation biofuels
  biofuel, 109f
  from biomass, 108
    cellulosic ethanol, 109
    DMF, 111–114
    EMF, 110–111
    long chain hydrocarbons, 114–119
Secondary building units (SBUs), 2
Silicoaluminophosphate (SAPO-*n*), 14–15
Sol–gel route, 25
Sonogashira reaction, 45–47
SSU. *See* Structural sub unit (SSU)
Stille reaction coupling, 47–48
Structural sub unit (SSU), 2
Structure-directing agents (SDAs), 3
Sulphides oxidation, 39–40
Supercritical carbon dioxide medium (SCC medium), 9–10
Sustainable catalysis systems
  properties of ionic liquids and potentials, 62–65
    density and viscosity, 64
    liquid range, 62–64
    polarity, 64–65
  utilization of ionic liquids, 65, 83–88
    CO_2 transformation to cyclocarbonates, 80–83
    dissolution and conversion of biomass, 74–80
    enzymatic reactions in ionic liquids, 70–74
    reactions catalysed by acidic functionalized ionic liquids, 84–87
    reactions catalyzed by basic functionalized ionic liquids, 87–88
    transition metal catalysis reactions, 66–70
Suzuki–Miyuara cross-coupling reaction, 44, 45f
SVLAC. *See* Laccases from *S. viridosporus* T7A (SVLAC)
Sylvan diesel process, 114–116

T

TAL. *See* Triacetic acid lactone (TAL)
Tertiary butyl hydroperoxide (TBHP), 36–40
Tetrahydrofuran (THF), 102–103
TH. *See* Thiophene (TH)
THF. *See* Tetrahydrofuran (THF)
Thiophene (TH), 13
Three-dimensional network structure (3D network structure), 25
Ti-MMM-1. *See* Titanium-containing micro-mesoporous materials (Ti-MMM-1)
Titanium-based meso/microcomposite, 13

Titanium-containing micro-mesoporous materials (Ti-MMM-1), 13
TOF. *See* Turnover frequency (TOF)
Tonnes per annum (TPA), 161
Top-down method, 6–7
Total reductive sugar (TRS), 77
TPA. *See* Tonnes per annum (TPA)
Transesterification, 33–34
Transition metal catalysis reactions, 66–70
  ionic liquids
    hydrogenation, 67–68
    oxidation in, 68–70
Transition metal compounds, 189–191
Transportation fuels, LA to, 118–119
Triacetic acid lactone (TAL), 159–160
TRS. *See* Total reductive sugar (TRS)
Turnover frequency (TOF), 24–25

U

Ullmann reaction, 48–49

V

Value-added chemicals, biomass to, 101
Vanillic catabolic pathway, 141
Vapour phase treatment (VPT), 16–18
Viscosity, 64
VPT. *See* Vapour phase treatment (VPT)

W

Water, 114–115
Water-gas shift (WGS), 180, 184
Wild-type *Streptomyces coelicolor* (SCWT), 142–143

X

X-ray absorption near edge structure (XANES), 133–134
XRD. *See* X-ray diffraction (XRD)

Z

Zeolite, 2–3, 4f
Zeolite-like molecular sieves, 2–3, 3f
Zeolite-template carbons (ZTCs), 150–151

Kolibri 360 GmbH
Ferdinand-Jühlke-Straße 7
99095 Erfurt, Deutschland
produktsicherheit@kolibri360.de